Probabilistic Boolean Networks

Probabilistic Boolean Networks

The Modeling and Control of Gene Regulatory Networks

Ilya Shmulevich

Institute for Systems Biology
Seattle, Washington

Edward R. Dougherty

Texas A&M University
College Station, Texas

Translational Genomics Research Institute
Phoenix, Arizona

siam. Society for Industrial and Applied Mathematics • Philadelphia

10 9 8 7 6 5 4 3 2 1

Library of Congress Cataloging-in-Publication Data

Shmulevich, Ilya, 1969-

Probabilistic boolean networks : the modeling and control of gene regulatory networks / Ilya Shmulevich, Edward R. Dougherty.

p. cm.

Includes bibliographical references and index.

ISBN 978-0-898716-92-4

1. Genetic regulation—Computer simulation. I. Dougherty, Edward R. II. Society for Industrial and Applied Mathematics. III. Title.

[DNLM: 1. Gene Regulatory Networks. 2. Models, Genetic. 3. Models, Statistical. QU 470 S558p 2010]

QH450.S56 2010

572.8′65—dc22

2009034643

This book is dedicated to the memory
of Norbert Wiener,
the father of modern translational science.

Contents

Preface

It was around the period of World War II that Arturo Rosenblueth and Norbert Wiener were taking the first steps in the direction of systems medicine. They formed an interesting pair: Rosenblueth, a physiologist at the Harvard Medical School, and Wiener, the father of modern engineering in the United States. For this book, their conception of science is salient. They wrote, "The intention and the result of a scientific inquiry is to obtain an understanding and a control of some part of the universe." [1] For them, as a research team, the part of the universe was physiology. An appreciation of their words is important. Understanding is not some vague, subjective explanation, but rather the precision of mathematical systems needed for the representation of relationships between measurable quantities and future predictions based on those relationships. Control is the ability to change physical behavior in a manner concomitant with the mathematical system representing the relevant phenomena.

Rosenblueth and Wiener take an active view of science: it is to change the world. In contemporary terminology, rather than science, one might say that they were describing translational science. "Translational science transforms a scientific mathematical model, whose purpose is to provide a predictive conceptualization of some portion of the physical world, into a model characterizing human intervention (action) in the physical world. Whereas the pure scientist typically tries to minimize human interference, translational science extends science to include conceptualization of human-originated action in the physical world and thereby raises epistemological issues relating to the knowledge of this intentional intervention into the natural order. Scientific knowledge is translated into practical knowledge by expanding a scientific system to include inputs that can be adjusted to affect the behavior of the system and outputs that can be used to monitor the effect of the external inputs and feed back information on how to adjust the inputs." [2] It is this translational scientific view that Wiener brought into line with modern science during his illustrious career. In perhaps the greatest transformation of engineering epistemology since antiquity, Wiener fundamentally altered the way human beings perceive scientifically based action in the world. Teaming with Rosenblueth, he brought that transformation into medicine.

Thinking of Wiener, this book should be read in two ways. First, considering the specific definitions, theorems, and equations, it discusses a particular dynamical model for gene regulatory networks—probabilistic Boolean networks (PBNs). It covers basic model properties, inference of model parameters from data, and intervention in the model to increase the likelihood of the network being in desirable states. Taking a wider perspective, one can view the PBN model as a vehicle in which to elucidate the therapeutic goals of translational genomics. The PBN model is rather general and includes both deterministic functional aspects and probabilistic characteristics inherent to the modeling of complex systems. Therefore, it is well suited to serve as a mathematical framework to study basic issues dealing with systems-based genomics, specifically, the relevant aspects of stochastic, nonlinear dynamical systems. These include long-run dynamical properties and how these correspond to therapeutic goals, the effect of complexity on model inference and the resulting consequences of model uncertainty, altering network dynamics via structural intervention, such as perturbing gene logic, optimal control of regulatory networks over time, limitations imposed on the ability to achieve optimal control owing to model complexity, and the effects of asynchronicity.

We do not know what models will ultimately be adopted for specific applications, but we do know that basic translational issues considered in this book will have to be confronted no matter what model is used. In translational science, we are faced with the problem of controlling complex systems in an environment of uncertainty—just the framework in which Wiener pioneered. We need to understand how the inevitable issues manifest themselves in the modeling and control of gene regulatory networks so that therapeutic control strategies can be derived and, even prior to that, so that appropriate experiments can be carried out to gain sufficient modeling information.

We have tried to unify the different strands of research that have been pursued over the last eight years and continue to be pursued. Issues such as inference, network reduction, constrained control, greedy control, and asynchronicity are just beginning to be studied. Moreover, how these are resolved in practice will depend on close interaction between biologists, physicians, mathematicians, and engineers. Only in that way will the phenomena, medical outcome, and model be brought into a coherent whole to fulfill the promise of translational science. We hope that this book provides a step in that direction.

[1] Rosenblueth, A., and N. Wiener, "The role of models in science." *Philosophy of Science,* 12, 316-321, 1945.

[2] Dougherty, E. R., "Translational science: epistemology and the investigative process," *Current Genomics*, 10 (2), 102-109, 2009.

Acknowledgments

The authors extend their appreciation to the numerous colleagues and students who contributed to developments discussed in this book. Although there are too many to mention explicitly, we would like to at least mention the names of several whose contributions play key roles in the present book: Michael M. Bittner, Marcel Brun, Aniruddha Datta, Babak Faryabi, Ivan Ivanov, Seungchan Kim, Peter Krawitz, Harri Lähdesmäki, Steve Marshall, Ranadip Pal, Xiaoning Qian, Golnaz Vahedi, Yufei Xiao, and Wei Zhang. We would also like to acknowledge those organizations that have supported the research behind this book: National Science Foundation, National Human Genome Research Institute, National Institute of General Medical Sciences, National Cancer Institute, Translational Genomics Research Institute, Institute for Systems Biology, W. M. Keck Foundation, University of Texas M. D. Anderson Cancer Center, and Texas A&M University. We thank Babak Faryabi and Xiaoning Qian for proofreading the manuscript and making numerous valuable suggestions. We are also grateful to our editor at SIAM, Elizabeth Greenspan, for her constant encouragement, patience, and support. IS would like to extend his sincere thanks to The Helen Riaboff Whiteley Center at the University of Washington Friday Harbor Laboratories, which provided a superbly peaceful and productive environment to work on this book. IS also thanks his family, Andrei, Anna, and Janette, for their continued love and support. Finally, let us note that we look back fondly at our days together with Wei Zhang in the Cancer Genomics Core Laboratory of the M. D. Anderson Cancer Center, where the basic ideas behind PBNs, as well as the more general mathematical concept of intervention in gene regulatory networks as a translational model for therapy, were germinated.

Chapter 1

Boolean Networks

Boolean networks are a class of discrete dynamical systems that can be characterized by the interactions over a set of Boolean variables. Random Boolean networks (RBNs), which are ensembles of random network structures, were first introduced by Stuart Kauffman in 1969 as a simple model class for studying dynamical properties of gene regulatory networks at a time when the structure of such networks was largely unknown. The idea behind such an approach is to define an ensemble of Boolean networks such that it fulfills certain known features of biological networks and then study random instances of these networks to learn more about general properties of such networks. This iterative and highly interdisciplinary process depends, on the one hand, on the input of new biological findings and might guide, on the other hand, the research of the experimentalist. For example, the now classical Kauffman ensembles in which every gene takes its inputs from exactly 2 other genes turned out to be "critical," lying at a phase transition between an ordered and a chaotic phase, in regard to the propagation of perturbations, and contributed to the hypothesis that living systems may be poised at the "edge of chaos"—a hypothesis for which experimental evidence has been mounting (Ramö *et al.*, 2006; Shmulevich *et al.*, 2005; Nykter *et al.*, 2008).

There have been hundreds of publications, primarily in the physics, mathematics, and theoretical biology literature, that have focused on the ensemble approach for studying genetic networks in which the center of attention lies not in the entities comprising a system but qualities possessed by the system as a whole—much like the concept of entropy in physics. We will not attempt to review or even summarize the numerous results and insights gained from such studies as the vast amount of material deserves an entire book on this subject, and we limit ourselves only to mentioning that such approaches have shed light on fundamental principles of complex living systems at all scales of organization, including adaptability and evolvability, robustness, coordination of complex behaviors, storage of information, and the relationships between the structure of such complex systems and their dynamical behavior. Further, we refer the reader to several excellent review articles that cover the ensemble properties of Boolean networks (Aldana *et al.*, 2002; Drossel, 2007), as well as Kauffman's book (Kauffman, 1993), which covers a much broader range of topics but contains much material, albeit somewhat dated, on Boolean networks. Our aim in this chapter is to give only the necessary definitions and concepts pertaining to Boolean networks to the extent that they are necessary for studying probabilistic Boolean networks. We will also review several examples of genetic networks using the Boolean network formalism.

Consider a directed graph where the vertices represent genes and the directed edges represent the actions of genes, or rather their products, on other genes. For example, directed edges from genes A and B into gene C indicate that A and B jointly act on C. The specific mechanism of action is not represented in the graph structure itself, so an additional representation is necessary. One of the simplest representation frameworks assumes that genes are binary-valued entities, indicating that they can be in one of two possible states of activity (e.g., ON or OFF) at any given point in time, and that they act on each other by means of rules represented by Boolean functions. Thus, in the above example, gene C is determined by the output of a Boolean function whose inputs are A and B. The underlying directed graph merely represents the input-output relationships. Let us now present this idea more formally.

A *Boolean network*, originally introduced by Kauffman (1969a, 1969b, 1974, 1993) is defined by a set of nodes (genes) $\{x_1, \ldots, x_n\}$ and a list of Boolean functions $\{f_1, f_2, \ldots, f_n\}$. Each gene $x_i \in \{0,1\}$ $(i = 1, \ldots, n)$ is a binary variable whose value at time $t+1$ is completely determined by the values of genes $x_{j_1}, x_{j_2}, \ldots, x_{j_{k_i}}$ at time t by means of a Boolean function $f_i : \{0,1\}^{k_i} \to \{0,1\}$. That is, there are k_i regulatory genes assigned to gene x_i that determine the "wiring" of that gene. Thus, we can write

$$x_i(t+1) = f_i(x_{j_1}(t), x_{j_2}(t), \ldots, x_{j_{k_i}}(t)). \tag{1.1}$$

In a *random Boolean network*, the functions f_i, sometimes called couplings (Aldana *et al.*, 2002), are selected randomly as are the genes that are used as their inputs. This is the basis of the ensemble approach that we mentioned above.

Each x_i represents the state (expression) of gene i, where $x_i = 1$ represents the fact that gene i is expressed and $x_i = 0$ means it is not expressed. Such a seemingly crude simplification of gene expression has ample justification in the experimental literature (Bornholdt, 2008). Indeed, consider the fact that many organisms exhibit an amazing determinism of gene activity under specific experimental contexts or conditions, such as *E. coli* under temperature change (Richmond *et al.*, 1999). The determinism is apparent despite the prevalent molecular stochasticity and experimental noise inherent to measurement technologies such as microarrays. Furthermore, accurate mathematical models of gene regulation that capture kinetic-level details of molecular reactions frequently operate with expressed molecular concentrations spanning several orders of magnitude, either in a saturation regime or in a regime of insignificantly small concentrations, with rapid switch-like transitions between such regimes (Davidich and Bornholdt, 2008a). Further, even higher organisms, which are necessarily more complex in terms of genetic regulation and heterogeneity, exhibit remarkable consistency when gene expression is quantized into two levels; for example, different subtypes of human tumors can be reliably discriminated in the binary domain (Shmulevich and Zhang, 2002).

In a Boolean network, a given gene transforms its inputs (regulatory factors that bind to it) into an output, which is the state or expression of the gene itself at the next time point. All genes are assumed to update synchronously in accordance with the functions assigned to them, and this process is then repeated. The artificial synchrony simplifies computation while preserving the qualitative, generic properties of global network dynamics (Kauffman, 1993). Although the synchronous updating scheme has been known to be biologically unrealistic, it has been applied in most analytical studies so far, as it is the only one that yields deterministic state transitions. Under asynchronous updating, many attractors of the synchronous scheme become unstable and decay into other dynamically

Table 1.1. *Truth tables of the functions in a Boolean network with 5 genes. The indices j_1, j_2, and j_3 indicate the input connections for each of the functions.*

	f_1	f_2	f_3	f_4	f_5
	0	0	0	0	0
	1	1	1	1	0
	1	1	1	-	0
	1	0	0	-	0
	0	0	1	-	0
	1	1	1	-	0
	1	1	0	-	0
	1	1	1	-	1
j_1	5	3	3	4	5
j_2	2	5	1	-	4
j_3	4	4	5	-	1

stable cycles. However, the very definition of an attractor, except for fixed points, remains difficult. Further, asynchronous updating is not *per se* biologically more realistic and has to be motivated carefully in every case not to fall victim to artifacts (Chaves *et al.*, 2005). Additionally, recent research indicates that some molecular control networks are so robustly designed that timing is not a critical factor (Braunewell and Bornholdt, 2006), that time ordering in the emergence of cell-fate patterns is not an artifact of synchronous updating in the Boolean model (Alvarez-Buylla *et al.*, 2008), and that simplified synchronous models are able to reliably reproduce the sequence of states in biological systems.

It is clear that the dynamics of a synchronous Boolean network are completely determined by Eq. (1.1)—a determinism from which we will depart upon introducing probabilistic Boolean networks. Let us give an example.

Consider a Boolean network consisting of 5 genes $\{x_1,\ldots,x_5\}$ with the corresponding Boolean functions given by the truth tables shown in Table 1.1. The *maximum connectivity* $K = \max_i k_i$ is equal to 3 in this case. Note that $x_4(t+1) = f_4(x_4(t))$ is a function of only one variable and is an example of autoregulation.

The dynamics of this Boolean network are shown in Fig. 1.1. Since there are 5 genes, there are $2^5 = 32$ possible states that the network can be in. Each state is represented by a circle and the arrows between states show the transitions of the network according to the functions in Table 1.1. It is easy to see that because of the inherent deterministic directionality in Boolean networks as well as only a finite number of possible states, certain states will be revisited infinitely often if, depending on the initial starting state, the network happens to transition into them. Such states are called *attractors* and the states that lead into them, including the attractors themselves, comprise their *basins of attraction*. For example, in Fig. 1.1, the state (00000) is an attractor and together with the 7 other (transient) states that eventually lead into it comprise its basin of attraction.

The attractors represent the *fixed points* of the dynamical system, thus capturing the system's long-term behavior. The attractors are always cyclical and may consist of more than one state. Starting from any state on an attractor, the number of transitions necessary for the system to return to it is called the *cycle length*. For example, the attractor (00000) has cycle length 1 while the states (11010) and (11110) comprise an attractor of length 2.

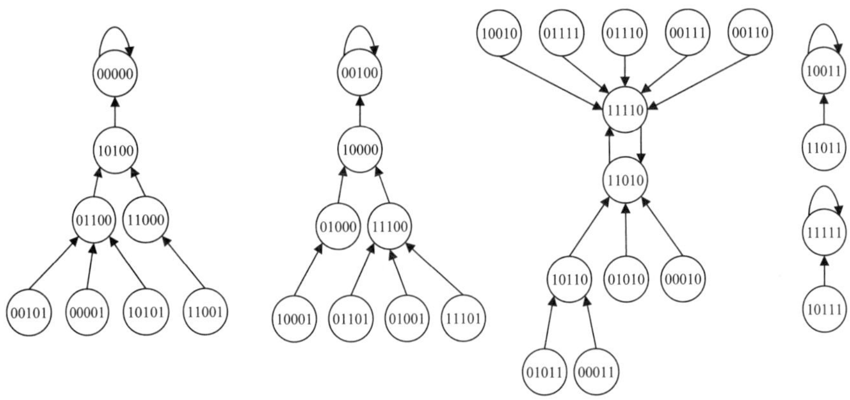

Figure 1.1. *The state transition diagram for the Boolean network defined in Table* 1.1 *(Shmulevich et al.,* 2002d*).*

Real genetic regulatory networks are highly stable in the presence of perturbations to the genes. Within the Boolean network formalism, this means that when a minimal number of genes transiently change value (say, by means of some external stimulus), the system typically transitions into states that reside in the same basin of attraction from which the network eventually flows back to the same attractor. Generally speaking, large basins of attraction correspond to higher stability. Such stability of networks in living organisms allows the cells to maintain their functional state within their environment.

Although in developmental biology, epigenetic, heritable changes in cell determination have been well established, it is now becoming evident that the same type of mechanisms may be also responsible in carcinogenesis and that gene expression patterns can be inherited without the need for mutational changes in DNA (MacLeod, 1996). In the Boolean network framework, this can be explained by so-called *hysteresis*, which is a change in the system's state caused by a stimulus that is not changed back when the stimulus is withdrawn (Huang, 1999). Thus, if the change of some particular gene does in fact cause a transition to a different attractor, the network will often remain in the new attractor even if that gene is switched off. Thus, the attractors of a Boolean network also represent a type of memory of the dynamical system (Huang, 1999). Indeed, the structure of the state space of a Boolean network, in which every state in a basin of attraction is associated with the corresponding attractor to which the system will ultimately flow, represents a type of associative memory. Interestingly, the amount of information that such a system is able to store or, equivalently, the complexity of the classification process that it implements (i.e., every "input" state in which the system may be placed gets associated with the corresponding "output" attractor), grows only with system size (number of nodes) when the system is poised at a critical phase transition between an ordered and a disordered phase (in the sense that a perturbation to one node gets propagated to only one other node on average), suggesting that critical systems are informationally optimal (Krawitz and Shmulevich, 2007). We will revisit criticality when we discuss the notions of influence of a gene on other genes and the sensitivity of Boolean functions to such one-gene perturbations. Further, when we discuss model inference in Chapter 3, it will turn out that the assumption of criticality can be statistically advantageous when learning a network model from data.

1.1 Cell Types and Cellular Functional States

Boolean networks reflect the nature of complex adaptive systems in that they are "systems composed of interacting agents described in terms of rules" (Holland, 1995). A central concept in dynamical systems is that of *structural stability*, which is the persistent behavior of a system under perturbation. Structural stability formally captures the idea of behavior that is not destroyed by small changes to the system. This is most certainly a property of real genetic networks, since the cell must be able to maintain homeostasis in metabolism or its developmental program in the face of external perturbations and stimuli. Boolean networks naturally capture this phenomenon as the system usually "flows" back into the attractors when some of the genes are perturbed. Real gene regulatory networks exhibit spontaneous emergence of ordered collective behavior of gene activity. Moreover, recent findings provide experimental evidence for the existence of attractors in real regulatory networks (Huang and Ingber, 2000; Huang *et al.*, 2005; Chang *et al.*, 2008). At the same time, many studies have shown (e.g., Wolf and Eeckman,1998) that dynamical system behavior and stability of equilibria can be largely determined from regulatory element organization. This suggests that there must exist certain generic features of regulatory networks that are responsible for the inherent robustness and stability. In addition, since there are many different cell types in multicellular organisms, despite the fact that each cell contains exactly the same DNA content, the cellular "fate" is determined by which genes are expressed.

This was the insight pursued by Kauffman in his pioneering studies of genetic regulatory networks (Kauffman, 1993). As explained above, the idea was to generate RBNs with certain properties and then systematically study the effects of these properties on the global dynamical behavior of the networks. For example, RBNs were studied with varying average connectivity, bias, which is the probability that a function outputs the value 1, and different classes of Boolean functions, such as canalizing functions. A Boolean function $f:\{0,1\}^n \to \{0,1\}$ is called *canalizing* in its ith input if there exist y and z such that for all $(x_1,x_2,\ldots,x_n) \in \{0,1\}^n$ with $x_i = y$, we have $f(x_1,x_2,\ldots,x_n) = z$. We will return to these questions shortly.

Kauffman's intuition was that the attractors in the Boolean networks should correspond to cellular types. This interpretation is quite reasonable if cell types are characterized by stable recurrent patterns of gene expression (Jacob and Monod, 1961), and today a gene expression profile of a cell is often regarded as a sufficient definition of its state.

Another interpretation of the attractors in Boolean networks is that they correspond to cellular states, such as *proliferation* (cell cycle), *apoptosis* (programmed cell death), and *differentiation* (execution of tissue-specific tasks). This highly appealing view was expounded by Huang (1999, 2001) with substantial biological justification. Such an interpretation can provide new insights into cellular homeostasis and cancer progression, the latter being characterized by a disbalance between these cellular states. For instance, if a (structural) mutation occurs, resulting in a reduction of the probability of the network entering the apoptosis attractor(s), then the cells will be less likely to undergo apoptosis and will exhibit uncontrolled growth. Similarly, an enlargement of the basins of attraction for the proliferation attractor would hyperstabilize it and result in hyperproliferation, typical of tumorigenesis. Such an interpretation need not be at odds with the interpretation that attractors represent cellular types. To the contrary, these views are complementary to each other, since for a given cell type, different cellular functional states must exist and be determined by the collective behavior of gene activity. Thus, one cell type can comprise several "neighboring" attractors each of which corresponds to different cellular functional states.

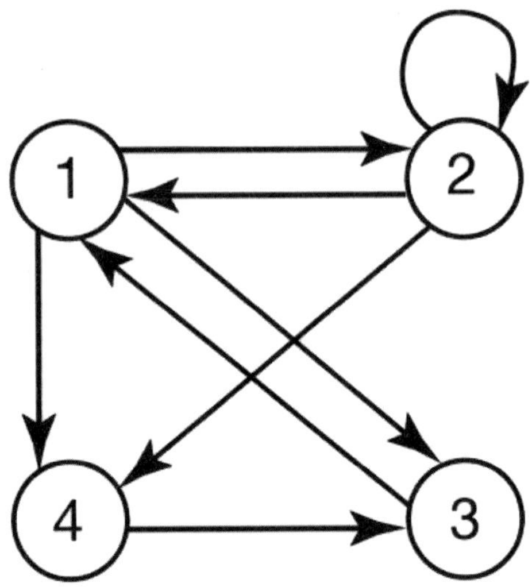

Figure 1.2. *The connectivity graph of a Boolean network of four nodes.*

1.2 Relevant Nodes

Given a particular starting state, the attractor to which the system will eventually flow starting from this state will be determined entirely by so-called *relevant nodes*. Some nodes of a network will always take on the same value regardless of the attractor on which the network finally settles down. These nodes establish the *frozen core* of a network. They are either updated by a constant Boolean function and are *frozen* to a fixed value right from the beginning or they freeze as the system transitions through its state space, in which case they are called *clamped*. Other nodes may have different values on different attractors, but their values are of no importance for determining the attractor to which the system will flow. These nodes are therefore called *nonfrozen* but *irrelevant*. Finally, there are the nonfrozen *relevant* nodes that can be recursively defined as influencing other relevant nodes. Relevant nodes ultimately have an effect on their own state. The concept of classifying the nodes of a network into clamped, irrelevant, and relevant, regarding their importance for the attractor dynamics, is simple yet powerful for the understanding of Boolean network dynamics (Bastolla and Parisi, 1998a,b; Socolar and Kauffman, 2003; Drossel, 2007).

The example network, which is illustrated in Fig. 1.2 and whose connectivity matrix, Boolean functions, and transition matrix are given in Tables 1.2–1.4, has two attractors: a fixed point, $(x_1x_2x_3x_4) = (1101)$, and an attractor of length two, $(0101) \rightleftarrows (1011)$. Only nodes 1 and 2 are relevant, whereas node 3 is nonfrozen but irrelevant and node 4 is a clamped node that freezes to the value $x_4 = 1$ as soon as the system reaches an attractor.

Algorithmically, relevant nodes can often be identified by an iterative procedure: First, links of frozen nodes to other nodes are cut, as this kind of input is not *variable* but *constant*. Second, so-called *degenerate* links have to be identified and removed from the connectivity matrix. In the example network, a closer look at the Boolean function of node 1 reveals that it actually depends only on its first variable, which is node 2. Therefore link $3 \to 1$ can be removed from the connectivity matrix. Third, nodes that do not influence other nodes are irrelevant and can be removed from the network.

Table 1.2. *Connectivity matrix of the example network.*

$x_i \backslash i$	1	2	3	4
x_{i_1}	2	1	1	1
x_{i_2}	3	2	4	2

Table 1.3. *Boolean functions of the example network.*

x_{i_1}	x_{i_2}	f_1	f_2	f_3	f_4
0	0	0	1	0	0
0	1	0	0	1	1
1	0	1	1	1	1
1	1	1	1	0	1

Table 1.4. *Transition matrix of the example network. The attractors are shown in bold.*

$\mathbf{x}$				$F(\mathbf{x})$			
x_1	x_2	x_3	x_4	f_1	f_2	f_3	f_4
0	0	0	0	0	1	0	0
0	0	0	1	0	1	1	0
0	0	1	0	0	1	0	0
0	0	1	1	0	1	1	0
0	1	0	0	1	0	0	1
0	**1**	**0**	**1**	**1**	**0**	**1**	**1**
0	1	1	0	1	0	0	1
0	1	1	1	1	0	1	1
1	0	0	0	0	1	1	1
1	0	0	1	0	1	0	1
1	0	1	0	0	1	1	1
1	**0**	**1**	**1**	**0**	**1**	**0**	**1**
1	1	0	0	1	1	1	1
1	**1**	**0**	**1**	**1**	**1**	**0**	**1**
1	1	1	0	1	1	1	1
1	1	1	1	1	1	0	1

This procedure is repeated until no further nodes can be removed from the network. In our example network, node 3 can be removed after cutting link $3 \to 1$. In the second round, node 4 can be removed, which leaves us with the two relevant nodes, 1 and 2. Some nodes identified by this method, however, are actually not relevant, but *freeze* to a constant value, as the following example illustrates: $f_1 = x_1 \vee x_2$, $f_2 = x_1 \vee \overline{x_2}$, where the symbol $\vee$ indicates disjunction and the overbar indicates negation. In summary, relevant nodes are the only ones that are important for determining the attractor structure of a Boolean network.

1.3 Network Properties and Dynamics

We now establish several concepts and give the necessary definitions that will be used in the remainder of the book. Although, as mentioned, we do not plan to discuss the ensemble approach, we will find it useful in some cases to make use of random Boolean functions when defining certain parameters. An important parameter characterizing a random function f_i is the *bias* p, which is the probability that the function takes on the value 1. A K-input random Boolean function with bias p can be generated by flipping a p-biased coin 2^K times and thereby filling in its truth table. If $p = 0.5$, then the function is said to be *unbiased*. We can also refer to an entire RBN as being p-biased or unbiased, referring to the corresponding property of each of its randomly generated functions.

Let us consider the "wiring" (input-output relationships between genes) in a Boolean network in terms of the corresponding directed graph. The in-degree of a gene is the number of inputs to the Boolean function governing its behavior. The simplest scenario is when each gene has the same in-degree K, implying that the in-degree distribution (probability mass function) has all of its mass at K and zero elsewhere (a delta function). This is the original definition, proposed by Kauffman (1969a). Other distributions are also possible, such as the power-law ("scale-free" distribution, which may be more reflective of the heterogeneous nature of biological systems (Oosawa and Savageau, 2002; Aldana and Cluzel, 2003). We examine only the simplest case where $k_i = K$ for all $i = 1,\ldots,n$ and refer the reader to Shmulevich and Dougherty (2007) for a more general treatment. Using only the bias p and the in-degree K of a random Boolean function, we will show analytically how the tendency of a Boolean function to propagate perturbations to one of its input variables can be determined and, following from that, how sensitive a Boolean network is to small perturbations to its state.

Generally, a dynamical system is said to behave in an *ordered* or *robust* manner when similar initial conditions will lead to similar trajectories in the phase (state) space. A system is called *chaotic* when even infinitesimal differences will cause an increasing divergence in the ensuing trajectories. In Boolean networks, the difference between two states $\mathbf{x}$ and $\mathbf{x}'$ is usually quantified in terms of the normalized Hamming distance η:

$$\eta(\mathbf{x},\mathbf{x}') = \frac{1}{n}\sum_{i=1}^{n} \left|x_i - x_i'\right|. \tag{1.2}$$

With this metric, the evolution of the overlap between two random states, $a(t) = 1 - \eta(\mathbf{x}(t),\mathbf{x}'(t))$, can be measured over time.

In the second half of the 1980s, Derrida, Flyvbjerg, and others began to study the propagation of perturbations in Boolean networks with a mean field approach that had already successfully been applied in the theory of spin glasses (Derrida and Pomeau, 1986; Flyvbjerg, 1988). Instead of considering the precise effects of all nodes on their linked nodes in a certain network instance, the mean field approach deals only with the average effect on a random node. If the overlap between two states, $\mathbf{x}$ and $\mathbf{x}'$, is $a(t)$ at a given time t, the probability that a random node i will receive the same input signal in the next step is $a(t)^K$. If the Boolean function f_i assigned to the node is generated with bias p, then the probability that two random input vectors yield the same function value is $P = p^2 + (1-p)^2$. The expected overlap of the two states after one time step is given by

$$a(t+1) = a(t)^K + P(1-a(t))^K. \tag{1.3}$$

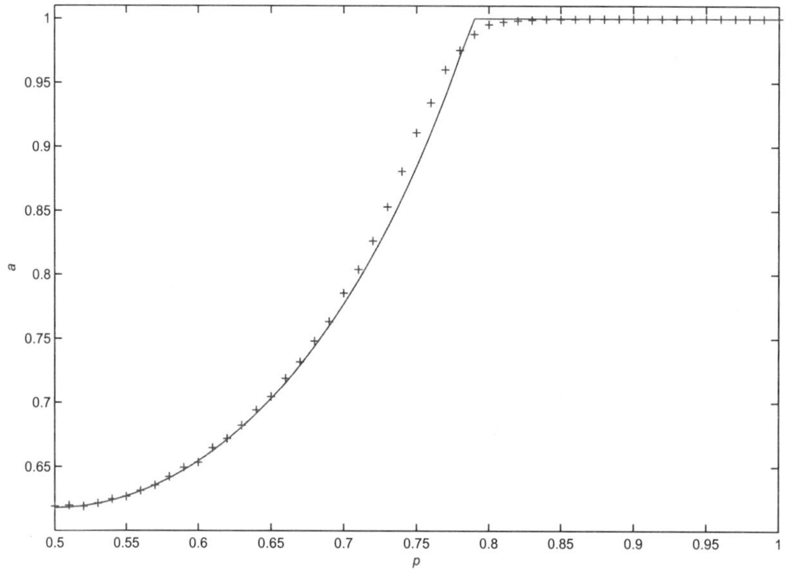

Figure 1.3. *The averaged overlap a, of two random states* $\mathbf{x}, \mathbf{x}'$ *with* $\eta(\mathbf{x}, \mathbf{x}') = 0.01$ *in network ensembles with* $n = 1000$, $K = 3$ *and* $0.5 \le p \le 1$, *after* 200 *time steps is indicated by* (+). *The solid line shows the annealed approximation (Eq.* (1.4)*).*

In the limit of t and $n \to \infty$, $a(t)$ asymptotically approaches the fixed point a^*, which obeys

$$a^* = 1 - (1 - (a^*)^K)2p(1-p). \tag{1.4}$$

The critical transition curve between the ordered phase, where the overlap approaches $a^* = 1$, and a chaotic phase, where the overlap approaches a value $a^* < 1$, is given by the relation

$$2Kp(1-p) = 1. \tag{1.5}$$

Figure 1.3 shows the averaged overlap a (+) of two random states of Hamming distance $\eta(\mathbf{x}, \mathbf{x}') = 0.01$ after 200 time steps in network ensembles of $n = 1000$ nodes and connectivity $K = 3$ as a function of the bias p, as well as the results of the annealed approximation (solid line, Eq. (1.4)). In ensembles with $p \ge 0.7887$ the initial perturbation dies out and the overlap approaches $a = 1$. Indeed, the value $p = 0.7887$ is the solution of the critical phase transition curve in Eq. (1.5) when $K = 3$.

We are now ready to introduce several key definitions that are useful for characterizing the strength of interactions among the variables in a Boolean network and relate these to the critical phase transition curve discussed above. In a Boolean function, some variables have a greater influence over the output of the function than other variables. To formalize this concept, let $f : \{0,1\}^K \to \{0,1\}$ be a Boolean function of K variables $x_1, \ldots, x_K$. Let

$$\frac{\partial f(\mathbf{x})}{\partial x_j} = f\left(\mathbf{x}^{(j,0)}\right) \oplus f\left(\mathbf{x}^{(j,1)}\right)$$

be the partial derivative of f with respect to x_j, where

$$\mathbf{x}^{(j,k)} = \left(x_1,\ldots,x_{j-1},k,x_{j+1},\ldots,x_K\right), \quad k = 0,1.$$

Clearly, the partial derivative itself is a Boolean function that specifies whether a change in the jth input causes a change in the original function f.

Now, the activity of variable x_j in function f is defined as

$$\alpha_j^f = \frac{1}{2^K} \sum_{\mathbf{x}\in\{0,1\}^K} \partial f(\mathbf{x})/\partial x_j. \tag{1.6}$$

Note that although the vector $\mathbf{x}$ consists of K components (variables), the jth variable is fictitious in $\partial f(\mathbf{x})/\partial x_j$. Recall that a variable x_j is fictitious in f if $f(\mathbf{x}^{(j,0)}) = f(\mathbf{x}^{(j,1)})$ for all $\mathbf{x}^{(j,0)}$ and $\mathbf{x}^{(j,1)}$. For a K-variable Boolean function f, we can form its activity vector $\alpha^f = [\alpha_1^f,\ldots,\alpha_K^f]$. It is easy to see that $0 \le \alpha_j^f \le 1$ for any $j = 1,\ldots,K$. In fact, we can consider α_j^f to be a probability that toggling the jth input bit changes the function value when the input vectors $\mathbf{x}$ are distributed uniformly over $\{0,1\}^K$. Since we are in the binary setting, the activity is also the expectation of the partial derivative with respect to the uniform distribution: $\alpha_j^f = E[\partial f(\mathbf{x})/\partial x_j]$. Under an arbitrary (not necessarily uniform) distribution over $\{0,1\}^K$, α_j^f is sometimes referred to as the *influence* of variable x_j on the function f (Kahn *et al.*, 1988). We note in passing that the influence of variables has a spectral interpretation in the context of harmonic analysis on the hypercube and is a powerful tool in computational learning theory and circuit complexity.

Another important quantity is the *sensitivity* of a Boolean function f, which measures how sensitive the output of the function is to changes in the inputs (this concept was introduced by Cook *et al.* (1986) under the name of critical complexity). The sensitivity $s^f(\mathbf{x})$ of f on vector $\mathbf{x}$ is defined as the number of Hamming neighbors of $\mathbf{x}$ on which the function value is different than on $\mathbf{x}$ (two vectors are Hamming neighbors if they differ in only one component; that is, the Hamming distance between them is equal to 1). That is,

$$\begin{aligned} s^f(\mathbf{x}) &= |\{i \in \{1,\ldots,K\} : f(\mathbf{x}\oplus e_i) \neq f(\mathbf{x})\}| \\ &= \sum_{i=1}^{K} 1\left[f(\mathbf{x}\oplus e_i) \neq f(\mathbf{x})\right], \end{aligned}$$

where e_i is the unit vector with 1 in the ith position and 0's everywhere else and $1[A]$ is an indicator function that is equal to 1 if and only if A is true. The *average sensitivity* s^f is defined by taking the expectation of $s^f(\mathbf{x})$ with respect to the distribution of $\mathbf{x}$. Under the uniform distribution, the average sensitivity is equal to the sum of the activities:

$$\begin{aligned} s^f = E\left[s^f(\mathbf{x})\right] &= \sum_{i=1}^{K} E\left[1\left[f(\mathbf{x}\oplus e_i) \neq f(\mathbf{x})\right]\right] \\ &= \sum_{i=1}^{K} \alpha_i^f. \end{aligned} \tag{1.7}$$

Therefore, s^f is a number between 0 and K and tells us how likely a function value is to change under a random change in one of its inputs.

Intuition tells us that highly biased functions (i.e., p far away from 0.5) should have low average sensitivity, since such a function with K inputs would be unlikely to change much between neighboring vectors on the K-dimensional hypercube. Indeed, consider a random Boolean function with bias p. The truth table of such a function is a 2^K-length vector of independent and identically distributed (i.i.d.) Bernoulli(p) random variables. Therefore, the probability that two Hamming neighbors are different is equal to $2p(1-p)$, since one can be 1 (with probability p) and the other 0 (with probability $1-p$), and vice versa. Since this is the same for all Hamming neighbors, all expected activities should be equal. That is, for each $i = 1, \ldots, K$,

$$E[\alpha_i^f] = 2p(1-p),$$

where the expectation is taken with respect to the distribution of the truth table of the function f. Thus, using Eq. (1.7), the expectation of the average sensitivity is

$$\begin{aligned} E\left[s^f\right] &= \sum_{i=1}^{K} E[\alpha_i^f] = \sum_{i=1}^{K} 2p(1-p) \\ &= K2p(1-p). \end{aligned} \tag{1.8}$$

This confirms our intuition about the relationship between bias and (expected) average sensitivity. Note that an unbiased function ($p = 0.5$) has (maximum) expected average sensitivity equal to $K/2$.

Let us now consider a RBN network with n genes with connectivity K and bias p. Two states that are Hamming neighbors should, on average, be succeeded at the next time moment by two states that are separated by a Hamming distance equal to $K2p(1-p)$, by the definition of average sensitivity and Eq. (1.8). If $K2p(1-p) > 1$, meaning that minimal perturbations are amplified, then the system is chaotic, and if $K2p(1-p) < 1$, when minimal perturbations are attenuated, then the system is ordered. Solving this equation for K at the phase transition, we obtain the critical connectivity K_c in terms of the bias p:

$$K_c = [2p(1-p)]^{-1}, \tag{1.9}$$

which is identical to Eq. (1.5). Similarly, we can solve for the critical bias p_c in terms of K. The critical curve given in Eq. (1.9) is shown in Fig. 1.4. All values of p and K that are above the critical curve yield chaotic network behavior, while values below the curve result in ordered behavior. Equation (1.9) was derived by Derrida and Pomeau (1986) using the so-called *annealed* approximation, where at each time step, a new set of Boolean functions is selected randomly.[1] Other approaches based on damage spreading (Luque and Solé, 1997), Lyapunov exponents (Luque and Solé, 2000) so-called stable core (Flyvbjerg, 1988), and others have been used. The reader can consult Aldana *et al.* (2002) for a review.

It is an appealing feature of the sensitivity parameter that it is entirely determined by the Boolean functions in a given network (Shmulevich and Kauffman, 2004). The sensitivity of a Boolean model of a biological network and the activities of its nodes are easy to

[1] This is in contrast to the *quenched* model, where the functions remain constant over time.

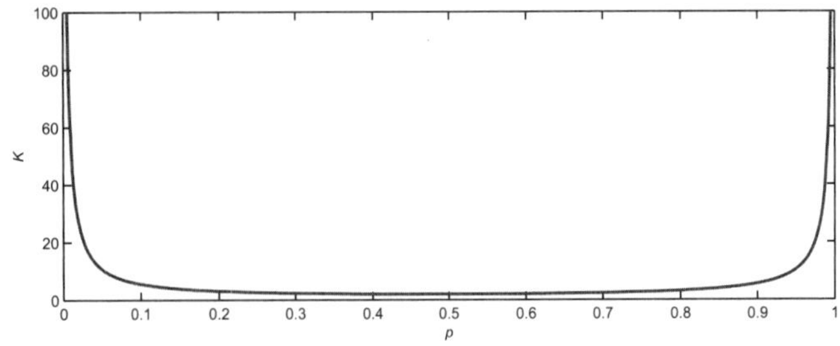

Figure 1.4. *The critical curve relating the bias p to the connectivity K in a RBN. All values of p and K above the curve result in chaotic network behavior, whereas all values below the curve yield ordered behavior.*

compute and may help conceptualize the dynamic characteristics of its elements. As such, it can be a very useful tool for analytically determining the effect of particular function structures, representing rules of genetic regulatory interactions, on the dynamics of the network. Let us give one example pertaining to canalizing functions, showing how the average sensitivity can be used to gain insights into how the abundance of such functions in a network affects the propagation of perturbations throughout the network.

Recall that a canalizing function is a type of Boolean function in which at least one of the input variables, called a canalizing variable, is able to determine the function output regardless of the values of the other variables. For example, the function $f(x_1,x_2,x_3) = x_1 \vee (x_2 \wedge x_3)$, where the symbols $\vee$ and $\wedge$ denote the Boolean disjunction and conjunction, respectively, is a canalizing function, since setting x_1 to 1 guarantees that the function value is 1 regardless of the value of x_2 or x_3. On the other hand, the exclusive OR function $f(x_1,x_2) = x_1 \oplus x_2$ is not a canalizing function, since the values of both variables always need to be known in order to determine the function output.

It has been known for quite some time that canalizing functions play a role in preventing chaotic behavior (Kauffman, 1969a; Kauffman, 1993; Stauffer, 1987; Lynch, 1995). By increasing the percentage of canalizing functions in a Boolean network, one can move closer toward the ordered regime and, depending on the connectivity and the distribution of the number of canalizing variables, cross the phase transition boundary (Kauffman, 2000; Kauffman *et al.*, 2004). In fact, there is strong evidence that canalizing functions are abundantly utilized in higher vertebrate gene regulatory systems (Kauffman, 1993). A recent large-scale study of the literature on transcriptional regulation in eukaryotes demonstrated a high bias towards canalizing rules (Harris *et al.*, 2002).

Let us look more closely at the reasons why canalizing functions promote ordered behavior, making use of the average sensitivity. We would thus expect that canalizing functions should have lower average sensitivity than noncanalizing functions. Moreover, we would expect that for canalizing functions, the importance of canalizing variables is higher than that of noncanalizing variables, as measured by their activities. The following discussion, adapted from Shmulevich and Kauffman (2004), illustrates this.

Recall that a function $f : \{0,1\}^K \to \{0,1\}$ is said to be canalizing if there exist an $i \in \{1,\ldots,K\}$ and $u,v \in \{0,1\}$ such that for all $x_1,\ldots,x_K \in \{0,1\}$, if $x_i = u$, then $f(x_1,\ldots,x_K) = v$. The input variable x_i is called the canalizing variable with canalizing value u and canalized value v. A canalizing function f can be written either as

$$f(x_1,\ldots,x_K) = x_i^q \vee g(x_1,\ldots,x_{i-1},x_{i+1},\ldots,x_K)$$

or

$$f(x_1,\ldots,x_K) = x_i^q \wedge g(x_1,\ldots,x_{i-1},x_{i+1},\ldots,x_K),$$

where $q \in \{0,1\}$. Here, $x_i^1 = x_i$ and $x_i^0 = x_i'$, where x_i' is the complement or negation of x_i. Also, recall that $\wedge$ takes precedence over $\vee$ so that writing $a \vee b \wedge c$ is the same as $a \vee (b \wedge c)$.

Let $f(x_1,\ldots,x_K)$ be a random canalizing function of the form

$$f(x_1,\ldots,x_K) = x_1 \vee g(x_2,\ldots,x_K),$$

where g is chosen randomly from the set of all $2^{2^{K-1}}$ Boolean functions. Without loss of generality, we are supposing that the first variable, x_1, is a canalizing variable. Furthermore, the discussion for other types of canalizing functions (e.g., $f(x_1,\ldots,x_K) = x_1 \wedge g(x_2,\ldots,x_K)$) would be nearly identical. Our first aim is to characterize the activities of each of the variables. The activities themselves are also random variables by virtue of f being random. It is clear that the activity of variables $x_2,\ldots,x_K$ should behave identically in the probabilistic sense if $g(x_2,\ldots,x_K)$ is a random unbiased function. Consequently, it will suffice to examine the activity of variable x_2, with the other variables behaving identically.

Let us first compute α_1^f – the activity of x_1 in f. First, we have

$$\begin{aligned}\frac{\partial f}{\partial x_1} &= (0 \vee g(x_2,\ldots,x_K)) \oplus (1 \vee g(x_2,\ldots,x_K)) \\ &= g(x_2,\ldots,x_K) \oplus 1 \\ &= g'(x_2,\ldots,x_K).\end{aligned}$$

Now, since g is a random unbiased function (i.e., $p = 1/2$), the expected activity of the canalizing variable x_1 is equal to

$$\begin{aligned}E\left[\alpha_1^f\right] &= E\left[2^{-(K-1)} \cdot \sum_{\mathbf{x}\in\{0,1\}^{K-1}} g'(x_2,\ldots,x_K)\right] \\ &= 2^{-(K-1)} \cdot \sum_{\mathbf{x}\in\{0,1\}^{K-1}} E\left[g'(x_2,\ldots,x_K)\right] \\ &= 2^{-(K-1)} \cdot \sum_{\mathbf{x}\in\{0,1\}^{K-1}} \frac{1}{2} \\ &= \frac{1}{2}.\end{aligned}$$

Now let us consider the expected activity of variable x_2. We have

$$\begin{aligned}
\frac{\partial f}{\partial x_2} &= \left(x_1 \vee g\left(\mathbf{x}^{(2,0)}\right)\right) \oplus \left(x_1 \vee g\left(\mathbf{x}^{(2,1)}\right)\right) \\
&= \left(x_1 \vee g\left(\mathbf{x}^{(2,0)}\right)\right) \wedge \left(x_1 \vee g\left(\mathbf{x}^{(2,1)}\right)\right)' \\
&\quad \vee \left(x_1 \vee g\left(\mathbf{x}^{(2,0)}\right)\right)' \wedge \left(x_1 \vee g\left(\mathbf{x}^{(2,1)}\right)\right) \\
&= \left(x_1 \vee g\left(\mathbf{x}^{(2,0)}\right)\right) \wedge \left(x_1' \wedge g'\left(\mathbf{x}^{(2,1)}\right)\right) \\
&\quad \vee \left(x_1' \wedge g'\left(\mathbf{x}^{(2,0)}\right)\right) \wedge \left(x_1 \vee g\left(\mathbf{x}^{(2,1)}\right)\right) \\
&= x_1' \wedge g\left(\mathbf{x}^{(2,0)}\right) \wedge g'\left(\mathbf{x}^{(2,1)}\right) \vee x_1' \wedge g'\left(\mathbf{x}^{(2,0)}\right) \wedge g\left(\mathbf{x}^{(2,1)}\right) \\
&= x_1' \wedge \left(g\left(\mathbf{x}^{(2,0)}\right) \oplus g\left(\mathbf{x}^{(2,1)}\right)\right) \\
&= x_1' \wedge \frac{\partial g}{\partial x_2},
\end{aligned}$$

where in the second equality we used the fact that $a \oplus b = a \wedge b' \vee a' \wedge b$, in the third equality we used de Morgan's identity $(a \vee b)' = a' \wedge b'$, in the fifth equality we again used the definition of $\oplus$, and in the last equality we used the definition of partial derivative. The expected activity of variable x_2 is now equal to

$$E\left[\alpha_2^f\right] = E\left[2^{-(K-1)} \cdot \sum_{\mathbf{x} \in \{0,1\}^{K-1}} x_1' \wedge \frac{\partial g}{\partial x_2}\right].$$

Note that $\partial g\left(x_2, \ldots, x_K\right) / \partial x_2$ is a Boolean function of $K-2$ variables and the above summation is taken over all $\mathbf{x} = (x_1, x_3, \ldots, x_K)$. Let us break up this summation into parts, corresponding to $x_1 = 0$ and $x_1 = 1$. We have

$$\begin{aligned}
E\left[\alpha_2^f\right] &= 2^{-(K-1)} \cdot \left[\sum_{\mathbf{x}^{(1,0)} \in \{0,1\}^{K-1}} E\left[1 \wedge \frac{\partial g}{\partial x_2}\right] + \sum_{\mathbf{x}^{(1,1)} \in \{0,1\}^{K-1}} E\left[0 \wedge \frac{\partial g}{\partial x_2}\right]\right] \\
&= 2^{-(K-1)} \cdot \sum_{\mathbf{x}^{(1,0)} \in \{0,1\}^{K-1}} E\left[\frac{\partial g}{\partial x_2}\right].
\end{aligned}$$

Since g is a random unbiased function, so is $\partial g / \partial x_2$. This means that the probability that a random function g differs on $\mathbf{x}^{(j,0)}$ and $\mathbf{x}^{(j,1)}$ is equal to $1/2$. Thus,

$$E\left[\alpha_2^f\right] = 2^{-(K-1)} \cdot \sum_{\mathbf{x}^{(1,0)} \in \{0,1\}^{K-1}} \frac{1}{2}$$

and since there are exactly 2^{K-2} different vectors $\mathbf{x}^{(1,0)} = (0, x_3, \ldots, x_K)$,

$$E\left[\alpha_2^f\right] = 2^{-(K-1)} \cdot \frac{1}{2} \cdot 2^{K-2} = \frac{1}{4}.$$

Thus, the expected activity of all noncanalizing variables is equal to $1/4$. The expected activity vector is then equal to $E[\alpha^f] = (\frac{1}{2}, \frac{1}{4}, \ldots, \frac{1}{4})$ and the expected sensitivity, by Eq. (1.7), is equal to $E[s(f)] = \frac{1}{2} + \frac{1}{4} \cdot (K-1) = (K+1)/4$. The situation for two or more canalizing variables is analogous. Note that a random unbiased function is expected to have each activity equal to $1/2$, which means the expected average sensitivity is equal to $K/2$, as already mentioned. Therefore, for $K > 1$ a canalizing function with just one canalizing variable is expected to have a lower average sensitivity, implying that the network constructed from such functions should be more ordered than an unbiased random network. We will make use of the concept of average sensitivity when we discuss network inference in the context of regularization (Chapter 3).

One global network parameter that pertains to the dynamics of the system is the recently introduced *basin entropy* (Krawitz and Shmulevich, 2007). A Boolean network classifies its state space into several basins of attraction in essence by assigning a unique attractor to every state. The weight w_ρ of basin ρ is the number of states belonging to that basin (including the attractor states, which are by definition part of the basin), divided by the total number of states (2^n). For example, in the network whose state transition diagram is shown in Fig. 1.1, the weights of the basins are $[8,8,12,2,2]/32$. The basin entropy,

$$h = -\sum_{\rho=1}^{m} w_\rho \ln w_\rho, \tag{1.10}$$

where m is the total number of basins/attractors, serves as a measure for the complexity of this classification process. Equivalently, the basin entropy measures the uncertainty of a system that is started in a random state. Surprisingly, basin entropy increases only in critical network ensembles, whereas it approaches a finite limit in the ordered and highly chaotic phase (Krawitz and Shmulevich, 2007). This suggests that the informationally optimal partition of the state space is achieved when the system is operating at the critical boundary between the ordered and disordered phases. For biological networks this means that their informational complexity will grow only with size when their elements are organized in a critical manner.

1.4 Boolean Models of Biological Networks

Biological networks can often be modeled as logical circuits from well-known local interaction data in a straightforward way. This is clearly one of the advantages of the Boolean network approach. Though logical models may sometimes appear obvious and simplistic, compared to detailed kinetic models of biomolecular reactions, they may help to understand the dynamic key properties of a regulatory process. Further, a Boolean network model can be formulated as a coarse-grained limit of the more detailed differential equations model for a system (Davidich and Bornholdt, 2008a). They may also lead the experimentalist to ask new questions and to test them first *in silico*. The following three network models exemplify how Boolean networks may serve as useful *in silico* tools from cell differentiation processes to cell signaling.

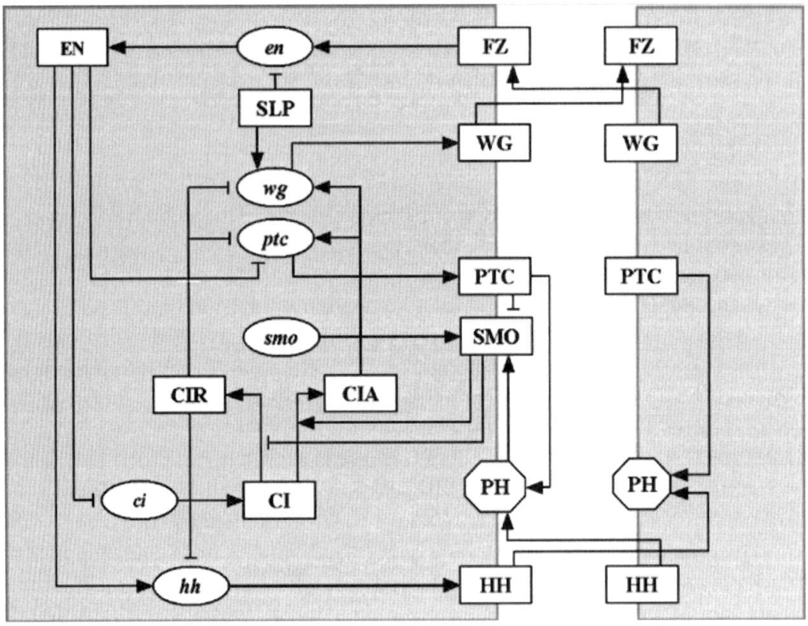

Figure 1.5. *The network of interactions between the segment polarity genes. The gray background layers illustrate two neighboring cells, indicating that some interactions in this network are intercellular (Albert and Othmer,* 2003*).*

1.4.1 The segment polarity network of the fruit fly

The segment polarity genes represent the last step in the hierarchical cascade of gene families initiating the segmented body of the fruit fly *Drosophila melanogaster.* The stable maintenance of the segment polarity gene expression pattern is a crucial requirement in the further development of the embryo. The dynamics of the best characterized segment polarity genes have been studied in different modeling approaches in order to understand the stability and robustness of this pattern formation (Chaves *et al.*, 2005, 2006; Albert and Othmer, 2003; Dassow *et al.*, 2000). Albert and Othmer suggested a Boolean model of gene and protein interactions which is depicted in Fig. 1.5. This model correctly reproduces the characteristic expression patterns observed in the wild-type development. Further, distinct *knock down* or *overexpression* phenotypes can be simulated through the model by fixing the state of a particular gene. An analysis of the robustness of the wild-type steady states under asynchronous updating schemes reveals that the model is very sensitive to temporal variations, whereas the wild-type development is stable as soon as certain differences in the temporal order of transcriptional and translational processes are incorporated into the updating procedure.

1.4.2 Control of the cell cycle

Proliferating eukaryotic cells perform a series of highly regulated and coordinated actions collectively referred to as the cell cycle. As cancer development is associated with a loss of control over the cell cycle, an understanding of its regulatory dynamics is of paramount importance. The integration of the biological and mostly qualitative knowledge about interacting genes and proteins into a Boolean network model allows the analysis of the dynamic regulatory properties. Fauré *et al.* (2006) studied the effects of different updating schemes on the dynamics of a highly abstracted mammalian cell-cycle control model (Fig. 1.6). Under the simplistic synchronous updating scheme the network gives rise to two different attractors, a stable state, matching the quiescent cell state (G0) when growth factors

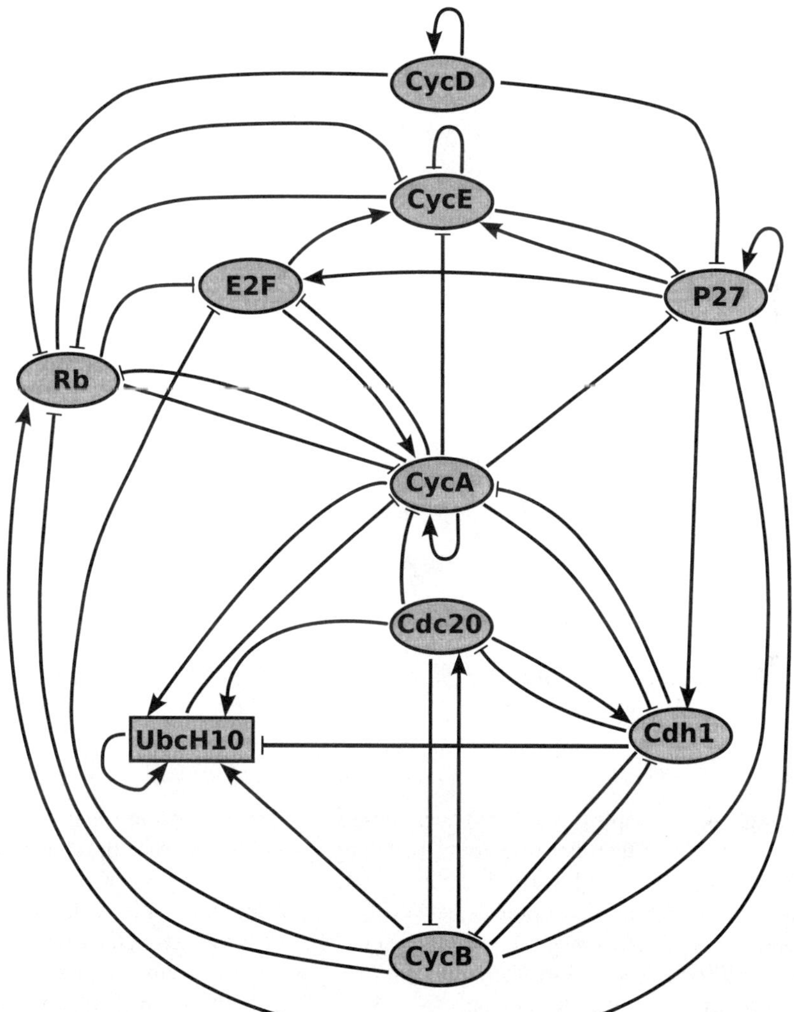

Figure 1.6. *Logical regulatory graph for the mammalian cell-cycle network. Each neode represents the activity of a key regulatory element, whereas the edges represent cross-regulations. Blunt arrows stand for inhibitory effects and normal arrows for activations (Fauré et al., 2006).*

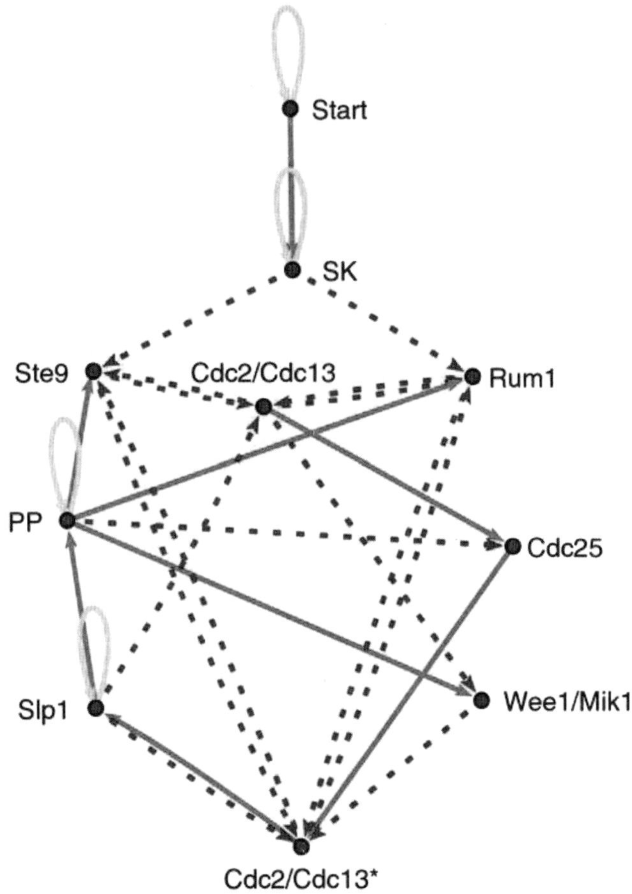

Figure 1.7. *Network model of the fission yeast cell-cycle regulation. Solid arrows represent activation while dashed arrows represent inhibition (Davidich and Bornholdt,* 2008b*).*

are lacking, and a complex dynamical cycle representing the cell cycle when the cyclin D complex is present. The order of activity switching (off or on) matches the available data, as well as the time plots published in the literature.

Another example of a Boolean network model of the cell cycle is that of the fission yeast *Schizosaccharomyces pombe* shown in Fig. 1.7, demonstrated by Davidich and Bornholdt (2008b). The cell-cycle system for this organism is well understood in terms of established differential equation models (Novak and Tyson, 1997), which are able to make predictions for wild-type and some known mutant cells, but is also quite different from that of the yeast *Saccharomyces cerevisiae*, for which a Boolean network model has also been studied (Li *et al.*, 2004).

Figure 1.8 shows the state space topology of the 10-gene network of the fission yeast cell cycle, with the thick arrows showing the state transitions that exactly match the corresponding biological time sequence in the cell-cycle control network, from the excited G1

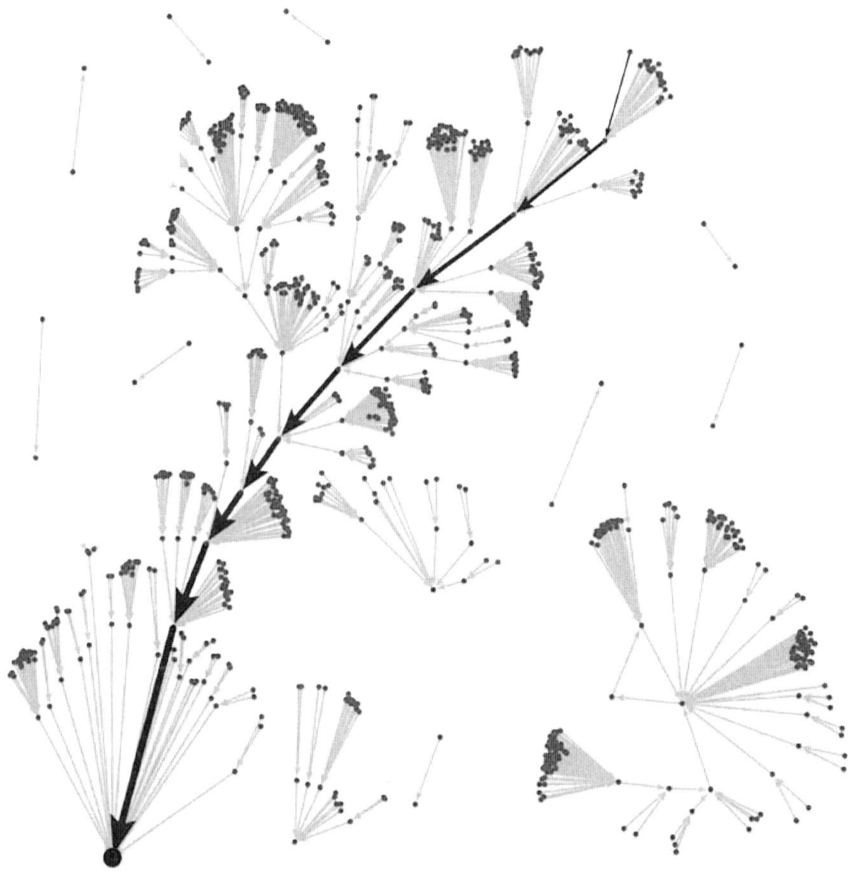

Figure 1.8. *State space of the* 1024 *possible network states of the fission yeast network. Each circle corresponds to one specific network state with each of the* 10 *proteins being in one specific activation state (active/inactive). The largest attractor tree corresponds to all network states flowing to the* G1 *fixed point (Davidich and Bornholdt,* 2008b*).*

state through S and G2 to the M phase and finally back to the stationary G1 state. This largest attractor attracts 73% of the entire state space, and the biological target state (G1) is robust to most perturbations.

An interesting biological difference between the cell-cycle network of *Saccharomyces cerevisiae* and that of *Schizosaccharomyces pombe* is that the former relies mostly on transcriptional regulation whereas the latter makes use of post-translational regulation, implying remarkably different time scales. A surprising conclusion that can be drawn from this study is that synchronous Boolean networks, which *ipso facto* neglect the timing of the biomolecular interactions, are nonetheless extremely robust against neglecting such different time scales.[2]

[2]Robustness here refers to the insensitivity of the dynamical behavior of the system, in this case against neglecting different time scales of regulation.

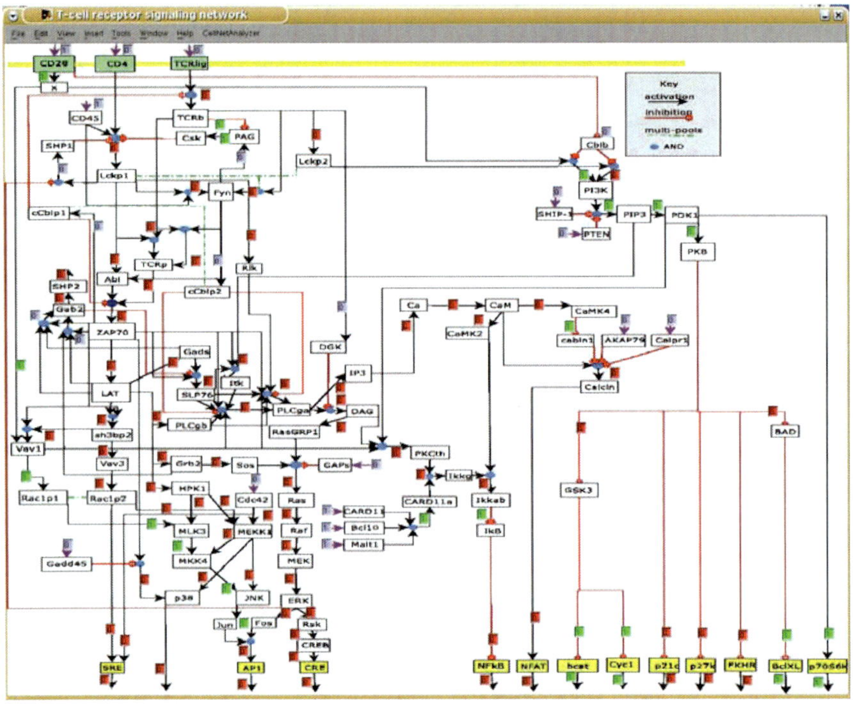

Figure 1.9. *Logical model of T cell signaling from the TCR and its coreceptors to the key transcription factors AP-1, NFAT, and NFκB, which determine T cell activation and function (Saez-Rodriguez et al., 2007).*

1.4.3 T-cell receptor signaling

T lymphocytes play a key role in the regulation of the adaptive immune system, as they are able to differentiate between self- and foreign-, potentially dangerous, antigens. T cells recognize foreign peptides that are presented by antigen presenting cells by means of the T cell receptor (TCR) and costimulatory molecules (CD4/8/28). This stimulus initiates a signaling cascade within the T cell that eventually influences its activation status. The T cell response must be tightly regulated, as a response to a pathogen that is too weak endangers the whole organism, whereas an overshooting reaction may lead to autoimmune disorders. Saez-Rodriguez *et al.* (2007) constructed a large Boolean model (comprising 94 nodes and 123 interactions) of this signaling network from local interaction knowledge (Fig. 1.9). In addition to reproducing known global behaviors, the model also predicted unexpected signaling events after antibody-mediated perturbation of CD28, which were then experimentally validated.

It is interesting to note that the directed graph of the network contains a core strongly connected component comprising only 33 nodes (out of a total of 94 nodes), meaning that for any two nodes in this core, there is a path between them; at the same time, the rest of the network that is not contained in the strongly connected component mainly consists of relatively simple input and output layers that feed into and out of this core. It then follows that the majority, if not all, of the nodes in this core are the relevant nodes in a Boolean network, discussed earlier, which are entirely responsible for determining the attractor structure of the network.

1.5 Discretization

Although the Boolean network models discussed thus far have been constructed "by hand" from known genetic regulatory interactions or from specific targeted experiments, a data-driven approach for model construction consists of inferring the model structure and rules of interactions from global measurement data, most commonly gene expression measurements using microarray technology. Since such measurements are continuous in nature, meaning that gene expressions are represented as real numbers, the inference of Boolean or other discrete model classes (e.g., ternary) requires discretization of the data as a necessary prior step. The most common discretization is binary (0 or 1) or ternary (typically -1, 0, 1). For example, a large number of Bayesian network approaches assume discrete data (see Shmulevich and Dougherty (2007) for a more extended discussion).

Although discretization is often justified from a conceptual point of view, in the sense that biological systems often exhibit switch-like on/off behavior, as discussed above, there are also pragmatic reasons for quantizing the data. Lower levels of model complexity mean less computation and lower data requirements for model identification. In a sense, this is an engineering form of Occam's razor in which the pragmatics of the problem imply some minimal level of necessary complexity. Additionally, quantization provides a certain level of robustness to noise in the observed data and may even improve accuracy of classification and simplify the obtained models (Pfahringer, 1995; Dougherty *et al.*, 1995). In particular, quantization has been shown to significantly improve error rates in microarray-based classification (Mircean *et al.*, 2004).

There are a number of approaches that can be used to discretize gene expression data. Generally speaking, this problem falls within the well-studied area of quantization developed in the signal processing community and relates to such topics as coding, compression, and rate distortion theory (Gersho and Gray, 1992). In the context of quantization, Snoussi (1989) suggested the following framework of which the Boolean formalization is a special case. Suppose a gene product x can act in different circumstances with different characteristic thresholds. The threshold refers to the limiting case of a sigmoidal curve whose critical transition point can vary depending on these circumstances. It is then natural to associate a multilevel discrete variable $\mathbf{x}$ with this gene product such that each value of the discrete variable corresponds to where in relation to the threshold x falls. Thus, given thresholds $\tau_1 < \tau_2 < \cdots$, the discrete variable is defined as $\mathbf{x} = \psi(x) = r_k$ for $\tau_k < x \leq \tau_{k+1}$. The values of the thresholds τ_k can either come from prior knowledge or be chosen automatically using the measurement data. For example, Pal *et al.* (2005a) employed two thresholds, thereby ternarizing the data, on the basis of induction or repression of genes under irradiated conditions of cell lines relative to the value of the genes under nonirradiated conditions.

In the field of signal processing, the operation $\psi(x)$ is referred to as *scalar quantization*. This process of representing continuous-valued signals by a set of finitely many values results in approximating the data at the expense of precision. However, as already mentioned, it can also have beneficial effects of enhancing inference precision on account of reduced model complexity and mitigating the effects of noise. Since quantization is a many-to-one mapping, it is *lossy* in the sense that the original signal cannot be recovered.

There are several ways one could determine the thresholds τ_k in an optimal way. The most popular method is the Lloyd–Max quantizer (Gersho and Gray, 1992), also known as k-means in the pattern recognition community. Given the number of quantization levels L, the optimality consists of selecting the τ_k and r_k, $k = 1,\ldots,L$, in order to minimize the mean square quantization error

$$\begin{aligned} D &= E[(x-r_k)^2] \\ &= \sum_{k=1}^{L} \int_{\tau_{k-1}}^{\tau_k} (x-r_k)^2 p(x)dx \end{aligned} \tag{1.11}$$

for a given input probability density function $p(x)$. Differentiating the error in Eq. (1.11) with respect to r_k and τ_k gives us the necessary conditions that must be satisfied by the optimal quantizer. Thus, $\partial D/\partial r_k = 0$ gives us

$$r_k = \frac{\int_{\tau_{k-1}}^{\tau_k} xp(x)dx}{\int_{\tau_{k-1}}^{\tau_k} p(x)dx}, \tag{1.12}$$

which can be interpreted as the expectation of x, or center of mass of the density function, given that we are in the interval $(\tau_{k-1}, \tau_k]$. Similarly, $\partial D/\partial \tau_k = 0$ results in

$$\tau_k = \frac{r_k + r_{k+1}}{2}, \tag{1.13}$$

which means that the optimal decision levels must be midway between the quantization levels. In general, Eqs. (1.12) and (1.13) do not yield closed-form solutions but can be solved by the following iterative algorithm. The algorithm starts with arbitrary positions of the quantization values and then iterates the following two steps until convergence:

1. For each data point, find and associate with it the nearest quantization value.
2. For each quantization interval, find the average of the data points associated with it and assign this average to be the new quantization value.

This iterative algorithm is expected to reach a local optimum. Unfortunately, it can be sensitive to the initial quantization values (seeds). To address this problem, one approach is to run the algorithm a fixed number of times and return the solution with the lowest quantization error D. Another approach is to run a preliminary quantization on 10% of the data and use the solution of that run as an initial condition.

There also exist several approaches specifically for binarizing gene expression data. One such approach, described in Shmulevich and Zhang (2002), first uses an optimization-based procedure for normalizing gene expression data as a necessary step prior to binarization and then determines an optimal threshold by a simple sorting procedure that selects the first sufficiently large "jump" in the sorted data as the threshold. Additionally, a global threshold needs to be employed so that genes that are not expressed above a reasonable level of detection are not subjected to spurious binarization. Such a global "noise floor" threshold can be determined from negative control spots on the array or other similar criteria.

Having binarized the data, the obvious metric of similarity between gene profiles becomes the Hamming distance (i.e., the number of genes that "disagree" in terms of their on/off values), which as an added benefit essentially avoids ambiguity in the choice of similarity measures and reflects the notion of similarity that biologists use in their everyday language and scientific literature. Using the Hamming distance in conjunction with multidimensional scaling, it was shown that a good degree of separation between different tumor subtypes (in two different data sets comprised of gliomas and sarcomas) can be achieved entirely in the binary domain (Shmulevich and Zhang, 2002).

Another approach, proposed by Zhou *et al.* (2003a), applies a binarization procedure based on a multiplicative model for expression up-regulation. Taken as a collection, the measurements for a particular gene, across a set of microarrays, compose a distribution of values that are modeled as a mixture of two distributions, one corresponding to lack of up-regulation and the other to up-regulation. The parameters of these two distributions are estimated from the expression data for the gene, and then data are quantized according to the modeled distributions. The mixture model depends on whether one is using ratios or the intensities directly. We note in passing that the question of whether ratios or absolute intensities should be used is highly nontrivial and depends on the purpose for which the microarray is being used. This problem has been investigated by Attoor *et al.* (2004) in the context of microarray-based classification.

Let us now discuss a particular application of quantization of gene expression data to the analysis of multivariate logical relationships among genes in a data set arising from radiation studies using the well-known NCI 60 anticancer drug screen cell lines (Pal *et al.*, 2005a). The relevance of this example is in demonstrating that logical relations based on coarse quantization closely reflect corresponding relations in the continuous (nonquantized) data and that such relations are typically not overly sensitive to the choice of thresholds used for quantization. The discussion will require us to first introduce a key concept that will be used throughout the remainder of the book—the coefficient of determination (CoD). We will return to it in greater detail when we discuss model inference in Chapter 3.

1.5.1 Coefficient of determination

The motivation behind the use of the coefficient of determination is the need to measure the strength of the relation between a set of predictor genes and a target gene, rather than identifying pairwise genetic coregulative responses to a particular stimulus. Indeed, pairwise measures such as correlation do not address the fundamental problem of determining sets of genes whose actions and interactions drive the cell's decision to set the transcriptional level of a particular gene.

The CoD, long used in the context of linear regression, was introduced in the context of optimal nonlinear filter design (Dougherty *et al.*, 2000) and then was used for inferring multivariate relationships between genes (Kim *et al.*, 2000a, 2000b). Such relationships, referred to as *predictors*, are the basic building blocks of a rule-based network. For the case of Boolean networks, a predictor for each gene is simply a Boolean function. In essence, relative to a given target gene and set of predictor genes, the CoD measures the relative increase in predictive capability using the predictor-gene expressions as opposed to predicting the target-gene expression based only on knowledge of the target gene's isolated behavior across the data set.

The method allows incorporation of knowledge of other conditions relevant to the prediction, such as the application of particular stimuli, or the presence of inactivating gene mutations, as predictive elements affecting the expression level of a given gene. Using the CoD, one can find sets of genes related in a multivariate matter to a given target gene.

Let us briefly discuss the CoD in the context of Boolean networks. Let x_i be a *target* gene that we wish to predict by observing some other genes $x_{i_1}, x_{i_2}, \ldots, x_{i_k}$. Also, suppose $f\left(x_{i_1}, x_{i_2}, \ldots, x_{i_k}\right)$ is an optimal predictor of x_i relative to some error measure ε. For example, in the case of mean-square error (MSE) estimation, it is well known that the optimal predictor is the conditional expectation of x_i given $x_{i_1}, x_{i_2}, \ldots, x_{i_k}$ (Scharf, 1991). Let ε_{opt} be the optimal error achieved by f. Then, the CoD for x_i relative to $x_{i_1}, x_{i_2}, \ldots, x_{i_k}$ is defined as

$$\theta = \frac{\varepsilon_0 - \varepsilon_{\text{opt}}}{\varepsilon_0}, \tag{1.14}$$

where ε_0 is the error of the best (constant) estimate of x_i in the absence of any conditional variables. That is, ε_0 is the error arising when using the best estimate of the target gene expression level given only statistics relating to the target gene itself, without using any information concerning other genes. If a predictor set can perfectly predict the target, then $\varepsilon_{\text{opt}} = 0$ and $\theta = 1$; at the other extreme, if a predictor set provides no additional information about the target, then $\varepsilon_{\text{opt}} = \varepsilon_0$ and $\theta = 0$. Thus, it is easily seen that the CoD must be between 0 and 1 ($0 \leq \theta \leq 1$) and measures the relative decrease in error from estimating x_i via f rather than by just the best constant estimate. Note that the definition of the CoD is general and applies to all model classes, not just Boolean networks.

In practice, the CoD must be estimated from training data with designed approximations being used in place of f. Those sets of (predictive) genes that yield the highest CoD, compared to all other sets of genes, are the ones used to construct the optimal predictor of the target gene. This procedure is applied to all target genes, thus estimating all the functions in a network. The method is computationally intensive and massively parallel architectures have been employed to handle large gene sets (Suh *et al.*, 2002).

The CoD technique has at least three advantages over standard correlation analysis. First, as already stated, the CoD can be applied to multiple predictors, thereby giving it the ability to discern multivariate intergene relationships. Second, the CoD can discover both linear and nonlinear relationships, whereas the correlation coefficient addresses only linear relationships. For instance, if gene x has the expression pattern $(0,0,0,1,1,0)$ across six conditions and gene y has the corresponding expression pattern $(1,1,-1,0,0,-1)$, then the relation $x = 1 - y^2$ is picked up by the CoD, with $\theta = 1$, but not picked up by the correlation coefficient which in this case is zero. A third advantage of the CoD is that, whereas the correlation coefficient is independent of the order of the genes, the CoD for x predicting y can be substantially different from that of y predicting x. In the example just given, the CoD of y predicting x is equal to 1, whereas the CoD for x predicting y is equal to $1/2$.

Let us now return to the question of robustness of discovered functional relations between genes relative to threshold selection. To define a measure of robustness, suppose that based on a given threshold we obtain the optimal predictor $z = f_{t_0}(x,y)$, where t_0 is the threshold, z is the discrete (say, binary or ternary) value of the target gene, and x and y are the discrete values of the predictor genes. The error of this best predictor determines the CoD, say, $\theta(x,y;z;t_0)$. If we change the threshold, then two things might happen. First, the error might change, thereby changing the CoD. Second, a different predictor may be optimal, thereby changing not only the CoD but the predictor function as well. Since the main interest is in finding functional relationships in gene networks, let us focus on the first possibility; i.e., we keep the function $f_{t_0}(x,y)$ and evaluate its performance when the discrete values have been obtained by different thresholds. To do this, we can define the

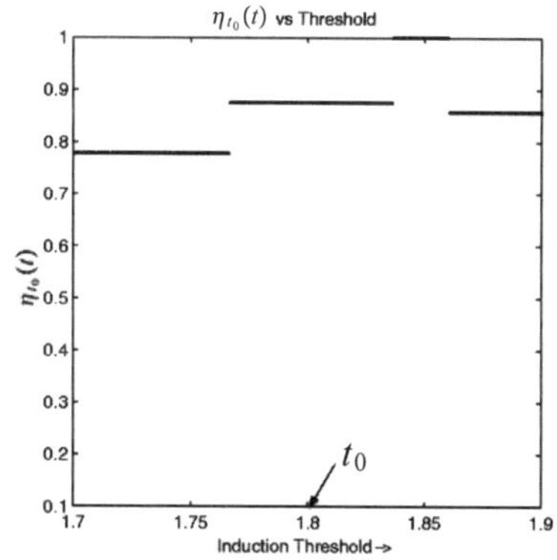

Figure 1.10. *The coefficient of threshold robustness $\eta_{t_0}(t)$ plotted as a function of the induction threshold shows stability with respect to small changes in the threshold values around t_0 (Pal et al.,* 2005a*).*

coefficient of threshold robustness by

$$\eta_{t_0}(t) = \frac{\varepsilon_0(t) - \varepsilon_{f_{t_0}}(t)}{\varepsilon_0(t)},$$

where $\varepsilon_0(t)$ is the error for the best predictor of z at threshold t given no observations and $\varepsilon_{f_{t_0}}(t)$ is the error of predicting z by $f_{t_0}(x,y)$ at threshold t. $\eta_{t_0}(t)$ measures the degree to which the best estimate for the transcriptional activity of a target gene is improved at threshold t by predicting it using the optimal predictor for threshold t_0 using the predictor genes thresholded at t, relative to the best estimate at t in the absence of any knowledge of the predictors. Note that by definition $\eta_{t_0}(t_0) = \theta(x,y;z;t_0)$ and that $\eta_{t_0}(t) \leq \theta(x,y;z;t)$. We judge the relationship $z = f_{t_0}(x,y)$ to be robust relative to threshold t_0 when $\eta_{t_0}(t)$ is stable for t near t_0. Noting that it is possible to have $\eta_{t_0}(t) > \eta_{t_0}(t_0)$, $f_{t_0}(x,y)$ is robust relative to threshold t_0 when $\eta_{t_0}(t)$ does not fall much below $\eta_{t_0}(t)$ for t near t_0.

With these definitions in hand, Pal *et al.* (2005a) applied the CoD analysis to ternarized data from the NCI 60 cell lines to identify relationships between genes responsive to ionizing radiation. To assess how the logical relations obtained depend on the thresholds, they plotted $\eta_{t_0}(t_0)$ in the neighborhood of t_0 to examine the relevance of the logical function $f_{t_0}(x,y)$ relative to the continuous data. It was found that for many cases, the coefficient of threshold robustness exhibits stability with respect to small changes in threshold values (see Fig. 1.10 for an example). Having established this robustness to quantization thresholds, the CoD analysis found a number of logical relationships, such as OR, AND, and XOR logic among many genes, with coefficient of determination values that significantly exceeded those expected by chance, using randomly generated data from the same distribution as for the actual data.

Chapter 2

Structure and Dynamics of Probabilistic Boolean Networks

Having discussed the structural and dynamical properties of Boolean networks, as well as the discretization of data and the basic inferential tool of coefficient of determination, we are now ready to take the next step and begin considering probabilistic or stochastic generalizations of this model class. This will afford us great flexibility and power in terms of making inferences from data, which necessarily contains uncertainty, as well as in terms of understanding the dynamical behavior of biological networks, particularly in relation to their structure. As we shall see, Boolean networks (or other deterministic discrete model classes) will become special cases of their probabilistic counterparts.

We will also see shortly that much of the theory of probabilistic Boolean networks, particularly pertaining to their dynamics, can be framed within the theory of Markov chains, affording us the many benefits inherent to this mature and well-established theory as well as with the existence of many analytical and numerical tools. Before embarking on our treatment of probabilistic Boolean networks, therefore, it is worth posing a fundamental question regarding the extent to which Markov chains are able to mimic biological regulation. This was addressed by Kim *et al.* (2002) in a paper intended to answer this very question.

2.1 Markov Chains as Models of Biological Regulation

It is easy to see that because of the determinism of a synchronous Boolean network, referring to the fact that from any given state the network transitions to one and only one successor state, its dynamics can be represented as a first-order homogeneous Markov chain whose state transition matrix is binary. In other words, every row of the state transition matrix, corresponding to a particular starting state, will contain only one entry equal to 1 in the position corresponding to its successor state (indicating a probability of 1 of transitioning to that state) and zeros in all other positions. Consider a more general approach of constructing a finite state Markov chain whose transitions (between states) are determined by state dependent multivariate conditional probabilities between gene expression levels, based on gene expression data. In that setting, mathematical tools that allow estimation of steady-state (long-run) behavior in biological systems would be useful for examining two ubiquitous aspects of biological system dynamics.

The first is homeostasis, the ability of cells to maintain their ongoing processes within the narrow ranges compatible with survival, and the second is a switchlike functionality that allow cells to rapidly transition between metastable states. In a Boolean network, long-run behavior is captured by the attractors. When in the vicinity of an attractor, a Boolean network that is not chaotic dynamically flows toward the attractor, representing the stability of the biological system that continuously restores its state back to the stable state after mild perturbations. On the other hand, a complex self-stabilizing system would not be expected to reach a desired target stable state starting from any arbitrary state; in a Boolean network, this can happen only when the starting state is in the basin of attraction of the target attractor states. In a biological system, this property affords the system its decision-making capacity (recall the basin entropy, which measures the degree of uncertainty a system possesses as regards its ultimate target attractor state).

This difficulty of reaching a target state from a state fairly distant from it provides a challenge in terms of modeling system dynamics with Markov chains. If a model is constructed to examine what happens with rule sets extracted from biological observations, the approach of examining the steady-state behavior achieved after many initializations from random states is unlikely to produce behavior similar to the biological system being modeled. Indeed, as is well known from the process of development (Davidson, 2006), the rules of interaction that govern transcriptional regulation of a particular gene at each step during the developmental process could be very different. Thus, an accurate model could not be expected to use any single simple set of rules to transition the system from the many unordered states produced by random initialization to the highly ordered target states from which the rules were abstracted. Furthermore, if an accurate model of the biological system that switches between relatively stable states is produced and that model is subjected to perturbations to allow it to reach the stable steady-state distributions favored by the transition rules derived from biological observations, then the model would be expected to have a significant probability of occupying steady states similar to each of the steady states from which the rules were derived. Kim *et al.* (2002) attempted to determine whether the kinds of biological behavior described above could be captured in a Markov chain model using a small network based on microarray data observations of human melanoma samples. The steady-state behavior of the constructed Markov chain was then compared to the initial observations.

Consider a Markov chain model that contains n nodes, each representing a gene that is assigned a ternary value: overexpressed (1), equivalently expressed (0), and under-expressed (-1) relative to a control. Thus, there are a total of 3^n states in the Markov chain. For capturing the dynamics of the network, the "wiring rule" is such that the expression state of each gene at step $t+1$ is predicted by the expression levels of genes at step t in the same network. Kim *et al.* (2002) chose three predictor genes $x_i^{(t)}, x_j^{(t)}, x_k^{(t)}$ with the highest CoD value and used the state of these predictor genes at step t and the corresponding conditional probabilities, which were estimated from observed data, to derive the state of the target gene $x_l^{(t+1)}$ at step $t+1$, where $i, j, k, l \in \{1, \ldots, n\}$ and

$$\sum_{m=-1,0,1} P(x_l^{(t+1)} = m | x_i^{(t)}, x_j^{(t)}, x_k^{(t)}) = 1. \tag{2.1}$$

Since the number of observations in microarray studies is often limited, there may be some states that are never observed in the data. In such a case, we simply use the probabilities

$P(x_l^{(t+1)} = m)$. Assuming independence, the state transition probability between state $\mathbf{x}^{(t)}$ and state $\mathbf{x}^{(t+1)}$ is equal to

$$\prod_{l=1}^{n} P(x_l^{(t+1)} = m | x_i^{(t)}, x_j^{(t)}, x_k^{(t)}). \tag{2.2}$$

We will revisit this formulation when we discuss the relationships between probabilistic Boolean networks and dynamic Bayesian networks, the latter being essentially equivalent to the Markov chain models discussed herein.

The important point to convey here is that once we have the Markov chain, we can compute its steady-state distribution, if it exists. A Markov chain is said to have a steady-state distribution if there exists a probability distribution $\pi = (\pi_1, \pi_2, \ldots, \pi_M)$ such that for all states $i, j \in \{1, 2, \ldots, M\}$, $\lim_{r\to\infty} P_{ij}^r = \pi_j$, where P_{ij}^r is the r-step transition probability. If there exists a steady-state distribution, then regardless of the initial state, the probability of the Markov chain being in state i in the long run can be estimated by sampling the observed states in the simulation (by simply counting the percentage of time the chain spends in that state). Although we will return to these concepts when we discuss probabilistic Boolean networks, the existence of a steady-state distribution can be ensured if the Markov chain is ergodic, meaning that any state is accessible from any other state after an arbitrary number of steps; this can be achieved by adding random gene perturbations, with a small probability, which essentially ensures that the chain never gets "stuck" in any set of states. A random gene perturbation simply means that any given gene has a small probability of being randomly flipped to a different value—for example, in the ternary case, if a gene has the value 0, it can be randomly flipped either to a 1 or to a -1. Although we will not discuss it here, it is possible to analytically compute the state transition probabilities for the Markov chain given the conditional probabilities stated above and the perturbation probability (Kim *et al.*, 2002). Using an inferred network comprised of 10 genes, based on the coefficients of determination between each target gene and many possible predictors, Kim *et al.* (2002) estimated the steady-state distribution by sampling from the Markov chain once it has converged to the steady-state (later we will also discuss the criteria for convergence). The remarkable finding was that only a small number of states (out all of 3^{10} states) had significant probabilities and most of those states with high probability were observed in the data. Figure 2.1 shows the steady-state distribution of all possible states of the Markov chain with a perturbation probability $p = 0.001$.

Furthermore, it was found that more than 85% of those states with high steady-state probability that were not observed in the data were very close to the observed data in terms of Hamming distance (for ternary data, the Hamming distance was defined simply as a sum of genewise differences between states), with only 2 to 4 genes being different between the observed data states and the estimated steady states. It is worth mentioning that if the perturbation probability becomes too large, it destroys the structure of the steady-state distribution, since low probability states acquire higher probabilities of occupancy simply due to random chance. It should be mentioned that the high probability steady states can be viewed as attractors of the underlying network model, since after the chain arrives into those states, it stays there for a long time. Figure 2.2 shows the state transition diagram for several high probability states. Based on the transition rules inferred from the data, there were approximately 700 transient states that drained into these high probability "attractor" states. It is also interesting that only one of the 10 genes (the last one) is changing, while all the other nine remain unchanged, suggesting highly stable nearly frozen attractors.

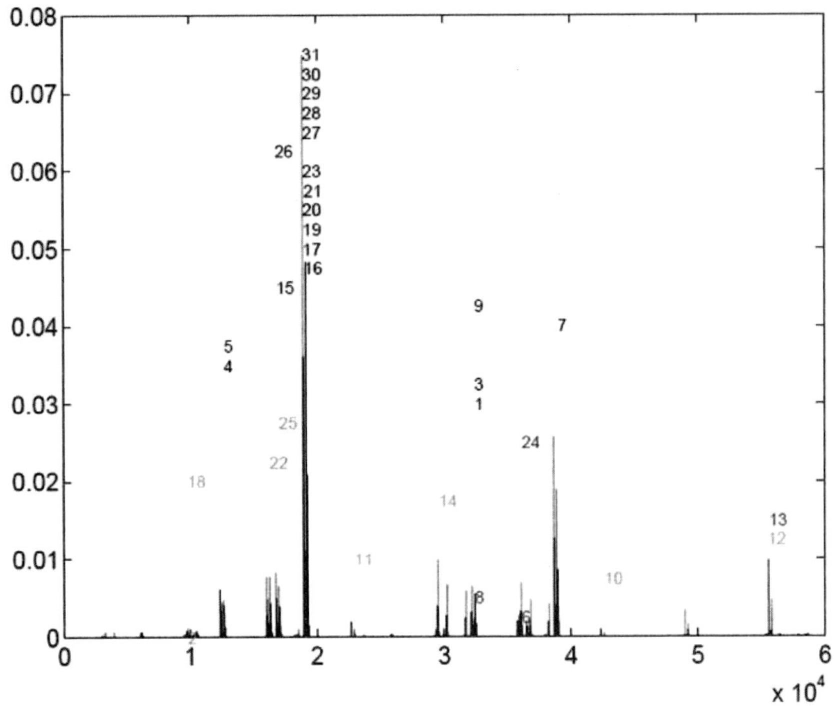

Figure 2.1. *The estimated steady-state distribution of all possible states of the Markov chain with perturbation probability* $p = 0.001$. *It can be seen that only a small number of states have significant steady-state probabilities, implying that the chain spends most of its time in those states (Kim et al.,* 2002*).*

In summary, the study by Kim *et al.* (2002) demonstrated the suitability of the Markov chain model to describe regulatory relationships between genes, since the model produced steady-state distributions that approximated the initial data observations and exhibited many properties associated with biological systems, to wit: the transition rules generated for the model produced localized stability, since initial states near the target states from which the model was built tended to stay in or near the target state demonstrating that the rules were sufficient to achieve self-stabilization and to guide the system to the target state; initial states far from the target state did not easily transition to the target state and required assistance in the form of random perturbation to get close enough to the target state to be self-stabilizing; the model rule sets inferred from the observations were sufficiently constraining to restrict the number of states seen in the steady state but sufficiently elastic to allow a collection of different states to be seen in the steady state. Taken together, these findings suggested that models incorporating rule-based transitions among states have the capacity to mimic biological regulation. With this, we now proceed to define the model class of probabilistic Boolean networks, which combines many aspects of the rule-based nature of Boolean networks and the probabilistic nature of Markov chains discussed thus far.

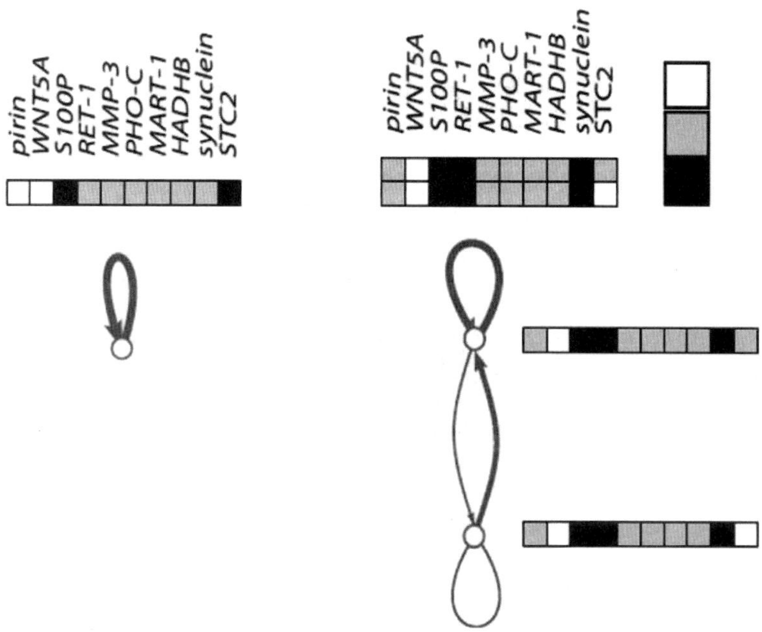

Figure 2.2. *The state transition diagram for several high probability states (Kim et al. 2002).*

2.2 Definition of Probabilistic Boolean Networks

Except in very restricted circumstances, gene expression data refute the determinism inherent to the Boolean network model, there typically being a number of possible successor states to any given state. Consequently, if we continue to assume the state at time $t+1$ is independent of the state values prior to time t, then, as we discussed in the previous section, the network dynamics are described by a Markov chain whose state transition matrix reflects the observed stochasticity. In terms of gene regulation, this stochasticity can be interpreted to mean that several predictor sets are associated with each gene and at any time point one of these predictor sets, along with a corresponding predictor function, is randomly chosen to provide the value of the gene as a function of the values within the chosen predictor set. It is this reasoning that motivated the original definition of a probabilistic Boolean network (PBN) in which the definition of a Boolean network was adapted in such a way that, for each gene, at each time point, a Boolean function (and predictor gene set) is randomly chosen to determine the network transition (Shmulevich *et al.*, 2002a, 2002d).

Rather than simply randomly assigning Boolean functions at each time point, one can take the perspective that the data come from distinct sources, each representing a "context" of the cell. From this perspective, the data derive from a family of deterministic networks and, in principle, the data could be separated into separate samples according to the contexts from which they have been derived. Given the context, the overall network would function as a Boolean network, its transition matrix reflecting determinism. If defined in this manner, a PBN is a collection of Boolean networks in which a constituent network governs gene activity for a random period of time before another randomly chosen constituent network takes over, possibly in response to some random event, such as an external stimulus or the action of a (latent) gene[3] outside the network. Since the latter is not part of the model, network switching is random. This model defines a "context-sensitive" PBN (Shmulevich *et al.*, 2002d; Brun *et al.*, 2005). The probabilistic nature of the constituent choice reflects the fact that the system is open, not closed, the idea being that changes between the constituent networks result from the genes responding to latent variables external to the model network.

We now formally define PBNs. Although we retain the terminology "Boolean" in the definition, this does not refer to the binary quantization assumed in standard Boolean networks but rather to the logical character of the gene predictor functions. In the case of PBNs, quantization is assumed to be finite but not necessarily binary. Formally, a PBN consists of a sequence $V = \{x_i\}_{i=1}^n$ of n nodes, where $x_i \in \{0,\ldots,d-1\}$, and a sequence $\{\mathbf{f}_l\}_{l=1}^m$ of vector-valued functions, defining constituent networks. In the framework of gene regulation, each element x_i represents the expression value of a gene. It is common to mix the terminology by referring to x_i as the ith gene. Each vector-valued function $\mathbf{f}_l = (f_l^{(1)}, f_l^{(2)} \ldots, f_l^{(n)})$ determines a constituent network, or *context*, of the PBN. The function $f_l^{(i)} : \{0,\ldots,d-1\}^n \to \{0,\ldots,d-1\}$ is a predictor of gene i whenever network l is selected. The number of quantization levels is denoted by d. At each updating epoch a decision is made whether to switch the constituent network. This decision depends on a binary random variable ξ: if $\xi = 0$, then the current context is maintained; if $\xi = 1$, then a constituent network is randomly selected from among all constituent networks according to the selection probability distribution $\{c_l\}_{l=1}^m$, $\sum_{l=1}^m c_l = 1$. The switching probability $q = P(\xi = 1)$ is a system parameter. If the current network is maintained, then the PBN behaves like a fixed network and synchronously updates the values of all the genes according to the current context. Note that, even if $\xi = 1$, a different constituent network is not necessarily selected because the "new" network is selected from among all contexts. In other words, the decision to switch is not equivalent to the decision to change the current network. If a switch is called for ($\xi = 1$), then, after selecting the predictor function $\mathbf{f}_l$, the values of genes are updated accordingly, that is, according to the network determined by $\mathbf{f}_l$. If $q < 1$, the PBN is said to be *context-sensitive*; if $q = 1$, the PBN is said to be *instantaneously random*, which corresponds to the original definition in (Shmulevich *et al.*, 2002a).

A PBN is defined so that the selection probabilities c_l, $l = 1,2,\ldots,m$, are not conditioned on the current network. The definition could have been different; that is, the selection probabilities could have been conditioned on the current network, in which case the definition would have required selection probabilities of the form c_{jl} equal to the probability of selecting network l given the current network is j (meaning that the selection probabilities themselves would constitute a Markov chain). This would have made the definition overly complicated and, more importantly, would have greatly increased the difficulty of inferring PBNs from data, as it would increase the number of parameters to be inferred. In particular, with conditional selection probabilities we could have had $c_{ll} = 0$, so that the current network could not be maintained. As we will see, PBNs have been considered with the condition that the current network cannot be continued, and this can be accomplished

[3]We use the term "gene" generically and understand it to be any biological molecule, e.g., protein, having regulatory potential.

using the definition that has been adopted by applying a conditional constraint. But, as defined in general, when $\xi = 1$, context selection is not conditioned on the current network.

Whereas a network switch corresponds to a change in a latent variable causing a structural change in the functions governing the network, a random perturbation corresponds to a transient value change that leaves the network wiring unchanged, as in the case of activation or inactivation owing to external stimuli such as stress conditions, small molecule inhibitors, etc. In a PBN with perturbation, there is a small probability p that a gene may change its value at each epoch. Perturbation is characterized by a random perturbation vector $\gamma = (\gamma_1, \gamma_2, \ldots, \gamma_n), \gamma_i \in \{0,1\}$ and $P(\gamma_i = 1) = p$, the perturbation probability. Thus, γ_i is a Bernoulli(p) random variable. If $\mathbf{x}(t)$ is the current state of the network and $\gamma(t+1) = \mathbf{0}$, then the next state of the network is given by $\mathbf{x}(t+1) = \mathbf{f}_l(\mathbf{x}(t))$; if $\mathbf{x}(t)$ is the current state and $\gamma(t+1) \neq \mathbf{0}$, meaning that at least one gene is to be perturbed, then $\mathbf{x}(t+1) = (x_1^\gamma(t+1), x_2^\gamma(t+1), \ldots, x_n^\gamma(t+1))$, where $x_i^\gamma(t+1) = x_i(t)$ if $\gamma_i(t+1) = 0$ and $x_i^\gamma(t+1)$ is uniformly distributed over the set $\{0,1,\ldots,d-1\} \backslash \{x_i(t)\}$ if $\gamma_i(t+1) = 1$. In the special case of a binary PBN,

$$\mathbf{x}(t+1) = 1(\gamma(t+1) = \mathbf{0})\mathbf{f}_l(\mathbf{x}(t)) + 1(\gamma(t+1) \neq \mathbf{0})(\mathbf{x}(t) \oplus \gamma(t+1)), \tag{2.3}$$

where $\oplus$ is componentwise XOR. The probability of no perturbation, in which case the next state is determined according to the current network function $\mathbf{f}_l$, is $(1-p)^n$ and the probability of a perturbation is $1-(1-p)^n$. The perturbation model captures the realistic situation where the activity of a gene undergoes a random alteration.

Two quantization levels have thus far been used in practice. If $d = 2$ (binary), then the constituent networks are Boolean networks with 0 or 1 meaning OFF or ON, respectively, as discussed in the previous chapter. The case $d = 3$ (ternary) arises when we consider a gene to be down-regulated (0), up-regulated (2), or invariant (1). This situation commonly occurs with gene expression microarrays, where a ratio is taken between the expression values of two samples cohybridized to the same array with two different fluorescent colors. Since basic properties, inference methodologies, and regulatory analyses that apply for binary PBNs usually apply to nonbinary PBNs, except where indicated we will focus on binary PBNs.

According to the definition of a PBN, there is a set $\mathbf{F} = \{\mathbf{f}_1, \mathbf{f}_2, \ldots, \mathbf{f}_m\}$ of network functions and the current network function, $\mathbf{f}$, is a random vector selected from $\mathbf{F}$ according to the distribution $\{c_l\}_{l=1}^m$ when a switch is called for ($\xi = 1$); that is, $P(\mathbf{f} = \mathbf{f}_l) = c_l$. The random vector function $\mathbf{f}$ is of the form $\mathbf{f} = (f^{(1)}, f^{(2)}, \ldots, f^{(n)})$, where $f^{(i)}$ is the predictor for gene i. In general, the selection of the component functions is not independent, meaning that it is not necessary to have

$$P(f^{(i)} = f_a^{(i)}, f^{(k)} = f_b^{(k)}) = P(f^{(i)} = f_a^{(i)})P(f^{(k)} = f_b^{(k)}). \tag{2.4}$$

A PBN is said to be *independent* if the random variables $f^{(1)}, f^{(2)}, \ldots, f^{(n)}$ are independent, so that the preceding equation and more general multivariate probabilities involving $f^{(1)}, f^{(2)}, \ldots, f^{(n)}$ decompose into products of unconditional probabilities. If the PBN is not independent, then these multivariate probabilities need to be expressed in terms of conditional probabilities.

A special case occurs when there exist function classes $F_1, F_2, \ldots, F_n$ with $F_i = \{f_1^{(i)}, f_2^{(i)}, \ldots, f_{r(i)}^{(i)}\}$ and $\mathbf{F} = F_1 \times F_2 \times \cdots \times F_n$. In this case $\mathbf{F}$ contains $m = \prod_{i=1}^n r(i)$ vector functions $\mathbf{f}_1, \mathbf{f}_2, \ldots, \mathbf{f}_m$, each determining a context and each of the form $(f_{j_1}^{(1)}, f_{j_2}^{(2)}, \ldots, f_{j_n}^{(n)})$

with $f_{j_i}^{(i)} \in F_i$. Assuming that the predictors are drawn independently from the function classes (so that the PBN is independent) and

$$P(f^{(i)} = f_j^{(i)}) = c_j^{(i)}, \tag{2.5}$$

the context selection probabilities are given by

$$P(\mathbf{f} = (f_{j_1}^{(1)}, f_{j_2}^{(2)}, \ldots, f_{j_n}^{(n)})) = \prod_{i=1}^{n} P(f^{(i)} = f_{j_i}^{(i)}) = \prod_{i=1}^{n} c_{j_i}^{(i)}. \tag{2.6}$$

As with Boolean networks, attractors play a major role in the study of PBNs. By definition, the attractor cycles of a PBN consist of the attractor cycles of the constituent networks, and their basins are likewise defined. Whereas in a Boolean network two attractor cycles cannot intersect, attractor cycles from different contexts can intersect in a PBN.

2.3 Dynamics: State Transition Probabilities

The dynamics of a PBN can be described via its associated Markov chain. In the general formulation of context-sensitive PBNs, there arises a possibility of confusion as regards the term "state." On the one hand, the word state refers to the state of the network, in the sense that we have been using it for Boolean networks, comprised of the activities of all of the genes in the network at a given time. This common terminology extends beyond Boolean networks to other model classes, such as systems of differential equations. On the other hand, Markov chains are also comprised of states, which in the special case of instantaneously random PBNs are one and the same (i.e., the state of the Markov chain is the vector of activities of all of the genes). Let us refer to this vector of activities as a GAP, which stands for "gene activity profile." The ambiguity arises when we discuss the more general case of context-sensitive PBNs, for which the states of the corresponding Markov chain consist of all (context, GAP) ordered pairs, the reason being that the probability of moving from one GAP to another depends on the context (i.e., on the constituent network). As stated above, in the special case of an instantaneously random PBN, the state space of the Markov chain reduces to the GAPs. We will use the term GAP when it is necessary to avoid this ambiguity, but will prefer to use the commonly accepted term state when no such ambiguity arises, particularly when referring to the network state rather than the state of the Markov chain. Derivation of the transition probability for a context-sensitive PBN depends on the recognition that four mutually exclusive events can occur at any time point:

1. The context is unchanged and the current network function is applied.
2. The context is unchanged and there is a random perturbation applied to the current GAP.
3. The context is changed and the new network function is applied.
4. The context is changed and there is a random perturbation applied to the current GAP.

According to the definition of a PBN, there are two mutually exclusive ways in which the context may remain unchanged: (a) $\xi = 0$, which means that no change is possible, or (b) $\xi = 1$ and the current network is selected.

Let z denote an arbitrary state of the Markov chain, and let $z_1 = (\kappa_1, \mathbf{x}_1)$ and $z_2 = (\kappa_2, \mathbf{x}_2)$ be two states, κ_1 and κ_2 being contexts and $\mathbf{x}_1$ and $\mathbf{x}_2$ being GAPs. We derive the probability,

$$P(z_2|z_1) = P(z(t+1) = (\kappa_2, \mathbf{x}_2) \,|\, z(t) = (\kappa_1, \mathbf{x}_1)), \tag{2.7}$$

of transitioning from $z_1 = z(t)$ to $z_2 = z(t+1)$. First, the transition probability can be split, depending on whether the context remains the same or is changed:

$$\begin{aligned} &P(z_2|z_1) \\ &= P(z(t+1) = (\kappa_2, \mathbf{x}_2) \,|\, z(t) = (\kappa_1, \mathbf{x}_1), \kappa_2 = \kappa_1) P(\kappa_1 = \kappa_2) \\ &+ P(z(t+1) = (\kappa_2, \mathbf{x}_2) \,|\, z(t) = (\kappa_1, \mathbf{x}_1), \kappa_2 \neq \kappa_1) P(\kappa_1 \neq \kappa_2). \end{aligned} \tag{2.8}$$

As pointed out, the context can remain the same in two mutually exclusive ways, depending on whether $\xi = 0$ or $\xi = 1$. Thus,

$$P(\kappa_2 = \kappa_1) = (1-q) + q c_{\kappa_1}. \tag{2.9}$$

The context will change from κ_1 to κ_2 if and only if a switch is called for and κ_2 is selected, so that

$$P(\kappa_2 \neq \kappa_1) = q\, c_{\kappa_2}. \tag{2.10}$$

In both cases, a transition from GAP $\mathbf{x}_1$ to GAP $\mathbf{x}_2$ may occur either according to the constituent network at the instant $t+1$ or through an appropriate number of perturbations. For $i, j = 1, 2, \ldots, n$, let $G_l(i, j) = 1$ if GAP j can be reached from GAP i in one transition of the constituent network l and $G_l(i, j) = 0$ otherwise.[4] Let $\eta(\mathbf{x}_1, \mathbf{x}_2)$ be the Hamming distance between GAPs $\mathbf{x}_1$ and $\mathbf{x}_2$. We have

$$\begin{aligned} &P(z(t+1) = (\kappa_2, \mathbf{x}_2) \,|\, z(t) = (\kappa_1, \mathbf{x}_1), \kappa_2 = \kappa_1) \\ &= 1(\kappa_2 = \kappa_1)((1-p)^n G_{\kappa_1}(\mathbf{x}_1, \mathbf{x}_2) \\ &+ (1-p)^{(n-\eta(\mathbf{x}_1,\mathbf{x}_2))} p^{\eta(\mathbf{x}_1,\mathbf{x}_2)} 1(\eta(\mathbf{x}_1, \mathbf{x}_2) \neq 0)) \end{aligned} \tag{2.11}$$

and

$$\begin{aligned} &P(z(t+1) = (\kappa_2, \mathbf{x}_2) \,|\, z(t) = (\kappa_1, \mathbf{x}_1), \kappa_2 \neq \kappa_1) \\ &= 1(\kappa_2 \neq \kappa_1)((1-p)^n G_{\kappa_2}(\mathbf{x}_1, \mathbf{x}_2) \\ &+ (1-p)^{(n-\eta(\mathbf{x}_1,\mathbf{x}_2))} p^{\eta(\mathbf{x}_1,\mathbf{x}_2)} 1(\eta(\mathbf{x}_1, \mathbf{x}_2) \neq 0)) \end{aligned} \tag{2.12}$$

The first parts of Eqs. (2.11) and (2.12) correspond to the probability of transition from GAP $\mathbf{x}_1$ to GAP $\mathbf{x}_2$ according to the predictor functions defined by the constituent network at the instant $t+1$. If the transition between two GAPs occurs due to perturbation, then its probability is presented as the second parts of these two equations.

By replacing the terms of Eq. (2.8) with their equivalents from Eqs. (2.9), (2.10), (2.11), and (2.12), we obtain that the probability of transition from any state $z_1 = (\kappa_1, \mathbf{x}_1)$

[4] $G_l(i, j)$ is essentially the binary state transition matrix of the constituent Boolean network.

to any state $z_2 = (\kappa_2, \mathbf{x}_2)$ is given by

$$\begin{aligned} &P(z_2|z_1) \\ &= 1(\kappa_2 = \kappa_1)((1-q)+q c_{\kappa_1})[(1-p)^n G_{\kappa_1}(\mathbf{x}_1,\mathbf{x}_2) \\ &+(1-p)^{(n-\eta(\mathbf{x}_1,\mathbf{x}_2))} p^{\eta(\mathbf{x}_1,\mathbf{x}_2)} 1(\eta(\mathbf{x}_1,\mathbf{x}_2) \neq 0)] \\ &+1(\kappa_2 \neq \kappa_1) q\, c_{\kappa_2}[(1-p)^n G_{\kappa_2}(\mathbf{x}_1,\mathbf{x}_2) \\ &+(1-p)^{(n-\eta(\mathbf{x}_1,\mathbf{x}_2))} p^{\eta(\mathbf{x}_1,\mathbf{x}_2)} 1(\eta(\mathbf{x}_1,\mathbf{x}_2) \neq 0)]. \end{aligned} \tag{2.13}$$

The preceding expression applies directly for instantaneously random PBNs by setting $q = 1$.

The basic definition of a PBN makes the assumption that, given a switch is allowed ($\xi = 1$), the context selection is independent of the current context. A change in this independence assumption has been considered in the literature under which the current context cannot be selected. Under this conditioning, given context κ_1, the selection probability for context κ_2 becomes

$$c_{\kappa_2|\kappa_1} = \frac{c_{\kappa_2}}{\sum_{l=1,l\neq\kappa_1}^{m} c_l}. \tag{2.14}$$

Equations (2.9) and (2.10) become $P(\kappa_2 = \kappa_1) = 1-q$ and $P(\kappa_2 \neq \kappa_1) = q$, respectively, and the transition probability becomes

$$\begin{aligned} &P(z_2|z_1) \\ &= 1(\kappa_2 = \kappa_1)(1-q)[(1-p)^n G_{\kappa_1}(\mathbf{x}_1,\mathbf{x}_2) \\ &+(1-p)^{(n-\eta(\mathbf{x}_1,\mathbf{x}_2))} p^{\eta(\mathbf{x}_1,\mathbf{x}_2)} 1(\eta(\mathbf{x}_1,\mathbf{x}_2) \neq 0)] \\ &+1(\kappa_2 \neq \kappa_1) q\, c_{\kappa_2|\kappa_1}[(1-p)^n G_{\kappa_2}(\mathbf{x}_1,\mathbf{x}_2) \\ &+(1-p)^{(n-\eta(\mathbf{x}_1,\mathbf{x}_2))} p^{\eta(\mathbf{x}_1,\mathbf{x}_2)} 1(\eta(\mathbf{x}_1,\mathbf{x}_2) \neq 0)]. \end{aligned} \tag{2.15}$$

Unless otherwise noted, we will assume that network selection is not conditioned by the current network.

Example 2.1 Suppose we are given an independent PBN without perturbation consisting of three genes $V = (x_1, x_2, x_3)$ and the function classes $F_1 = \{f_1^{(1)}, f_2^{(1)}\}$, $F_2 = \{f_1^{(2)}\}$, and $F_3 = \{f_1^{(3)}, f_2^{(3)}\}$. Let the functions be given by the following truth tables.

$x_1x_2x_3$	$f_1^{(1)}$	$f_2^{(1)}$	$f_1^{(2)}$	$f_1^{(3)}$	$f_2^{(3)}$
000	0	0	0	0	0
001	1	1	1	0	0
010	1	1	1	0	0
011	1	0	0	1	0
100	0	0	1	0	0
101	1	1	1	1	0
110	1	1	0	1	0
111	1	1	1	1	1
$c_j^{(i)}$	0.6	0.4	1	0.5	0.5

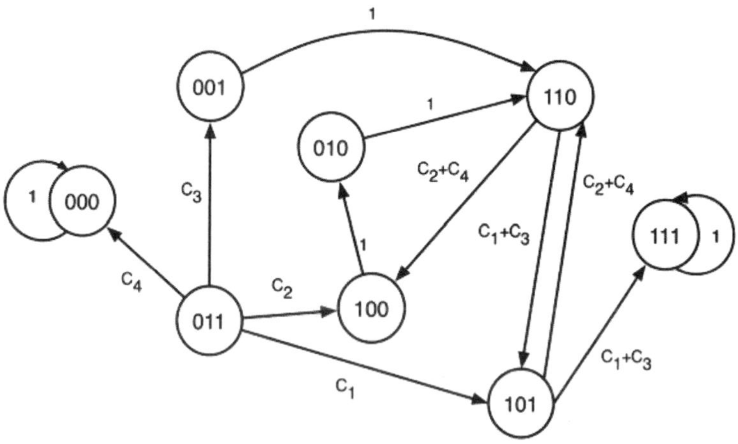

Figure 2.3. *State transition diagram corresponding to the state transition matrix in Eq.* (2.16).

Since there are 2 functions for gene x_1, 1 function for gene x_2, and 2 functions for gene x_3, there are $r = 2 \times 1 \times 2 = 4$ possible vector functions $\mathbf{f}_1, \mathbf{f}_2, \mathbf{f}_3, \mathbf{f}_4$ with associated probabilities $c_1 = 0.3$, $c_2 = 0.3$, $c_3 = 0.2$, and $c_4 = 0.2$. To see this, if, for example, $\mathbf{f}_1 = (f_1^{(1)}, f_1^{(2)}, f_2^{(3)})$, then multiplying their probabilities as in Eq. (2.6) gives $c_1 = 0.6 \times 1 \times 0.5 = 0.3$. Thus, the state transition matrix $\mathbf{A}$ is given by

$$\mathbf{A} = \begin{bmatrix} 1 & 0 & 0 & 0 & 0 & 0 & 0 & 0 \\ 0 & 0 & 0 & 0 & 0 & 0 & 1 & 0 \\ 0 & 0 & 0 & 0 & 0 & 0 & 1 & 0 \\ c_4 & c_3 & 0 & 0 & c_2 & c_1 & 0 & 0 \\ 0 & 0 & 1 & 0 & 0 & 0 & 0 & 0 \\ 0 & 0 & 0 & 0 & 0 & 0 & c_2+c_4 & c_1+c_3 \\ 0 & 0 & 0 & 0 & c_2+c_4 & c_1+c_3 & 0 & 0 \\ 0 & 0 & 0 & 0 & 0 & 0 & 0 & 1 \end{bmatrix}. \tag{2.16}$$

Let us consider one of the entries in matrix $\mathbf{A}$ to clarify its construction. Suppose we wish to compute the probability of the state transition $(1,1,0) \to (1,0,0)$, which corresponds to the entry $\mathbf{A}_{7,5}$ (the indexing starts with 1). To do this, we need to use the row corresponding to $(x_1, x_2, x_3) = (1,1,0)$ in the network truth table given above. Then, we look for possible combinations of the predictors for each of the three genes that will give us the values $(1,0,0)$. By direct inspection, we can see that either $(f_1^{(1)}, f_1^{(2)}, f_2^{(3)})$ or $(f_2^{(1)}, f_1^{(2)}, f_2^{(3)})$ result in $(1,0,0)$. The two possible combinations (i.e., constituent networks) have probabilities c_2 and c_4, and so the transition probability is equal to $c_2 + c_4$. All other entries in $\mathbf{A}$ are computed similarly. The state transition diagram corresponding to this matrix is shown in Fig. 2.3. For example, the 7th row of matrix $\mathbf{A}$ corresponds to $(1,1,0)$ and it can be seen that the only possible transitions are to $(1,0,0)$ or $(1,0,1)$, corresponding to columns 5 and 6, respectively. ■

The ability to determine the state transition probabilities, and therefore compute the entire state transition matrix, allows us to determine the steady-state distribution of the model and therefore determine its long-run behavior. Importantly, it will allow us to assess the effect of various parameter changes in the model (whether they are continuous, such as p, c_l, or q, or discrete, such as different functions or wiring) on the steady-state behavior. Before we discuss how the steady-state distribution can be computed or numerically estimated, let us first briefly examine the criteria for the existence of steady-state distributions.

2.4 The Existence of Steady-State Distributions

When considering the long-run behavior of a Markov chain, it is useful to consider equivalence classes within the set of states. This is especially true for genomic systems in which the state space can be extremely large and may be partitioned according to various subsystems. If an equivalence class is closed, meaning that no state outside the class is accessible from a state within the class, then for long-run analysis that class can be treated as an irreducible Markov chain in its own right: once inside the class, the system cannot leave it. Hence, we will consider long-run dynamics in terms of a single irreducible finite Markov chain.

A key property in the characterization of long-run behavior is periodicity. State i has period k if any return to state i must occur in multiples of k time steps. Since periodicity is a class property, an irreducible Markov chain can either be considered to be aperiodic or not. A homogeneous Markov chain with finite state space $S = \{1, 2, \ldots, M\}$ is said to possess a *stationary distribution* (or *invariant distribution*) if there exists a probability distribution $\pi = (\pi_1, \pi_2, \ldots, \pi_M)$ such that, for any $j \in S$ and for any number r of time steps,

$$\pi_j = \sum_{i=1}^{M} \pi_i P_{ij}^r, \tag{2.17}$$

where P_{ij}^r is the r-step transition probability. Hence, if the initial distribution is $\pi = (\pi_1, \pi_2, \ldots, \pi_M)$, then the probability of being in state i at time r is equal to π_i for all r and the Markov chain is a strictly stationary random process. The Markov chain is said to possess a *steady-state (limiting) distribution* if there exists a probability distribution $\pi = (\pi_1, \pi_2, \ldots, \pi_M)$ such that, for all states $i, j \in S$,

$$\lim_{r \to \infty} P_{ij}^r = \pi_j.$$

If there exists a steady-state distribution, then, regardless of the starting state, the probability of the Markov chain being in state i in the long run is π_i. In particular, for any initial distribution $D^0 = \left(D_1^0, D_2^0, \ldots, D_M^0\right)$, the state probability D_i^k approaches π_i as $k \to \infty$. Relative to the probability vector π, the vector D^k satisfies $\lim_{k \to \infty} D^k = \pi$. Every irreducible, finite-state, homogeneous Markov chain possesses a unique probability vector π, with $0 < \pi_i < 1$, providing the stationary distribution. If the chain is also aperiodic, then π also provides the steady-state distribution. Should the chain be only irreducible, and not necessarily aperiodic, then it may not possess a steady-state distribution. A more detailed treatment of the above concepts can be found in most textbooks on stochastic processes, such as Çınlar (1997).

If the chain has a steady-state distribution, we can answer the following question: In the long run, what is the probability that the chain is in state i? The answer does not depend on the initial state. Suppose the states are divided into two classes, C_1 and C_2. Then we

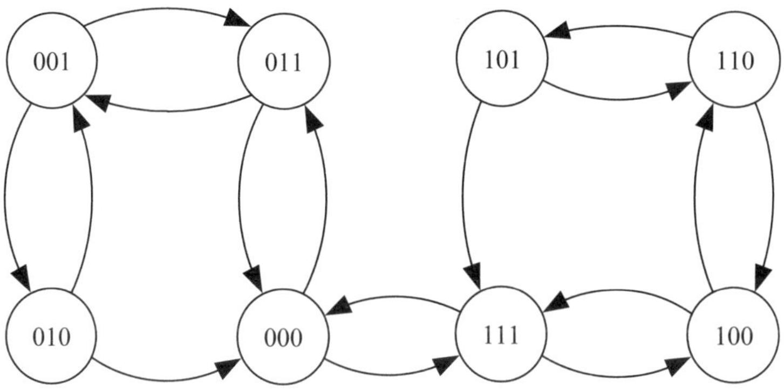

Figure 2.4. *State transition diagram of an independent PBN without a steady-state distribution (Shmulevich et al.,* 2002a*).*

can answer the following question without concern for the initial state: In the long run, what is the probability that the chain is in class C_1 (or C_2)? Such a question need not be answerable if there does not exist a steady-state distribution (if the chain is not aperiodic).

To illustrate lack of a steady-state distribution, let us consider a three-variable independent instantaneously random PBN. Since we are not concerned with the probabilities, but only the possible Boolean functions, we can use a simplified notation to list the possible functions. We use a table consisting of eight rows corresponding to the eight states and three columns corresponding to the possible values the Boolean functions can have for the three variables given the state determining the row. The entry * in the table means that the value of the predictor for that gene given the values of the genes in that row can be either 0 or 1. Consider the following function table:

$x_1x_2x_3$	$f^{(1)}$	$f^{(2)}$	$f^{(3)}$
000	*	1	1
001	0	1	*
010	0	0	*
011	0	0	*
100	1	1	*
101	1	1	*
110	1	0	*
111	*	0	0

So, for example, there are four possible predictors $f_1^{(1)}, f_2^{(1)}, f_3^{(1)}, f_4^{(1)}$ for the first gene. Similarly, there are 256 possible vector functions (network realizations) of the form $\mathbf{f} = (f^{(1)}, f^{(2)}, f^{(3)})$. The corresponding Markov diagram is given in Fig. 2.4. Every state has period 2, and therefore the PBN is not aperiodic. Thus, there does not exist a steady-state distribution. The requirement for a PBN to possess a steady-state distribution may be imposed so that the associated long-run questions may be posed. If so, then this imposes a constraint on the collections of Boolean functions. Certain sets of Boolean functions, such as the one just considered, are not permissible.

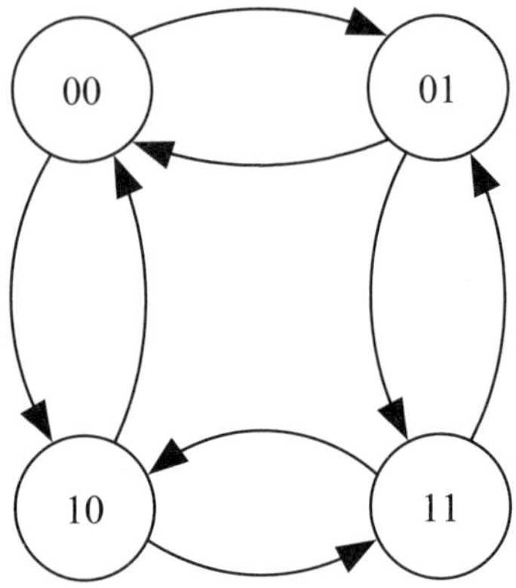

Figure 2.5. *Example of a dependent PBN not containing a steady-state distribution (Shmulevich et al.,* 2002a*).*

The previous example considered an independent PBN. The steady-state requirement can be even more constraining for dependent PBNs. Consider the following very simple PBN with only two network functions (realizations), $\mathbf{f}_1(x_1,x_2) = (f_1^{(1)}, f_1^{(2)}) = (\bar{x}_1, x_2)$ and $\mathbf{f}_2(x_1,x_2) = (f_2^{(1)}, f_2^{(2)}) = (x_1, \bar{x}_2)$. Since the PBN is dependent, the selection of the predictor for the first gene may not be viewed independently of the selection of the predictor for the second gene. These two possible network realizations imply, for instance, that if $f_1^{(1)} = \bar{x}_1$ is selected for the first gene, then $f_2^{(2)} = \bar{x}_2$ cannot be simultaneously selected for the second gene. That is, the probability that the network $\mathbf{f}$ takes on any realization other than the two given above, say $(f_1^{(1)}, f_2^{(2)})$, is zero. The corresponding Markov diagram is given in Fig. 2.5. This PBN is not aperiodic and does not possess a steady-state distribution. Note that the addition of the network function $\mathbf{f}_3(x_1,x_2) = (f_3^{(1)}, f_3^{(2)}) = (\bar{x}_1, \bar{x}_2)$ makes the PBN aperiodic.

It is easy to see from Eq. (2.13) that if the perturbation probability p is nonzero, then the Markov chain becomes irreducible, since there is a nonzero probability of transitioning from any state to any other state in any number of time steps. It may not be possible to transition from a state to itself in one time step (i.e., some of the diagonal entries of the state transition matrix may be zero) as there may not exist a context κ such that $G_\kappa(\mathbf{x}, \mathbf{x}) = 1$, but the chain can always return to the same state in any other number of steps by virtue of random perturbations, implying that the chain is also aperiodic and thus ergodic.

The fact that the Markov chain is ergodic implies that it possesses a steady-state distribution equal to the stationary distribution, which can be estimated empirically simply by running the network for a sufficiently long time and by collecting information about the proportion of time the process spends in each state (recall Fig. 2.1). The convergence rate, however, will surely depend on the perturbation probability p.

2.5 Steady-State Analysis of PBNs

As we discussed above, a key aspect of the analysis of the dynamics of PBNs is the determination of their steady-state (long-run) behavior. This is a crucial task in many contexts.

For instance, suppose we are interested in the long-term joint behavior of several selected genes. That is, we would like to obtain their limiting joint distribution. This information can supply answers to questions of the following type: "What is the probability that gene A will be expressed in the long run?" or "What is the probability that gene B and gene C will *both* be expressed in the long run?" Steady-state analysis is necessary for answering such questions.

When the network is not very large, it is possible to compute the entire state transition matrix of the corresponding Markov chain. However, it should be remembered that the size of this matrix is md^n, where m is the number of contexts, d is the number of discretization levels, and n is the number of genes. A variety of approaches using iterative, projection, decompositional, and other methods could potentially be used (Stewart, 1994). However, because of the exponential growth of the state space, the computation of the state transition matrix becomes prohibitive for matrix-based numerical analysis of even moderately sized networks.

For network sizes that do allow for the computation of the state transition matrix, there are several ways in which this matrix computation can be performed more efficiently than computing each entry one by one. First, in the general context-sensitive case, it can be seen from Eq. (2.13) that the transition matrix is essentially decoupled into a matrix that corresponds to the state transitions of the constituent Boolean networks (defined by the matrices $G_\kappa(\mathbf{x}_1,\mathbf{x}_2)$, where κ is either κ_1 or κ_2) and a matrix that corresponds to perturbations. In other words, the state transition matrix is a sum of the matrix without perturbation multiplied by $(1-p)^n$ and the perturbation matrix comprised of $(1-p)^{(n-\eta(\mathbf{x}_1,\mathbf{x}_2))}p^{\eta(\mathbf{x}_1,\mathbf{x}_2)}1(\eta(\mathbf{x}_1,\mathbf{x}_2)\neq 0)$.

For the special case of instantaneously random PBNs, with perturbations, the state transition matrix $P(\mathbf{x}_1,\mathbf{x}_2)$ is expressed as

$$P(\mathbf{x}_1,\mathbf{x}_2) = (1-p)^n A(\mathbf{x}_1,\mathbf{x}_2) + (1-p)^{n-\eta(\mathbf{x}_1,\mathbf{x}_2)} p^{\eta(\mathbf{x}_1,\mathbf{x}_2)} 1(\mathbf{x}_1 \neq \mathbf{x}_2), \tag{2.18}$$

where

$$A(\mathbf{x}_1,\mathbf{x}_2) = \sum_{j=1}^{m} c_j G_j(\mathbf{x}_1,\mathbf{x}_2). \tag{2.19}$$

Thus, as pointed out by Ching *et al.* (2007), since the perturbation matrix (the second term in the above expression, which we can denote as $\tilde{P}$) is the same for different networks as it depends only on the number of genes and the gene perturbation probability p, once this matrix is constructed, it can be used for all other networks thereon after.

Continuing with the instantaneously random case and focusing our attention only on $A(\mathbf{x}_1,\mathbf{x}_2)$, the part of the state transition matrix that is determined by the constituent Boolean networks, it can further be noticed that this matrix is sparse in practice and therefore much time would be used to compute the zero entries. One approach to speed up the computation, proposed by Zhang *et al.* (2007), is to consider only the nonzero entries. The idea is that each entry in the matrix (recall that we are not considering the perturbations because of the decoupling) is the sum of the probabilities of each of the constituent Boolean networks that are able to perform the one-step transition corresponding to the entry in the matrix. Not having to compute the zero entries in the state transition matrix reduces the computational complexity from $O(n \cdot m \cdot d^{2n})$ to $O(n \cdot m \cdot d^n)$, where n is the number of genes, m is the number of constituent networks, and d is the number of discretization levels (Ching *et al.*, 2007).

Having the state transition matrix at hand, now including the perturbation part, the computation of the steady-state distribution essentially amounts to computing the eigenvector corresponding to the maximum eigenvalue (which for a stochastic matrix is guaranteed to be equal to one, with all other eigenvalues being less than one). In other words, the steady-state distribution π, written here as a row vector, is the solution to the equation $\pi = \pi P$, where P is the state transition matrix (see Eq. (2.17)). This can be done using any number of numerical methods, a particularly simple and effective one being the power method (Stewart, 1994). Briefly, the power method is a simple algorithm that is effective for finding the eigenvector corresponding to the dominant eigenvalue in large matrices. Starting from an initial vector π_0, this algorithm is described by the iteration

$$\pi_{k+1} = \frac{\pi_k P}{\|\pi_k P\|}.$$

A subsequence of π_k is guaranteed to converge to the steady-state distribution, with the convergence rate determined by the ratio $|\lambda_2/\lambda_1|$, where λ_2 is the second dominant eigenvalue (recall that for the stochastic matrix P, $\lambda_1 = 1$). We note that the power method is used by Google to compute the page rank of webpages in their search engine.

A further savings in computation of the state transition matrix can be achieved by observing that in a realization of a PBN, many constituent Boolean networks can have very low probabilities of being selected. Thus, as proposed by Ching *et al.* (2007), one can consider only those Boolean networks whose probability is greater than a certain threshold. Let us examine the expected error in the steady-state distribution that would result by neglecting constituent Boolean networks whose probabilities of selection are below some threshold.

Suppose there are n_0 Boolean networks that are removed whose corresponding transition matrices are $(G_1, G_2, \ldots, G_{n_0})$ with selection probabilities $c_1, c_2, \ldots, c_{n_0}$, respectively. Then, after the removal of these n_0 Boolean networks followed by normalization, the state transition matrix becomes

$$\hat{A} = \frac{(1-p)^n}{1 - \sum_{i=1}^{n_0} c_i} \times \left(A - \sum_{i=1}^{n_0} c_i G_i \right) + \tilde{P}. \tag{2.20}$$

Let $\hat{\pi}$ be the steady-state probability vector for the linear system $\hat{\pi}\hat{A} = \hat{\pi}$, which, using Eq. (2.20), results in

$$\hat{\pi}\left((1-p)^n \left(A - \sum_{i=1}^{n_0} c_i G_i \right) + \left(1 - \sum_{i=1}^{n_0} c_i \right) \tilde{P} \right)$$
$$= \left(1 - \sum_{i=1}^{n_0} c_i \right) \hat{\pi}. \tag{2.21}$$

Therefore we have

$$
\begin{aligned}
&\|\hat{\pi} P - \hat{\pi}\|_\infty \\
&= \|\hat{\pi}((1-p)^n A + \tilde{P}) - \hat{\pi}\|_\infty \\
&= \left\| \hat{\pi}(1-p)^n \sum_{i=1}^{n_0} c_i G_i + \hat{\pi}\tilde{P} \sum_{i=1}^{n_0} c_i - \hat{\pi} \sum_{i=1}^{n_0} c_i \right\|_\infty \\
&= \left\| \sum_{i=1}^{n_0} c_i \hat{\pi}((1-p)^n G_i + \tilde{P} - I) \right\|_\infty \\
&\leq \sum_{i=1}^{n_0} c_i \hat{\pi} \|((1-p)^n G_i + \tilde{P} - I)\|_\infty .
\end{aligned}
\tag{2.22}
$$

We note that in each row of $G_k, k = 1, 2, \ldots, n_0$, there is one nonzero entry equal to one, since each such matrix corresponds to one Boolean network with deterministic state transitions. As the exact form of each A_k is unknown, we can assume that the position of the nonzero entry is uniformly distributed. Let

$$
y_i = \sum_{j=1}^{2^n} [(1-p)^n G_k + \tilde{P}]_{ij}, \qquad Y = \max\{y_1, y_2, \ldots, y_{2^n}\}.
$$

Since the term $(1-p)^n G_k + \tilde{P}$ is the state transition matrix with perturbation corresponding to the kth constituent Boolean network, $E[y_i] = 1$. Recall that the Chernoff bound states that if $Z_1, Z_2, \ldots, Z_m$ are independent Poisson trials such that for $i = 1, \ldots, m$, $P(Z_i) = p_i$, where $0 < p_i < 1$ (here, $P(\cdot)$ denotes probability), then for

$$
Z = \sum_{i=1}^{m} Z_i, \quad \mu = E[Z] = \sum_{i=1}^{m} p_i
$$

and $\delta > 2e - 1$,

$$
P(Z > (1+\delta)\mu) < 2^{-(1+\delta)\mu}.
$$

By letting $m = 2^n$, $\delta = 2n$, and $\mu = 1$, we have

$$
P(y_i > (1+2n)) < \frac{1}{2^{1+2n}},
$$

and since the event $\{Y > (1+2n)\} = \cup_{i=1}^{2^n} \{y_i > (1+2n)\}$,

$$
P(Y > (1+2n)) \leq \sum_{i=1}^{2^n} P(y_i > (1+2n)) < 2^n \times \frac{1}{2^{1+2n}} = \frac{1}{2^{1+n}}.
$$

We should note that n should be larger than 2 since $n > (2e-1)/2$ is assumed (e is the base of the natural logarithm). However, if $n \leq 2$, $Y \leq 1 + 2n$ always holds.

Thus, we have

$$
E[Y] < \max\left(1 + 2n, 2^n \times \frac{1}{2^{1+n}}\right) = 1 + 2n
$$

and

$$\begin{aligned}E[\|\hat{\pi}((1-p)^n G_k + \tilde{P}) - \hat{\pi}\|_\infty] &< (1+E[Y])\|\hat{\pi}\|_\infty \\ &< (2+2n)\|\hat{\pi}\|_\infty.\end{aligned}$$

Finally, from Eq. (2.22) and the above, the expected residual is bounded by

$$E[\|\hat{\pi}P - \hat{\pi}\|_\infty] < \sum_{i=1}^{n_0} c_i(2+2n)\|\hat{\pi}\|_\infty.$$

If $\|\pi\|_\infty$ is equal to or very close to $\|\hat{\pi}\|_\infty$, then

$$E\left[\frac{\|\hat{\pi}P - \hat{\pi}\|_\infty}{\|\pi\|_\infty}\right] < \sum_{i=1}^{n_0} c_i(2+2n).$$

Since this error estimate gives only an expected upper bound on the error, it cannot be applied in all cases to estimate n_0. From this analysis, we can see that the error bound can be determined by the total probability of the removed constituent Boolean networks and the number of genes in the PBN. Ching *et al.* (2007) also derived theoretical results on the probability distribution of the number of "neglected" Boolean networks given a threshold level, under the assumption that there are only two Boolean functions assigned to each gene.

2.5.1 Steady-state analysis via simulation

It is clear that given the state transition matrix of the Markov chain corresponding to a PBN, one can compute the steady-state distribution directly from it. However, as already mentioned, this is not always possible due to the exponential growth of the state transition matrix relative to the number of genes. One useful approach for determining such steady-state probabilities is to use Monte Carlo simulation methods. Very large state spaces are commonly encountered in Markov chain Monte Carlo (MCMC) methods for many applications, including Markov random field modeling in image processing (Winkler, 1995), where efficient simulation and estimation are routinely performed. Thus, Monte Carlo methods represent a viable alternative to numerical matrix-based methods for obtaining steady-state distributions. Informally speaking, this consists of running the Markov chain for a sufficiently long time until convergence to the stationary distribution is reached and observing the proportion of time the process spends in the parts of the state space that represent the information of interest, such as the joint stationary distribution of several specific genes. A key factor is convergence, which to a large extent depends on the perturbation probability p. In general, a larger p results in quicker convergence, but making p too large is not biologically meaningful.

In order for us to perform long-term analysis of the Markov chain corresponding to a PBN using Monte Carlo methods, we need to be able to estimate the convergence rate of the process. Only after we are sufficiently sure that the chain has reached its stationary distribution can we begin collecting information from it. Typical approaches for assessing convergence are based on the second-largest eigenvalue of the state transition matrix. Unfortunately, as mentioned above, for even a moderate number of genes, obtaining the eigenvalues of the transition matrix may be impractical. Thus, it is advantageous to be able to determine the number of iterations necessary until satisfactory convergence is reached. One approach for obtaining *a priori* bounds on the number of iterations is based on the so-called *minorization condition* for Markov chains (Rosenthal, 1995). This approach is discussed by Shmulevich *et al.* (2003) in the context of PBNs. However, their analysis indicates that even if one makes assumptions about the relative magnitudes of the perturbation and transition probabilities, one is not likely to obtain a useful bound on convergence via this approach. Another approach is to diagnose the convergence to the steady-state distribution, to which we turn now.

Diagnosing convergence In a practical situation, it is important to be able to empirically determine when to stop the chain and produce our estimates. For this purpose, there are a number of monitoring methods available (Cowles and Carlin, 1996; Robert, 1995). Consider, for example, the Kolmogorov–Smirnov test—a nonparametric test of stationarity that can be used to assess convergence.

When the chain is stationary, distributions $\pi^{(k_1)}$ and $\pi^{(k_2)}$ are the same for arbitrary times k_1 and k_2. Thus, given a sample $\mathbf{x}^{(1)},\ldots,\mathbf{x}^{(T)}$, we can compare the two halves: $\mathbf{x}^{(1)},\ldots,\mathbf{x}^{(T/2)}$ and $\mathbf{x}^{(T/2+1)},\ldots,\mathbf{x}^{(T)}$. In order to correct for non-i.i.d. (correlated) samples, we introduce a "batch size" G leading to the construction of two (quasi-) independent samples (Robert and Casella, 1999). We thus select subsamples $\mathbf{x}_1^{(G)},\mathbf{x}_1^{(2G)},\ldots$ and $\mathbf{x}_2^{(G)},\mathbf{x}_2^{(2G)},\ldots$ and use the Kolmogorov–Smirnov statistic with the lexicographical ordering to define the indicator:

$$K = \frac{1}{M}\max_{\eta}\left|\sum_{g=1}^{M} 1_{[0\cdots 0,\eta)}\left(\mathbf{x}_1^{(gG)}\right) - \sum_{g=1}^{M} 1_{[0\cdots 0,\eta)}\left(\mathbf{x}_2^{(gG)}\right)\right|, \tag{2.23}$$

where the maximum is over the state space, the vertices of an n-dimensional Boolean hypercube (for binary-valued networks). As $\sqrt{M}K$ has the cumulative distribution function $R(x) = 1 - \sum_{k=1}^{\infty}(-1)^{k-1}e^{-2k^2x^2}$ (Robert and Casella, 1999), the corresponding p-value can be computed for each T until it reaches a desired level.

This can also be used to assess convergence for a selected group of genes $j_1,\ldots,j_m$ by replacing state vectors $\mathbf{x}_1^{(gG)}$ and $\mathbf{x}_2^{(gG)}$ in Eq. (2.23) with just the vectors of their j_1th, $\ldots, j_m$th coordinates and modifying the domain of η into the hypercube over these coordinates only. For example, if only the distribution for the first gene is of interest, the maximum in Eq. (2.23) degenerates into just an absolute difference of the numbers of zeros in the first coordinate between the two samples.

Two-state Markov chain approach Another approach originally proposed by Raftery and Lewis (1992) can be useful in the context of PBNs (Shmulevich *et al.*, 2003). This method reduces the study of the convergence of the chain to the study of the convergence of a two-state Markov chain. Suppose that we are interested in knowing the steady-state probability of the event {Gene A is ON and Gene B is OFF}. Then, we can partition the state space into two disjoint subsets such that one subset contains all states on which the event occurs and the other subset contains the rest of the states. Consider the two "metastates" corresponding to these two subsets. Although the sequence of these metastates does not form a Markov chain in itself, it can be approximated by a first-order Markov chain if

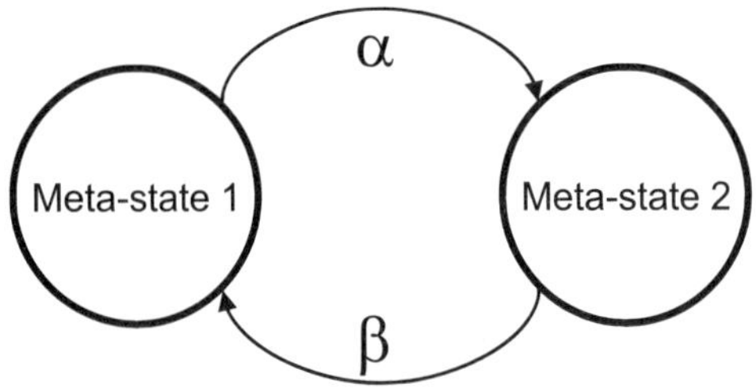

Figure 2.6. *An illustration of the two metastates in the approach of Raftery and Lewis* (1992), *with transition probabilities α and β.*

every k states from the original Markov chain are discarded (i.e., the chain is subsampled), if one assumes that the dependence in the sequence of metastates falls off fairly rapidly with lag. It turns out in practice that k is usually set to 1, meaning that nothing is discarded and the sequence of metastates is treated as a homogeneous Markov chain (see Raftery and Lewis (1992) for details) with transition probabilities α and β between the two metastates. This is illustrated in Fig. 2.6.

Using standard results for two-state Markov chains, it can be shown that the "burn-in" period (the number of iterations necessary to achieve stationarity) m_0 satisfies

$$m_0 \geq \log\left(\frac{\varepsilon\,(\alpha+\beta)}{\max(\alpha,\beta)}\right) / \log(1-\alpha-\beta). \tag{2.24}$$

For the examples that we will give below, we will set $\varepsilon = 0.001$. In addition, it can be shown that the minimum total number of iterations ν necessary to achieve a desired accuracy r (we use $r = 0.01$ in our examples below) is

$$\nu = \frac{\alpha\beta\,(2-\alpha-\beta)}{(\alpha+\beta)^3}\left(\frac{r}{\Phi\left(\frac{1}{2}(1+s)\right)}\right)^{-2}, \tag{2.25}$$

where $\Phi(\cdot)$ is the standard normal cumulative distribution function and s is a parameter that we set to 0.95 in our examples. For detailed explanations of the "precision" parameters ε, r, and s, see Raftery and Lewis (1992). The question becomes how to estimate the transition probabilities α and β, as these are unknown. The solution is to perform a test run from which α and β can be estimated and from which m_0 and ν can then be computed. Then, another run with the computed burn-in period m_0 and the total number of iterations N is performed and the parameters α and β are re-estimated from which m_0 and ν are recomputed. This can be done several times in an iterative manner until the estimates of m_0 and ν are smaller than the number of iterations already achieved. Let us now give an example of how this method can be used to determine the steady-state probabilities of several genes.

Using a human glioma gene expression data set described in Fuller *et al.* (1999), Shmulevich *et al.* (2003) constructed a small subnetwork consisting of 15 genes. The entire 597-gene network was inferred using the coefficient of determination, as will be described in Chapter 3 (see also Shmulevich *et al.* (2002a)). The algorithm for building the

Table 2.1. *Steady-state analysis of several pairs of genes.*

Tie-2	NFκB	%	Tie-2	TGFB3	%	TGFB3	NFκB	%
OFF	OFF	15.68	OFF	OFF	14.75	OFF	OFF	10.25
OFF	ON	41.58	OFF	ON	42.50	OFF	ON	12.47
ON	OFF	9.21	ON	OFF	7.96	ON	OFF	14.64
ON	ON	31.53	ON	ON	32.78	ON	ON	60.65

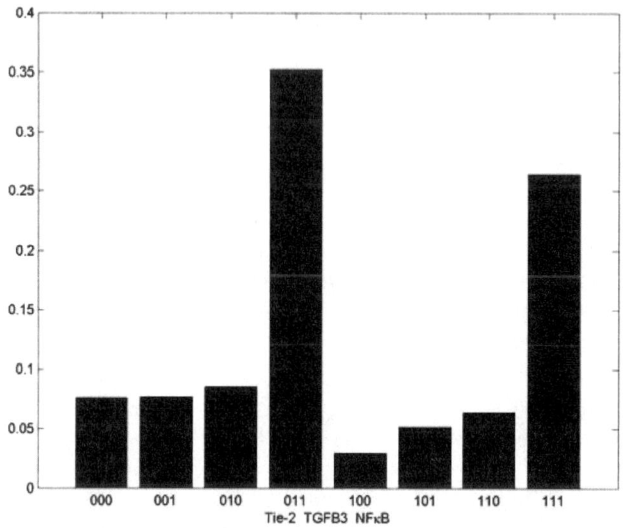

Figure 2.7. *Steady-state analysis of all three genes: Tie-2, TGFB3, and NFκB. The gene combinations are coded in binary. For example,* 010 *means that Tie-2 is OFF, TGFB3 is ON, and NFκB is OFF. The bars show the joint steady-state probabilities (Shmulevich et al.,* 2003*).*

subnetwork starting from so-called seed genes, which uses influences of genes and ensures that the subnetwork functions fairly autonomously from the rest of the genes, is described by Hashimoto *et al.* (2004).

The joint steady-state probabilities of several combinations of two genes were analyzed: Tie-2 and NFκB; Tie-2 and TGFB3; and TGFB3 and NFκB. For example, for Tie-2 and NFκB, the two-state Markov chain method described above, when applied to an initial run of 10,000 iterations, produced a burn-in period of $m_0 = 87$ and a total number of iterations of $\nu = 48,268$. The transition probabilities α and β were both approximately equal to 0.03. The perturbation probability p was set to 0.001. After running the network for another 38,268 steps, the recomputed values of m_0 and ν became 91 and 50,782, respectively. Running the network for yet another 3,000 iterations was sufficient for the given accuracy and the steady-state probabilities of these two genes could be determined. The steady-state probabilities for all pairs of considered genes are shown in Table 2.1 as percentages. Figure 2.7 shows the joint steady-state probabilities for all three of these genes using a bar graph.

Tie-2 is a receptor tyrosine kinase expressed on the endothelial cells. Its two ligands, angiopoietin 1 and 2, bind Tie-2 and regulate vasculogenesis (Sato *et al.*, 1993), an important process in embryonic development and tumor development. Other related regulators for vasculogenesis are VEGF and VEGF receptors, which are often overexpressed in the advanced stage of gliomas (Cheng *et al.*, 1996). Although no experimental evidence supports a direct transcriptional regulation of those regulators by the transcriptional factor NFκB, which is also frequently activated in glioma progression (Hayashi *et al.*, 2001) as predicted in this analysis, the results showed that NFκB, at least indirectly, may influence the expression of Tie-2 expression. Thus, it may not be surprising that when NFκB is on, Tie-2 is on about $31.53/(41.58+31.53) = 43\%$ of the time. Because Tie-2 is only one of the regulators for the important vasculogenesis in glioma progression, it is consistent that this analysis of long-term (steady-state) gene expression activities shows that about 40% of the time Tie-2 is on. In contrast, NFκB is on 73% of the time, implying that fewer redundancies exist for NFκB activity.

Interestingly, a similar relationship exists between Tie-2 and TGFB3, as can be seen by comparing the percentages in columns 3 and 6 of Table 2.1. This suggests that TGFB3 and NFκB are more directly linked, which is also shown in the last three columns of the table (60% of the time, they are both on). This relationship is very likely because TGFB1, a homologue of TGFB3, was shown to have a direct regulatory relationship with NFκB (Arsura *et al.*, 1996). Furthermore, a study by Strauch *et al.* (2003) showed that NFκB indeed regulates TGFB3 expression (and cell migration).

Steady-state analysis of PBNs is an important problem. For even $n = 20$ genes, working with $2^{20} \times 2^{20}$ matrices becomes cumbersome and quickly prohibitive for larger n. However, Monte Carlo techniques can be successfully used as long as we are sufficiently confident that the Markov chain corresponding to the PBN has converged to its equilibrium distribution. Moreover, Monte Carlo techniques exhibit favorable scaling behavior with respect to the number of genes. Despite the fact that the size of the state space grows exponentially with n, efficient steady-state analysis can still be carried out. We again note that MCMC methods on images containing $512 \times 512 = 2.6214 \times 10^5$ pixels, resulting in state spaces on the order of $10^{78,912}$ for binary models, can nonetheless be effectively performed.

Another important application of steady-state analysis concerns the computation of the steady-state probabilities of the attractors in PBNs. Clearly, the existence of the entire state transition matrix makes this problem straightforward, since the steady-state probabilities of the states on the attractors can either be pulled out from the entire steady-state distribution, if available, or estimated by simulation as discussed above. It is also possible to compute the steady-state probabilities of the attractors (and basins) from the PBN parameters directly, as shown by Brun *et al.* (2005).

2.5.2 Steady-state probabilities of attractors and basins

The attractors of a Boolean network characterize its long-run behavior. If, however, we incorporate random perturbation, then the network can escape its attractors. In this case, full long-run behavior is characterized by its steady-state distribution, as discussed above. Nonetheless, if the probability of perturbation is very small, the network will lie in its attractor cycles for a large majority of the time, meaning that attractor states will carry most of the steady-state probability mass. The amount of time spent in any given attractor depends on its basin. Large basins tend to produce attractors possessing relatively large steady-state mass.

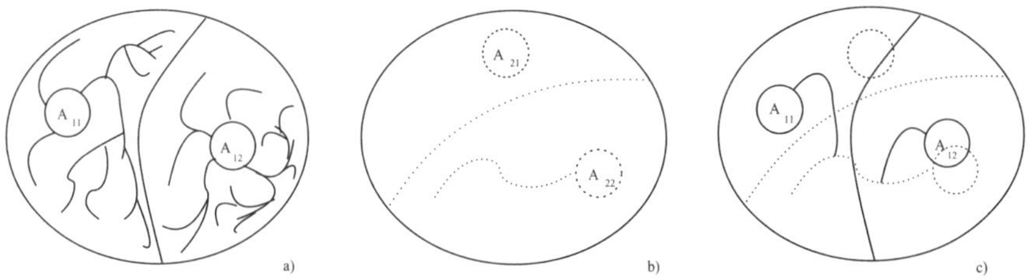

Figure 2.8. *An illustration of the behavior of a context-sensitive PBN (Brun et al., 2005).*

Let us now consider context-sensitive PBNs. So long as there is no functional switching (i.e., a new realization of a constituent network), the current Boolean-network realization of the PBN characterizes the activity of the PBN and it will transition into one of its attractor cycles and remain there until a switch occurs. When a network switch does occur, the present state becomes an initialization for the new realization and the network will transition into the attractor cycle whose basin contains the present state. It will remain there until another network switch. Recall that the attractor family of the PBN is defined to be the union of all the attractors in the constituent Boolean networks and that the attractors of a PBN need not be disjoint, although those corresponding to each constituent Boolean network must necessarily be disjoint.

Figure 2.8 shows an example of the behavior of a context-sensitive PBN relative to its attractors under a change of function. Part (a) shows the attractor cycles A_{11} and A_{12} for a network function $\mathbf{f}_1$, its corresponding basins, and some trajectories. Part (b) shows the attractor cycles A_{21} and A_{22} for a network function $\mathbf{f}_2$ and its corresponding basins. In part (c), we can see that if the system is using the function $\mathbf{f}_2$ and it makes a function change, to $\mathbf{f}_1$, then the future of the system depends on which part of the trajectory it is at the moment of the function change. In this example, for the particular trajectory shown with the dotted line toward the attractor A_{22}, the first part of the trajectory is in the basin corresponding to the attractor A_{11}, and the end of the trajectory is inside the basin corresponding to the attractor A_{12}. Therefore, if the change of function occurs before the system crosses the boundary between the basins, it will transition toward the attractor A_{11}. If the change of function occurs after it crosses the boundary, then it will transition toward the attractor A_{12}. In particular, we see that the attractor A_{22} lies completely inside the basin corresponding to the attractor A_{12}. In this case, if a change of function occurs when the system is inside the attractor A_{22}, it will always transition to the attractor A_{12}.

If one now incorporates perturbation into the PBN model, the stationary distribution characterizes the long-run behavior of the network. If both the switching and perturbation probabilities are very small, then the attractors still carry most of the steady-state probability mass. This property has been used to validate network inference from data (Kim *et al.*, 2002; Zhou *et al.*, 2004) and to formulate analytic expressions of the probabilities of attractor states (Brun *et al.*, 2005), to which we turn next.

As we have already seen, for a context-sensitive PBN, the state $\mathbf{X}_t$ at time t does not constitute a homogeneous Markov chain because the transition probabilities $P[\mathbf{X}_{t+1} = \mathbf{x}|\mathbf{X}_t = \mathbf{y}]$ depend on the vector function $\mathbf{f}$ selected at time t. Recall that instead of the GAPs $\mathbf{x}$ serving as the states of a Markov chain, we can represent the states $(\mathbf{x}, \mathbf{f})$ as states of a homogeneous Markov chain, $(\mathbf{X}_t, \mathbf{F}_t)$, with the following notation for the transition probabilities (see Eq. (2.7)):

$$P_{\mathbf{y},\mathbf{g}}(\mathbf{x},\mathbf{f}) = P[\mathbf{X}_{t_0+1} = \mathbf{x}, \mathbf{F}_{t_0+1} = \mathbf{f} | \mathbf{X}_{t_0} = \mathbf{y}, \mathbf{F}_{t_0} = \mathbf{g}]$$

for any time t_0. Here, we use the notation (GAP, context) for convenience. The steady-state probabilities $\pi(\mathbf{x})$ are the marginal probabilities of the steady-state probabilities $\pi(\mathbf{x},\mathbf{f})$. Let

$$P^{(t)}_{\mathbf{y},\mathbf{g}}(\mathbf{x},\mathbf{f}) = P[\mathbf{X}_{t_0+t} = \mathbf{x}, \mathbf{F}_{t_0+t} = \mathbf{f} | \mathbf{X}_{t_0} = \mathbf{y}, \mathbf{F}_{t_0} = \mathbf{g}],$$

for any time t_0, be the probability of transition from $(\mathbf{y},\mathbf{g})$ to $(\mathbf{x},\mathbf{f})$ in t steps. The steady-state distribution π of the Markov chain is now defined, for any starting state-function pair $(\mathbf{y},\mathbf{g})$, by $\pi(\mathbf{x},\mathbf{f}) = \lim_{t\to\infty} P^{(t)}_{\mathbf{y},\mathbf{g}}(\mathbf{x},\mathbf{f})$. We call *restart* the situation where there occurs a perturbation, a function change, or both of them at the same time. We define, for each time t, the random variables $\tau_p(t)$ and $\tau_c(t)$ that measure the time elapsed between t and the last perturbation and the time elapsed between t and the last function change, respectively. The random variable $\tau(t) = \min(\tau_c(t), \tau_p(t))$ measures the time elapsed between t and the last restart. The events $\tau_p(t) = 0$ and $\tau_c(t) = 0$ indicate the existence of perturbation and function change, respectively, at time t. They are determined by the probabilities p and q, being $P[\tau_p(t) = 0] = 1-(1-p)^n$ and $P[\tau_c(t) = 0] = q$, and are independent of the time t. For each value of t, if we assume that there has been at least one perturbation before time t, the distribution for $\tau(t)$ is given by

$$P[\tau(t) = j] = \frac{(1-b)b^j}{1-b^t} 1_{[0\le j\le t-1]}, \tag{2.26}$$

where $b = (1-q)(1-p)^n$.

Since the Markov chain is homogeneous, and the probabilities $\tau_p(t) = 0$ and $\tau_c(t) = 0$ are independent of t, for any time t we can split the transition matrix $P_{\mathbf{y},\mathbf{f}_l}(\mathbf{x},\mathbf{f}_k)$ by conditioning it on the existence of a perturbation and function change:

$$\begin{aligned}
&P_{\mathbf{y},\mathbf{f}_l}(\mathbf{x},\mathbf{f}_k) = P[\mathbf{X}_t = \mathbf{x}, \mathbf{F}_t = \mathbf{f}_k | \mathbf{X}_{t-1} = \mathbf{y}, \mathbf{F}_{t-1} = \mathbf{f}_l] \\
&= P[\mathbf{X}_t = \mathbf{x}, \mathbf{F}_t = \mathbf{f}_k | \mathbf{X}_{t-1} = \mathbf{y}, \mathbf{F}_{t-1} = \mathbf{f}_l, \tau_p(t) > 0, \tau_c(t) > 0] P[\tau_p(t) > 0, \tau_c(t) > 0] \\
&+ P[\mathbf{X}_t = \mathbf{x}, \mathbf{F}_t = \mathbf{f}_k | \mathbf{X}_{t-1} = \mathbf{y}, \mathbf{F}_{t-1} = \mathbf{f}_l, \tau_p(t) > 0, \tau_c(t) = 0] P[\tau_p(t) > 0, \tau_c(t) = 0] \\
&+ P[\mathbf{X}_t = \mathbf{x}, \mathbf{F}_t = \mathbf{f}_k | \mathbf{X}_{t-1} = \mathbf{y}, \mathbf{F}_{t-1} = \mathbf{f}_l, \tau_p(t) = 0, \tau_c(t) > 0] P[\tau_p(t) = 0, \tau_c(t) > 0] \\
&+ P[\mathbf{X}_t = \mathbf{x}, \mathbf{F}_t = \mathbf{f}_k | \mathbf{X}_{t-1} = \mathbf{y}, \mathbf{F}_{t-1} = \mathbf{f}_l, \tau_p(t) = 0, \tau_c(t) = 0] P[\tau_p(t) = 0, \tau_c(t) = 0].
\end{aligned}$$

Each condition corresponds to one of the four cases listed previously. Table 2.2 summarizes the probabilities of the conditional transition matrices for the events.

In Table 2.2, the last 3 events correspond to restarts of the system. The restart probability $P^*_{\mathbf{y},\mathbf{f}_l}(\mathbf{x},\mathbf{f}_k)$, for any t, is the probability of being in $(\mathbf{x},\mathbf{f}_k)$ after a restart, given that the system was in $(\mathbf{y},\mathbf{f}_l)$ before the restart:

$$P^*_{\mathbf{y},\mathbf{f}_l}(\mathbf{x},\mathbf{f}_k) = P[\mathbf{X}_t = \mathbf{x}, \mathbf{F}_t = \mathbf{f}_k | \mathbf{X}_{t-1} = \mathbf{y}, \mathbf{F}_{t-1} = \mathbf{f}_l, \tau(t) = 0].$$

Table 2.2. *Conditional transition probabilities for the four cases.*

Event A	Probability $P(A)$	$P[\mathbf{X}_t = \mathbf{x}, \mathbf{F}_t = \mathbf{f}_k \mid \mathbf{X}_{t-1} = \mathbf{y}, \mathbf{F}_{t-1} = \mathbf{f}_l, A]$
$[\tau_p(t) > 0, \tau_c(t) > 0]$	$(1-p)^n(1-q)$	$\mathbf{1}_{[\mathbf{x}=\mathbf{f}_l(\mathbf{y})]} \cdot \mathbf{1}_{[k=l]}$
$[\tau_p(t) > 0, \tau_c(t) = 0]$	$(1-p)^n q$	$P(\mathbf{F} = \mathbf{f}_k) \cdot \mathbf{1}_{[\mathbf{x}=\mathbf{f}_l(\mathbf{y})]}$
$[\tau_p(t) = 0, \tau_c(t) > 0]$	$(1-(1-p)^n)(1-q)$	$\frac{p^{\eta(\mathbf{x},\mathbf{y})}(1-p)^{n-\eta(\mathbf{x},\mathbf{y})}}{1-(1-p)^n} \cdot \mathbf{1}_{[\mathbf{x}\neq\mathbf{y}]} \cdot \mathbf{1}_{[k=l]}$
$[\tau_p(t) = 0, \tau_c(t) = 0]$	$(1-(1-p)^n)q$	$\frac{p^{\eta(\mathbf{x},\mathbf{y})}(1-p)^{n-\eta(\mathbf{x},\mathbf{y})}}{1-(1-p)^n} \cdot P(\mathbf{F} = \mathbf{f}_k) \cdot \mathbf{1}_{[\mathbf{x}\neq\mathbf{y}]}$

A direct way to compute the steady-state probabilities $\pi(A_{ki})$ is as the marginal probabilities of $\pi(A_{ki}, \mathbf{f}_l)$. We will first compute the steady-state probabilities $\pi(B_{ki}, \mathbf{f}_k)$ for the basins, then the steady-state probabilities $\pi(A_{ki}, \mathbf{f}_k)$ for the attractors, and finally the marginal steady-state probabilities $\pi(A_{ki})$.

The computation of the steady-state probabilities $\pi(A_{ki}, \mathbf{f}_k)$ will be split into three stages:

1. the computation of the steady-state probabilities $\pi(B_{ki}, \mathbf{f}_k)$ of the basins,
2. the computation of the conditional probabilities $\pi(A_{ki}, \mathbf{f}_k \mid B_{ki}, \mathbf{f}_k)$,
3. the approximation to the marginal steady-state probabilities $\pi(A_{ki})$.

Basins The next theorem represents the steady-state distribution of the basins as the solution of a system of linear equations in which the coefficients are perturbation-based probabilities. We will need the following definition. For any starting state-function pair $(\mathbf{h}, \mathbf{g})$, let

$$P^*_{B_{lv},\mathbf{f}_l}(B_{ki}, \mathbf{f}_k) = \lim_{t\to\infty} P[\mathbf{X}_t \in B_{ki}, \mathbf{F}_t = \mathbf{f}_k \mid \mathbf{X}_{t-1} \in B_{lv}, \mathbf{F}_{t-1} = \mathbf{f}_l, \mathbf{X}_0 = \mathbf{h}, \mathbf{F}_0 = \mathbf{g}, \tau(t) = 0]$$

be the steady-state probability of a restart in the basin B_{ki} using the function $\mathbf{f}_k$ after a restart, given that the system was inside the basin B_{lv} and using the function $\mathbf{f}_l$ before the restart.

Theorem 2.2 (Brun *et al.*, 2005).

$$\pi(B_{ki}, \mathbf{f}_k) = \sum_{l=1}^{r} \sum_{v=1}^{m_l} P^*_{B_{lv},\mathbf{f}_l}(B_{ki}, \mathbf{f}_k)\pi(B_{lv}, \mathbf{f}_l).$$

Proof. Because we are interested in $\pi(B_{ki}, \mathbf{f}_k) = \lim_{t\to\infty} P^{(t)}_{\mathbf{h}}(B_{ki}, \mathbf{f}_k)$, we can assume that for any t there has been a previous restart, and Eq. (2.26) applies. We can compute $P^{(t)}_{\mathbf{h}}(B_{ki}, \mathbf{f}_k)$ as a function of the probability $P[\tau(t) = j]$, the probability of reaching the basin B_{ki} and the function $\mathbf{f}_k$ at time $t - j$ (after the restart). For any starting state $\mathbf{h}$ and function $\mathbf{g}$,

$$\begin{aligned}
\pi(B_{ki},\mathbf{f}_k) &= \lim_{t\to\infty} P[\mathbf{X}_t \in B_{ki},\mathbf{F}_t = \mathbf{f}_k|\mathbf{X}_0 = \mathbf{h},\mathbf{F}_0 = \mathbf{g}] \\
&= \lim_{t\to\infty} \sum_{j=0}^{\infty} P\left[\mathbf{X}_t \in B_{ki},\mathbf{F}_t = \mathbf{f}_k|\mathbf{X}_0 = \mathbf{h},\mathbf{F}_0 = \mathbf{g},\tau(t) = j\right] P\left[\tau(t) = j\right] \\
&= \lim_{t\to\infty} \sum_{j=0}^{\infty}\sum_{l=1}^{r}\sum_{v=1}^{m_l} \\
&\quad P\left[\mathbf{X}_t \in B_{ki},\mathbf{F}_t = \mathbf{f}_k|\mathbf{X}_0 = \mathbf{h},\mathbf{F}_0 = \mathbf{g},\tau(t) = j,\mathbf{X}_{t-j-1} \in B_{lv},\mathbf{F}_{t-j-1} = \mathbf{f}_l\right] \\
&\quad \times P\left[\mathbf{X}_{t-j-1} \in B_{lv},\mathbf{F}_{t-j-1} = \mathbf{f}_l|\mathbf{X}_0 = \mathbf{h},\mathbf{F}_0 = \mathbf{g},\tau(t) = j\right] P\left[\tau(t) = j\right] \\
&= \lim_{t\to\infty} \sum_{j=0}^{\infty}\sum_{l=1}^{r}\sum_{v=1}^{m_l} \\
&\quad P\left[\mathbf{X}_{t-j} \in B_{ki},\mathbf{F}_{t-j} = \mathbf{f}_k|\mathbf{X}_0 = \mathbf{h},\mathbf{F}_0 = \mathbf{g},\tau(t) = j,\mathbf{X}_{t-j-1} \in B_{lv},\mathbf{F}_{t-j-1} = \mathbf{f}_l\right] \\
&\quad \times P\left[\mathbf{X}_{t-j-1} \in B_{lv},\mathbf{F}_{t-j-1} = \mathbf{f}_l|\mathbf{X}_0 = \mathbf{h},\mathbf{F}_0 = \mathbf{g},\tau(t) = j\right] P\left[\tau(t) = j\right],
\end{aligned}$$

where in the fourth equality we have used the fact that if, at time $t-j$, the system reaches the basin B_k and it is using the function $\mathbf{f}_k$, then it will stay there until time t, because the condition $\tau(t) = j$ says that there are no restarts between $t-j$ and t. Looking at Eq. (2.26) for $P\left[\tau(t) = j\right]$, we can see that for a t large enough, $1-b^t > 0.5$. Therefore, for each j, the expression inside the sum is bounded by $2\cdot(1-b)b^j$. Hence, by dominated convergence, the limit can be brought inside and

$$\begin{aligned}
\pi(B_{ki},\mathbf{f}_k) &= \sum_{l=1}^{r}\sum_{v=1}^{m_l}\sum_{j=0}^{\infty} P^*_{B_{lv},\mathbf{f}_l}(B_{ki},\mathbf{f}_k)\pi(B_{lv},\mathbf{f}_l)(1-b)b^j \\
&= \sum_{l=1}^{r}\sum_{v=1}^{m_l} P^*_{B_{lv},\mathbf{f}_l}(B_{ki},\mathbf{f}_k)\pi(B_{lv},\mathbf{f}_l),
\end{aligned}$$

where in the first equality we have used the fact that the probability of being in the basin at time $t-j-1$ is independent of the existence of a perturbation at time $t-j$. □

The expression for the steady-state distribution for the basins depends on the unknown quantities $P^*_{B_{lv},\mathbf{f}_l}(B_{ki},\mathbf{f}_k)$. The next step is to find an approximation for these quantities. For any starting state-function pair $(\mathbf{h},\mathbf{g})$, let

$$\pi^*(\mathbf{x},\mathbf{f}) = \lim_{t\to\infty} P[\mathbf{X}_{t-1} = \mathbf{x},\mathbf{F}_{t-1} = \mathbf{f}|\mathbf{X}_0 = \mathbf{h},\mathbf{F}_0 = \mathbf{g},\tau(t) = 0]$$

be the steady-state probability for $\mathbf{x}$, conditioned on times previous to perturbation. Then, for any set B,

$$\pi^*(\mathbf{x},\mathbf{f}|B,\mathbf{f}) = \lim_{t\to\infty} P[\mathbf{X}_{t-1} = \mathbf{x},\mathbf{F}_{t-1} = \mathbf{f}|\mathbf{X}_{t-1} \in B,\mathbf{F}_{t-1} = \mathbf{f},\mathbf{X}_0 = \mathbf{h},\tau(t) = 0].$$

Lemma 2.3.

$$P^*_{B_{lv},\mathbf{f}_l}(B_{ki},\mathbf{f}_k) = \sum_{\mathbf{x}\in B_{ki}}\sum_{\mathbf{y}\in B_{lv}} P^*_{\mathbf{y},\mathbf{f}_l}(\mathbf{x},\mathbf{f}_k)\pi^*(\mathbf{y},\mathbf{f}_l|B_{lv},\mathbf{f}_l).$$

Proof. For any starting state $\mathbf{h}$ and function $\mathbf{g}$,

$$\begin{aligned}
P^*_{B_{lv},\mathbf{f}_l}(B_{ki},\mathbf{f}_k) &= \lim_{t\to\infty} \sum_{\mathbf{x}\in B_{ki}} \\
&\quad P[\mathbf{X}_t = \mathbf{x}, \mathbf{F}_t = \mathbf{f}_k | \mathbf{X}_{t-1} \in B_{lv}, \mathbf{F}_{t-1} = \mathbf{f}_l, \mathbf{X}_0 = \mathbf{h}, \mathbf{F}_0 = \mathbf{g}, \tau(t) = 0] \\
&= \lim_{t\to\infty} \sum_{\mathbf{x}\in B_{ki}} \sum_{\mathbf{y}\in B_{lv}} \\
&\quad P[\mathbf{X}_t = \mathbf{x}, \mathbf{F}_t = \mathbf{f}_k | \mathbf{X}_{t-1} = \mathbf{y}, \mathbf{F}_{t-1} = \mathbf{f}_l, \mathbf{X}_0 = \mathbf{h}, \mathbf{F}_0 = \mathbf{g}, \tau(t) = 0] \\
&\quad \times P[\mathbf{X}_{t-1} = \mathbf{y} | \mathbf{X}_{t-1} \in B_{lv}, \mathbf{F}_{t-1} = \mathbf{f}_l, \mathbf{X}_0 = \mathbf{h}, \mathbf{F}_0 = \mathbf{g}, \tau(t) = 0] \\
&= \lim_{t\to\infty} \sum_{\mathbf{x}\in B_{ki}} \sum_{\mathbf{y}\in B_{lv}} P[\mathbf{X}_t = \mathbf{x}, \mathbf{F}_t = \mathbf{f}_k | \mathbf{X}_{t-1} = \mathbf{y}, \mathbf{F}_{t-1} = \mathbf{f}_l, \tau(t) = 0] \\
&\quad \times P[\mathbf{X}_{t-1} = \mathbf{y} | \mathbf{X}_{t-1} \in B_{lv}, \mathbf{F}_{t-1} = \mathbf{f}_l, \mathbf{X}_0 = \mathbf{h}, \mathbf{F}_0 = \mathbf{g}, \tau(t) = 0] \\
&= \sum_{\mathbf{x}\in B_{ki}} \sum_{\mathbf{y}\in B_{lv}} P^*_{\mathbf{y},\mathbf{f}_l}(\mathbf{x},\mathbf{f}_k)\pi^*(\mathbf{y},\mathbf{f}_l | B_{lv},\mathbf{f}_l),
\end{aligned}$$

where in the third equality we have removed the conditions $\mathbf{X}_0 = \mathbf{h}$ and $\mathbf{F}_0 = \mathbf{g}$ because the transition probability, under perturbation, is independent of the starting state-function pair. □

The only unknown value is now $\pi^*(\mathbf{y},\mathbf{f}_l | B_{lv},\mathbf{f}_l)$, the steady-state conditional probability of being in $\mathbf{y}$ given that the system is in B_{lv} using the function $\mathbf{f}_l$, and given that there will be a restart in the next point of time. If the probability of restart is small, then the average time between restarts is large, and, for approximation purposes, we can assume that the system has already reached an attractor previous to each restart. Therefore, we make the assumption that if there will be a restart in time t, and the system is in the basin B_{lv} at time $t-1$, then it is inside the attractor A_{lv} corresponding to the basin B_{lv} at time $t-1$. In such a case, for any starting state $\mathbf{h}$ and function $\mathbf{g}$, and for any time t, we have the approximation

$$\begin{aligned}
&P[\mathbf{X}_{t-1} = \mathbf{y} | \mathbf{X}_{t-1} \in B_{lv}, \mathbf{F}_{t-1} = \mathbf{f}_l, \mathbf{X}_0 = \mathbf{h}, \mathbf{F}_0 = \mathbf{g}, \tau(t) = 0] \\
&\quad \approx \frac{1}{|A_{lv}|}\mathbf{1}_{[\mathbf{y}\in A_{lv}]}
\end{aligned}$$

resulting in

$$\pi^*(\mathbf{y},\mathbf{f}_l | B_{lv},\mathbf{f}_l) \approx \frac{1}{|A_{lv}|}\mathbf{1}_{[\mathbf{y}\in A_{lv}]} \tag{2.27}$$

and

$$P^*_{B_{lv},\mathbf{f}_l}(B_{ki},\mathbf{f}_k) \approx \frac{1}{|A_{lv}|} \sum_{\mathbf{x}\in B_{ki}} \sum_{\mathbf{y}\in B_{lv}} P^*_{\mathbf{y},\mathbf{f}_l}(\mathbf{x},\mathbf{f}_k).$$

If we define the values $C_{ki,lv}$ by

$$C_{ki,lv} = \frac{1}{|A_{lv}|} \sum_{\mathbf{x}\in B_{ki}} \sum_{\mathbf{y}\in B_{lv}} P^*_{\mathbf{y},\mathbf{f}_l}(\mathbf{x},\mathbf{f}_k), \tag{2.28}$$

then

$$\pi(B_{ki},\mathbf{f}_k) \approx \sum_{l=1}^{r}\sum_{v=1}^{m_l} C_{ki,lv}\pi(B_{lv},\mathbf{f}_l). \tag{2.29}$$

Equations (2.28) and (2.29) give an approximation for the vector of steady-state probabilities $\pi(B_1),\ldots,\pi(B_m)$ as the stationary probability vector of the stochastic matrix $C_{ki,lv}$.

Attractors Once we have obtained the steady-state probabilities $\pi(B_{ki},\mathbf{f}_k)$, we can compute the steady-state probabilities $\pi(A_{ki},\mathbf{f}_k)$, given by

$$\pi(A_{ki},\mathbf{f}_k) = \pi(A_{ki},\mathbf{f}_k|B_{ki},\mathbf{f}_k)\pi(B_{ki},\mathbf{f}_k)$$

with

$$\pi(A_{ki},\mathbf{f}_k|B_{ki},\mathbf{f}_k) = \lim_{t\to\infty} P^{(t)}_{\mathbf{h},\mathbf{g}}(A_{ki},\mathbf{f}_k|B_{ki},\mathbf{f}_k).$$

To find the expression for $\pi(A_k)$ we need to prove the following.

Lemma 2.4. *For basin B_{ki}, starting state $\mathbf{h}$, starting function $\mathbf{g}$, fixed value $j \geq 0$, and $\mathbf{x} \in B_{ki}$,*

$$\begin{aligned}&\lim_{t\to\infty} P\left[\mathbf{X}_{t-j} = \mathbf{x}, \mathbf{F}_{t-j} = \mathbf{f}_k | \mathbf{X}_{t-j} \in B_{ki}, \mathbf{F}_{t-j} = \mathbf{f}_k, \mathbf{X}_0 = \mathbf{h}, \mathbf{X}_0 = \mathbf{g}, \tau(t) = j\right]\\ &= \frac{1}{\pi(B_{ki},\mathbf{f}_k)}\sum_{l=1}^{r}\sum_{v=1}^{m_l}\sum_{\mathbf{y}\in B_{lv}} P^*_{\mathbf{y},\mathbf{f}_l}(\mathbf{x},\mathbf{f}_k)\pi^*(\mathbf{y},\mathbf{f}_l|B_{lv},\mathbf{f}_l)\pi(B_{lv},\mathbf{f}_l).\end{aligned}$$

Proof. Let L be the limit in the lemma. We split the probability over all possible values for $\tau(t)$. For any $\mathbf{x} \in B_{ki}$,

$$\begin{aligned}L &= \lim_{t\to\infty} \frac{P\left[\mathbf{X}_{t-j} = \mathbf{x}, \mathbf{F}_{t-j} = \mathbf{f}_k | \mathbf{X}_0 = \mathbf{h}, \mathbf{X}_0 = \mathbf{g}, \tau(t) = j\right]}{P\left[\mathbf{X}_{t-j} \in B_{ki}, \mathbf{F}_{t-j} = \mathbf{f}_k | \mathbf{X}_0 = \mathbf{h}, \mathbf{X}_0 = \mathbf{g}, \tau(t) = j\right]}\\
&= \frac{1}{\pi(B_{ki},\mathbf{f}_k)} \lim_{t\to\infty} P\left[\mathbf{X}_{t-j} = \mathbf{x}, \mathbf{F}_{t-j} = \mathbf{f}_k | \mathbf{X}_0 = \mathbf{h}, \mathbf{X}_0 = \mathbf{g}, \tau(t) = j\right]\\
&= \frac{1}{\pi(B_{ki},\mathbf{f}_k)} \lim_{t\to\infty} \sum_{l=1}^{r}\sum_{v=1}^{m_l}\sum_{\mathbf{y}\in B_{lv}}\\
&\quad P\left[\mathbf{X}_{t-j} = \mathbf{x}, \mathbf{F}_{t-j} = \mathbf{f}_k | \mathbf{X}_{t-j-1} = \mathbf{y}, \mathbf{F}_{t-j-1} = \mathbf{f}_l, \mathbf{X}_0 = \mathbf{h}, \mathbf{X}_0 = \mathbf{g}, \tau(t) = j\right]\\
&\quad \times P\left[\mathbf{X}_{t-j-1} = \mathbf{y}, \mathbf{F}_{t-j-1} = \mathbf{f}_l | \mathbf{X}_{t-j-1} \in B_{lv}, \mathbf{F}_{t-j-1} = \mathbf{f}_l, \mathbf{X}_0 = \mathbf{h}, \mathbf{X}_0 = \mathbf{g}, \tau(t) = j\right]\\
&\quad \times P\left[\mathbf{X}_{t-j-1} \in B_{lv}, \mathbf{F}_{t-j-1} = \mathbf{f}_l | \mathbf{X}_0 = \mathbf{h}, \mathbf{X}_0 = \mathbf{g}, \tau(t) = j\right]\\
&= \frac{1}{\pi(B_{ki},\mathbf{f}_k)}\sum_{l=1}^{r}\sum_{v=1}^{m_l}\sum_{\mathbf{y}\in B_{lv}} P^*_{\mathbf{y},\mathbf{f}_l}(\mathbf{x},\mathbf{f}_k)\pi^*(\mathbf{y},\mathbf{f}_l|B_{lv},\mathbf{f}_l)\pi(B_{lv},\mathbf{f}_l),\end{aligned}$$

where in the second equality we have used the fact that the probability of the system being in the basin B_{lv} is independent of the time since the previous restart, or the time to the next restart (j in this case), and where the last equality results from calling $t' = t - j$, bringing the limit inside the summation to the products, and again using the fact that the transition probability at the restart is independent of the starting point $(\mathbf{h},\mathbf{g})$. □

Lemma 2.5. *If $\delta(\mathbf{x},A_k)$ is the number of iterations of $\mathbf{f}$ needed to reach the attractor A_k from the state $\mathbf{x}$, then for any $\mathbf{x}\in A_k$, $b<1$,*

$$\sum_{j=\delta(\mathbf{x},A_k)}^{\infty}(1-b)b^j = b^{\delta(\mathbf{x},A_k)}.$$

Proof.

$$\sum_{j=\delta(\mathbf{x},A_k)}^{\infty}(1-b)b^j = 1-\sum_{j=0}^{\delta(\mathbf{x},A_k)-1}(1-b)b^j = 1-(1-b)\frac{1-b^{\delta(\mathbf{x},A_k)}}{1-b}. \quad \square$$

Theorem 2.6 (Brun *et al.*, 2005).

$$\pi(A_{ki},\mathbf{f}_k) = \sum_{l=1}^{r}\sum_{v=1}^{m_l}\left[\sum_{\mathbf{x}\in B_{ki}}\sum_{\mathbf{y}\in B_{lv}} P^*_{\mathbf{y},\mathbf{f}_l}(\mathbf{x},\mathbf{f}_k)\pi^*(\mathbf{y},\mathbf{f}_l|B_{lv},\mathbf{f}_l)b^{\delta(\mathbf{x},A_{ki})}\right]\pi(B_{lv},\mathbf{f}_l). \tag{2.30}$$

Proof. We prove the theorem by splitting $P^{(t)}_{\mathbf{h},\mathbf{g}}(A_{ki},\mathbf{f}_k|B_{ki},\mathbf{f}_k)$ over the possible values for the random variable $\tau(t)$ and using the fact that the system cannot leave a basin between restarts. In the third equality below, we replace $\mathbf{X}_t\in B_{ki}$ by $\mathbf{X}_{t-j}\in B_{ki}$ because they are the same event: if the system reaches the basin B_{ki} at the time $t-j$, and is using the function $\mathbf{f}_k$, it will not leave the basin until time t (at least), because there is no restart between $t-j$ and t (j being the time between t and the last restart). Moreover, we split the event $\mathbf{X}_{t-j}\in B_{ki}$ over the possible starting states $\mathbf{x}\in B_{ki}$:

$$\begin{aligned}
\pi(A_{ki},\mathbf{f}_k) &= \pi(B_{ki},\mathbf{f}_k)\lim_{t\to\infty}P^{(t)}_{\mathbf{h},\mathbf{g}}(A_{ki},\mathbf{f}_k|B_{ki},\mathbf{f}_k)\\
&= \pi(B_{ki},\mathbf{f}_k)\\
&\quad\times\lim_{t\to\infty}\sum_{j=0}^{\infty}P\left[\mathbf{X}_t\in A_{ki},\mathbf{F}_t=\mathbf{f}_k|\mathbf{X}_t\in B_{ki},\mathbf{F}_t=\mathbf{f}_k,\mathbf{X}_0=\mathbf{h},\mathbf{F}_0=\mathbf{g},\tau(t)=j\right]\\
&\quad\times P[\tau(t)=j]\\
&= \pi(B_{ki},\mathbf{f}_k)\\
&\quad\times\lim_{t\to\infty}\sum_{j=0}^{\infty}\sum_{\mathbf{x}\in B_{ki}}\\
&\quad P\left[\mathbf{X}_t\in A_{ki},\mathbf{F}_t=\mathbf{f}_k|\mathbf{X}_{t-j}=\mathbf{x},\mathbf{F}_{t-j}=\mathbf{f}_k,\mathbf{X}_0=\mathbf{h},\mathbf{F}_0=\mathbf{g},\tau(t)=j\right]\\
&\quad\times P\left[\mathbf{X}_{t-j}=\mathbf{x}|\mathbf{X}_{t-j}\in B_{ki},\mathbf{F}_{t-j}=\mathbf{f}_k,\mathbf{X}_0=\mathbf{h},\mathbf{F}_0=\mathbf{g},\tau(t)=j\right]P[\tau(t)=j].
\end{aligned}$$

For each state $\mathbf{x}$, the system will reach the attractor A_{ki} ($f^j(\mathbf{x})\in A_{ki}$) if and only if the distance (trajectory length) between $\mathbf{x}$ and A_{ki} is smaller than j, that is, $j\geq\delta(\mathbf{x},A_{ki})$, so that we use the indicator function $\mathbf{1}_{[j \geq \delta(\mathbf{x},A_{ki})]}$:

$$\pi(A_{ki},\mathbf{f}_k) = \pi(B_{ki},\mathbf{f}_k)\sum_{j=0}^{\infty}\sum_{\mathbf{x}\in B_{ki}} \mathbf{1}_{[j \geq \delta(\mathbf{x},A_{ki})]}$$
$$\times \lim_{t\to\infty} P\left[\mathbf{X}_{t-j} = \mathbf{x}|\mathbf{X}_{t-j} \in B_{ki}, \mathbf{F}_{t-j} = \mathbf{f}_k, \mathbf{X}_0 = \mathbf{h}, \mathbf{F}_0 = \mathbf{g}, \tau(t) = j\right]$$
$$\times \lim_{t\to\infty} P[\tau(t) = j].$$

The proof is completed by using dominated convergence, because for a large t, $P[\tau(t) = j]$ is bounded by $2\cdot(1-b)b^j$. We move the limit inside the sum and then apply Lemmas 2.4 and 2.5:

$$\pi(A_{ki},\mathbf{f}_k) = \pi(B_{ki},\mathbf{f}_k)\sum_{j=0}^{\infty}\sum_{\mathbf{x}\in B_{ki}} \mathbf{1}_{[j \geq \delta(\mathbf{x},A_{ki})]}$$
$$\times \left[\frac{1}{pi(B_{ki},\mathbf{f}_k)}\sum_{l=1}^{r}\sum_{v=1}^{m_l}\sum_{\mathbf{y}\in B_{lv}} P^*_{\mathbf{y},\mathbf{f}_l}(\mathbf{x},\mathbf{f}_k)\pi^*(\mathbf{y},\mathbf{f}_l|B_{lv},\mathbf{f}_l)\pi(B_{lv},\mathbf{f}_l)\right](1-b)b^j$$
$$= \sum_{\mathbf{x}\in B_{ki}}\left[\sum_{l=1}^{r}\sum_{v=1}^{m_l}\sum_{\mathbf{y}\in B_{lv}} P^*_{\mathbf{y},\mathbf{f}_l}(\mathbf{x},\mathbf{f}_k)\pi^*(\mathbf{y},\mathbf{f}_l|B_{lv},\mathbf{f}_l)\pi(B_{lv},\mathbf{f}_l)\right]\sum_{j\geq\delta(\mathbf{x},A_{ki})}^{\infty}(1-b)b^j$$
$$= \sum_{l=1}^{r}\sum_{v=1}^{m_l}\left[\sum_{\mathbf{x}\in B_{ki}}\sum_{\mathbf{y}\in B_{lv}} P^*_{\mathbf{y},\mathbf{f}_l}(\mathbf{x},\mathbf{f}_k)\pi^*(\mathbf{y},\mathbf{f}_l|B_{lv},\mathbf{f}_l)b^{\delta(\mathbf{x},A_{ki})}\right]\pi(B_{lv},\mathbf{f}_l). \qquad \square$$

Finally, as $\pi^*(\mathbf{y},\mathbf{f}_l|B_{lv},\mathbf{f}_l)$ is unknown, if the probability $q(1-(1-p)^n)$ of a restart is small, then we can approximate $\pi(A_{ki},\mathbf{f}_k)$ by

$$\pi(A_{ki},\mathbf{f}_k) \approx \sum_{l=1}^{r}\sum_{v=1}^{m_l}\frac{1}{|A_{lv}|}\left[\sum_{\mathbf{x}\in B_{ki}}\sum_{\mathbf{y}\in B_{lv}} P^*_{\mathbf{y},\mathbf{f}_l}(\mathbf{x},\mathbf{f}_k)b^{\delta(\mathbf{x},A_{ki})}\right]\pi(B_{lv},\mathbf{f}_l). \tag{2.31}$$

Once we have the steady-state probabilities for the pairs $(A_{ki},\mathbf{f}_k)$, we are interested in the steady-state probabilities $\pi(A_{ki})$ of being in the attractor A_{ki} of the Boolean function $\mathbf{f}_k$. Because we can be inside this attractor A_{ki} even using a different function $\mathbf{f}_l$, with $l \neq k$, this probability is

$$\pi(A_{ki}) = \sum_{l=1}^{r}\pi(A_{ki},\mathbf{f}_l).$$

Equations (2.30) and (2.31) give the steady-state probability $\pi(A_{ki},\mathbf{f}_l)$ of being in the attractor A_{ki} and using the Boolean function $\mathbf{f}_l$, only when $l = k$. We need to compute $\pi(A_{ki},\mathbf{f}_l)$ for $l \neq k$. We will approximate these probabilities again assuming that the system is almost always inside some attractor; that is, A_{li}, $i = 1,2,\ldots,m_l$, is a partition of X. Using the approximation

$$\pi(A_{ki},\mathbf{f}_l|A_{lj},\mathbf{f}_l) \approx \frac{|A_{ki}\cap A_{lj}|}{|A_{lj}|} \tag{2.32}$$

we can approximate $\pi(A_{ki},\mathbf{f}_l)$ by

$$\begin{aligned}\pi(A_{ki},\mathbf{f}_l) &\approx \sum_{j=1}^{m_l} \pi(A_{ki},\mathbf{f}_l|A_{lj},\mathbf{f}_l)\pi(A_{lj},\mathbf{f}_l) \\ &\approx \sum_{j=1}^{m_l} \frac{|A_{ki}\cap A_{lj}|}{|A_{lj}|}\pi(A_{lj},\mathbf{f}_l).\end{aligned} \tag{2.33}$$

Equation (2.33) is still valid for $l=k$ because in this case $A_{ki}\cap A_{kj}=\emptyset$ for $i\neq j$. Therefore, the marginal steady-state probability for the attractors can be approximated by

$$\pi(A_{ik}) \approx \sum_{l=1}^{r}\sum_{j=1}^{m_l} \frac{|A_{ki}\cap A_{lj}|}{|A_{lj}|}\pi(A_{lj},\mathbf{f}_l). \tag{2.34}$$

Union of attractors Equation (2.34) provides representation of the steady-state probabilities for attractors. In the Boolean model, this at once provides representation for unions of attractors because of their disjointness. For PBNs, they need not be disjoint. Hence, we wish to apply the probability addition theorem, but first we need to obtain the steady-state probability of an intersection of attractors. Choosing one of the attractors to condition on, we obtain

$$\pi\left(\bigcap_{u=1}^{n} A_{k_u,i_u}\right) = \pi\left(\bigcap_{u=1}^{n-1} A_{k_u,i_u}|A_{k_n,i_n}\right)\pi(A_{k_n,i_n}).$$

We obtain the approximation

$$\pi\left(\bigcap_{u=1}^{n} A_{k_u,i_u}\right) \approx \frac{|\bigcap_{u=1}^{n} A_{k_u,i_u}|}{|A_{k_n,i_n}|}\pi(A_{k_n,i_n}). \tag{2.35}$$

The intersection is null if any two k_u are equal, and therefore the probability in this case is 0. Since conditioning in Eq. (2.35) can be with respect to any of the attractors in the intersection and since the accuracy of the approximation depends on approximation of the conditional probability, to avoid the unfortunate consequence of choosing a conditioning attractor for which the approximation is particularly bad, we take the average over all attractors in the intersection for our final approximation,

$$\pi\left(\bigcap_{u=1}^{n} A_{k_u,i_u}\right) \approx \frac{1}{n}\sum_{j=1}^{n} \frac{|\bigcap_{u=1}^{n} A_{k_u,i_u}|}{|A_{k_j,i_j}|}\pi(A_{k_j,i_j}),$$

where the attractor probabilities are defined in Eq. (2.34). The probability addition theorem yields

$$\pi\Big(\bigcup_{u=1}^{n} A_{k_u,i_u}\Big) = \sum_{u=1}^{n}\pi(A_{k_u,i_u}) + \sum_{v=2}^{n}(-1)^{v+1} \sum_{1\le u_1<u_2<u_3<\cdots<u_v\le n} \pi\Big(\bigcap_{w=1}^{v} A_{k_{u_w},i_{u_w}}\Big)$$

$$\approx \sum_{u=1}^{n}\pi(A_{k_u,i_u}) + \sum_{v=2}^{n}(-1)^{v+1}\frac{1}{n} \sum_{1\le u_1<u_2<u_3<\cdots<u_v\le n} \sum_{j=1}^{n} \frac{|\bigcap_{w=1}^{n} A_{k_{u_w},i_{u_w}}|}{|A_{k_{u_j},i_{u_j}}|}\pi(A_{k_{u_j},i_{u_j}}).$$

Brun *et al.* (2005) performed simulations with several artificially generated PBNs to compare the steady-state probabilities computed from the above expressions with those determined from the state transition matrix and found excellent concordance. When the probabilities p and q are large (e.g., 0.2), the approximation becomes poor, as the assumption in the Eq. (2.27) is violated. One particularly interesting example consisted of a PBN comprised of 8 genes and 576 Boolean functions with a total of 1404 pairs $(A_{ki}, \mathbf{f}_k)$, with an average of 2.43 attractors A_{ki} per function $\mathbf{f}_k$, and a maximum of 8 attractors in one function. The transition matrix is huge, with nearly 22 billion elements, and therefore cannot be employed. Instead, a Monte Carlo simulation was performed, as described in the previous section, and it was found that the results of the approximation were indeed very close to those obtained by the large simulation. Were the probabilities p and q significantly smaller than they were (0.001), the simulation would be problematic because, despite the Markov chain being ergodic, the convergence would be slow and the simulation would only rarely flip out of an attractor and therefore many state visitations would be rare, especially for those constituent networks that themselves possess low probability.

2.6 Relationships of PBNs to Bayesian Networks

Recall the discussion in section 2.1, where we considered the extent to which Markov chains are able to mimic biological regulation. The conditional probabilities for the expression of a target gene given the expressions of a predictor set (see Eqs. (2.1) and (2.2)) are, in fact, basic building blocks of Bayesian networks, which we now briefly review so as to establish the necessary framework for understanding the relationships between PBNs and Bayesian networks.

2.6.1 Bayesian networks

A Bayesian network is essentially a compact representation of a joint probability distribution (Pearl, 1988; Jensen, 2001; Lauritzen, 1996). This representation takes the form of a directed acyclic graph (DAG) in which the nodes (vertices) of the graph represent random variables and the directed edges or lack thereof represent conditional dependencies or independencies. The other ingredient in this model is the specification of a conditional probability distribution for each of the random variables (nodes). This distribution, which in the case of discrete-valued variables takes the form of a conditional probability table, specifies the probability of a *child* node taking on a particular value given the values of its *parent* nodes. For example, in the case of genetic networks, the values of the nodes can correspond to (possibly discretized) gene expression levels or other measurable events, including external conditions.

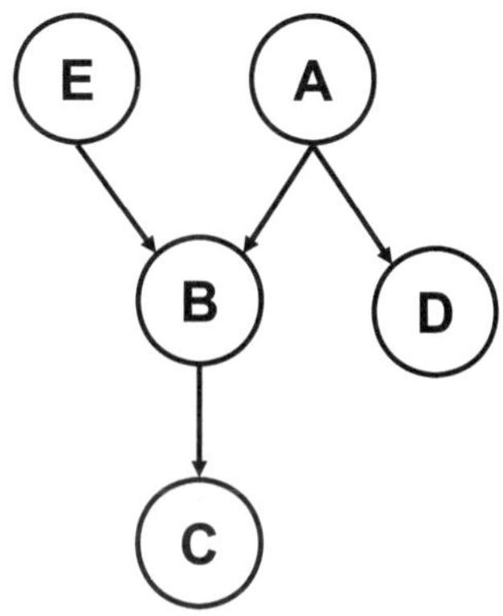

Figure 2.9. *An example of a Bayesian network consisting of five nodes.*

Specifically, consider a random vector $\mathbf{X} = (X_1, \ldots, X_n)$ representing some measurements, such as expressions of genes. A Bayesian network $B = (G, \Theta)$ is defined by (1) a DAG G whose vertices correspond to $X_1, \ldots, X_n$ and (2) a set Θ of local conditional probability distributions for each vertex X_i, given its parents in the graph. The graph encodes the so-called *Markov assumption*, which states that each variable X_i is independent of its nondescendants, given its parents in G. Using the chain rule of probabilities, any joint distribution satisfying the Markov assumption can be decomposed into a product of the local conditional probabilities. Let us denote by $\mathrm{Pa}(X_i)$ the Markovian parents of X_i in the graph G. Then,

$$P\{x_1, \ldots, x_n\} = \prod_{i=1}^{n} P\{x_i | \mathrm{Pa}(x_i)\}. \tag{2.36}$$

Consider the following example used by Friedman *et al.* (2000), shown in Fig. 2.9. The graph, consisting of five nodes, implies a number of conditional independencies. For example, the only nondescendant of B is D and the parents of B are A and E. Thus, according to the Markov assumption, B is independent of D, given both A and E. In other words,

$$P\{B|D, A, E\} = P\{B|A, E\}.$$

Similarly,

$$P\{D|B, C, E, A\} = P\{D|A\},$$

since B, C, and E are nondescendants of D and A is the only parent of D. There are other such independencies implied by this graph. The joint distribution, then, factors as

$$P\{A, B, C, D, E\} = P\{A\} P\{E\} P\{B|A, E\} P\{D|A\} P\{C|B\}. \tag{2.37}$$

Thus, a Bayesian network graphically represents a joint probability distribution in a compact and intuitive way.

To see why such a representation is more compact, consider the simplest case where nodes are binary-valued. Then, to represent the joint distribution of A, B, C, D, and E in the example above, one would need to store a conditional probability table (CPT) containing $2^5 = 32$ entries, one for each combination of the five variables. However, with the factored form given in Eq. (2.37), we need only one entry to store $P\{A\}$, one to store $P\{E\}$, four entries to store $P\{B|A,E\}$, and two entries for each of $P\{D|A\}$ and $P\{C|B\}$, for a total of 10 entries. As an arbitrary example, the CPT for $P\{B|A,E\}$ might look something like

A	E	$P\{B=1|A,E\}$
0	0	0.3
0	1	0.6
1	0	0.6
1	1	0.8

(2.38)

For large numbers of variables, the savings become much more appreciable. In general, for a binary-valued network with n nodes in which there are no more than k parents of any given node, the representation will require $O\left(n2^k\right)$ space as opposed to $O\left(2^n\right)$ for storing the entire joint probability distribution. For continuous-valued random variables, the CPT clearly cannot be used. One common choice is to use linear Gaussian conditional densities, which we will not cover here.

As we discussed above, each Bayesian network implies a set of independence statements. It turns out that two different Bayesian networks can imply identical sets of such independence statements. As a trivial example, consider $X \to Y$ and $Y \to X$. Clearly, in each case, there are no independencies and so, both networks, although different, are said to be *equivalent* because they imply the same independencies. This is an important issue because when one infers the network from measurement data (Shmulevich and Dougherty, 2007), one cannot distinguish between equivalent networks. One way around this difficulty is to perform targeted intervention experiments where genes are manipulated. This can help to narrow down the range of possible networks.

Bayesian networks, being representations of static probabilistic relationships, can naturally be extended to the temporal domain and be used to model stochastic processes, resulting in *dynamic Bayesian networks* (DBNs) (Murphy and Mian, 1999). Thus, DBNs generalize hidden Markov models and linear dynamical systems by representing the conditional dependencies and independencies between (state) variables over time. This added flexibility allows us to model feedback relationships, which we know are ubiquitous in genetic regulation, whereas standard Bayesian networks whose structure is a DAG cannot. It is typically assumed that the DBN model is time-invariant (graph structure and parameters do not change over time). Let us restrict our attention to first-order Markov processes in X, i.e., to processes whose transition probability obeys

$$P\{X(t)|X(0),X(1),\ldots,X(t-1)\} = P\{X(t)|X(t-1)\}.$$

A DBN that represents the joint probability distribution over all possible time series of variables in $\mathbf{X}$ consists of two parts: (i) an initial Bayesian network $B_0 = (G_0, \Theta_0)$ that defines the joint distribution of the variables in $\mathbf{X}(0)$, and (ii) a transition Bayesian network $B_1 = (G_1, \Theta_1)$ that specifies the transition probabilities $P\{\mathbf{X}(t)|\mathbf{X}(t-1)\}$ for all t. Boldface here indicates vector-valued random variables. Thus, a DBN is defined by a pair (B_0, B_1). The structure of DBNs is restricted in two ways. First, the DAG G_0 in the initial Bayesian network B_0 is assumed to have only "within-slice" connections, i.e., $\mathrm{Pa}(X_i(0)) \subseteq \mathbf{X}(0)$ for all $1 \le i \le n$. Second, because of the first-order Markovian assumption, we also constrain the variables in time slice $\mathbf{X}(t)$, $t > 0$, to have all their parents in slice $t-1$, i.e., $\mathrm{Pa}(X_i(t)) \subseteq \mathbf{X}(t-1)$ for all $1 \le i \le n$ and $t > 0$. An example of the basic building blocks of DBNs, B_0 and B_1, is shown in Fig. 2.10.

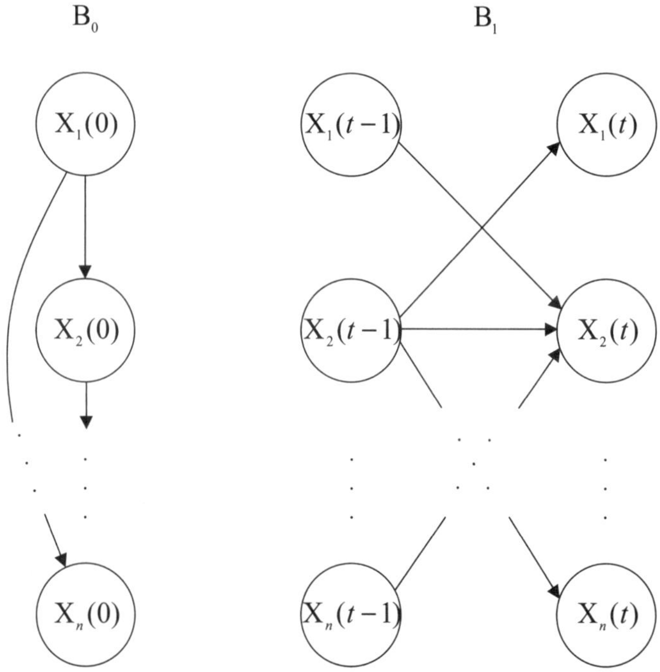

Figure 2.10. *An example of the basic building blocks of DBNs* (B_0, B_1)*.* B_0 *and* B_1 *are the initial and transition Bayesian networks, respectively (Lähdesmäki et al.,* 2006*).*

Using Eq. (2.36) and the assumptions on the initial and transition Bayesian networks discussed above, the joint probability over a finite set of random variables $\mathbf{X}(0) \cup \mathbf{X}(1) \cup \cdots \cup \mathbf{X}(T)$ can be expressed as

$$\begin{aligned} P\{\mathbf{X}(0), \mathbf{X}(1), \ldots, \mathbf{X}(T)\} &= P\{\mathbf{X}(0)\} \prod_{t=1}^{T} P\{\mathbf{X}(t) | \mathbf{X}(t-1)\} \\ &= \prod_{i=1}^{n} P\{X_i(0) | \operatorname{Pa}(X_i(0))\} \prod_{t=1}^{T} \prod_{j=1}^{n} P\{X_j(t) | \operatorname{Pa}(X_j(t))\}. \end{aligned}$$

We note that DBNs have also been used to infer genetic regulatory networks from gene expression data (Murphy and Mian, 1999; Kim *et al.*, 2003; Perrin *et al.*, 2003; Zou and Conzen, 2005). Bernard and Hartemink (2005) have used DBNs for learning dynamic models of transcriptional regulatory networks from gene expression data and transcription factor binding location data.

Let us now examine the structure of PBNs in relation to Bayesian networks. Rather than considering the formulation wherein a PBN consists of a set of constituent Boolean networks, each with a selection probability, let us turn our focus on the individual genes and the set of predictors (Boolean functions) assigned to them. For a node (gene) x_i, there corresponds a set $F_i = \{f_j^{(i)}\}_{j=1,\ldots,l(i)}$ of possible predictor functions along with their probabilities $c_1^{(i)},\ldots,c_{l(i)}^{(i)}$, $\sum_{j=1}^{l(i)} c_j^{(i)} = 1$, where $l(i)$ is the number of possible functions assigned to gene x_i. The constituent Boolean networks $\mathbf{f}_1,\mathbf{f}_2,\ldots,\mathbf{f}_m$, which we defined earlier, are of the form $\mathbf{f}_k = (f_{k_1}^{(1)}, f_{k_2}^{(2)},\ldots,f_{k_n}^{(n)})$, for $k = 1,2,\ldots,m$, $1 \leq k_i \leq l(i)$, and where $f_{k_i}^{(i)} \in F_i (i = 1,\ldots,n)$. Note the different interpretation of the notation $f_j^{(i)}$, being the jth possible predictor of gene i taking into account all possible constituent networks (and not the predictor of gene i when the jth network is selected). As there is no danger of ambiguity, we prefer to avoid complicating the notation to make this distinction. The selection probability $c_j^{(i)}$ of the individual predictor $f_j^{(i)}$ can be related to the selection probability c_k of the kth constituent network as $c_j^{(i)} = \sum_{k: f_{k_i}^{(i)} = f_j^{(i)}} c_k$.

We are interested in computing the conditional probability of gene X_i (a random variable) given its parents, namely, the set of all genes involved in predicting it. To make this more formal, let the set $\mathbf{x}_j^{(i)} \subseteq \{x_1,\ldots,x_n\}$, $j = 1,\ldots,l(i)$, denote the set of essential variables used by predictor $f_j^{(i)}$ for gene x_i. Then, the parent set

$$\mathrm{Pa}(x_i) = \bigcup_{j=1}^{l(i)} \mathbf{x}_j^{(i)} \tag{2.39}$$

is the set of all variables used to predict gene x_i. For convenience of notation, let us expand the domains of all predictors by adding fictitious variables so that all are functions of the variables in $\mathrm{Pa}(x_i)$. Computationally, this step is not necessary. Then, the conditional probability that $X_i = 1$, given the fact that predictor $f_j^{(i)}$ is used, is equal to

$$P\left\{X_i = 1 | f_j^{(i)} \text{ is used}\right\} = \sum_{\mathbf{x} \in \{0,1\}^{|\mathrm{Pa}(x_i)|}} D_i(\mathbf{x}) f_j^{(i)}(\mathbf{x}), \tag{2.40}$$

where $D_i(\mathbf{x})$ is the joint distribution over the variables in $\mathrm{Pa}(x_i)$. Because we're in the binary setting, Eq. (2.40) can also be interpreted as the conditional expectation of X_i given the predictor $f_j^{(i)}$. Computationally, we need only to consider the essential variables of $f_j^{(i)}$ and their joint distribution, which can be obtained from the entire joint distribution $D(\mathbf{x})$ by "integrating out" the fictitious variables. Moreover, Eq. (2.40) can be computed as a vector multiplication of the joint distribution and the truth table of the predictor.

Ultimately, we are interested in obtaining $P\{X_i = 1\}$. Since $c_j^{(i)} = P\{f_j^{(i)} \text{ is used}\}$,

$$\begin{aligned} P\{X_i = 1\} &= \sum_{j=1}^{l(i)} P\left\{X_i = 1 | f_j^{(i)} \text{ is used}\right\} \cdot c_j^{(i)} \\ &= \sum_{j=1}^{l(i)} c_j^{(i)} \cdot \sum_{\mathbf{x} \in \{0,1\}^{|\mathrm{Pa}(x_i)|}} D_i(\mathbf{x}) f_j^{(i)}(\mathbf{x}). \end{aligned} \tag{2.41}$$

Equation (2.41) provides the expression of the probability of the target gene X_i in terms of the probabilities of the predictors and the joint distribution of the target's parent genes. To establish the relationships further, we need to consider DBNs, since PBNs are inherently temporal. As shown by Lähdesmäki *et al.* (2006), PBNs and a certain subclass of DBNs can represent the same joint probability distribution over their common variables.

Now suppose that we are interested in the joint probability distribution of the variables expanded over a finite number of updating steps, say T steps, i.e., $\{\mathbf{X}(i) : 0 \leq i \leq T\}$. Given the initial state $\mathbf{x}(t-1)$, recall from Eq. (2.19) that the probability of moving to some state $\mathbf{x}(t)$ after one step of the network is

$$A(\mathbf{x}(t-1), \mathbf{x}(t)) = \sum_{j:\mathbf{f}_j(\mathbf{x}(t-1))=\mathbf{x}(t)} P\{\mathbf{F} = \mathbf{f}_j\} = \sum_{j=1}^{m} c_j G_j(\mathbf{x}(t-1), \mathbf{x}(t)), \tag{2.42}$$

where we are not taking into account perturbations. In an instantaneously random PBN, the network realizations are selected independently for each time instant and thus the joint probability distribution over all possible time series of length $T+1$ can be expressed as

$$\begin{aligned} &P\{\mathbf{X}(0) = \mathbf{x}(0), \ldots, \mathbf{X}(T) = \mathbf{x}(T)\} \\ &\quad = P\{\mathbf{X}(0) = \mathbf{x}(0)\} \prod_{t=1}^{T} A(\mathbf{x}(t-1), \mathbf{x}(t)), \end{aligned} \tag{2.43}$$

where $P\{\mathbf{X}(0) = \mathbf{x}(0)\}$ denotes the distribution of the first state.

Let us concentrate on independent PBNs and rewrite the state transition probabilities $A(\mathbf{x}(t-1), \mathbf{x}(t))$. The dependent PBN case is treated in detail by Lähdesmäki *et al.* (2006). Let $(\mathbf{x}(t))_i$, $1 \leq i \leq n$, denote the ith element of $\mathbf{x}(t)$, and $A(\mathbf{x}(t-1), (\mathbf{x}(t))_i)$ denote the probability that the ith element of $\mathbf{x}(t)$ will be $(\mathbf{x}(t))_i$ after one step of the network, given that the current state is $\mathbf{x}(t-1)$. Since all nodes are assumed to be independent, each node is updated independently, and Eq. (2.42) can be written as

$$A(\mathbf{x}(t-1), \mathbf{x}(t)) = \prod_{i=1}^{n} A(\mathbf{x}(t-1), (\mathbf{x}(t))_i). \tag{2.44}$$

Thus, given Eqs. (2.43) and (2.44), the joint probability distribution over all possible time series of length $T+1$ can be expressed as

$$\begin{aligned} &P\{\mathbf{X}(0) = \mathbf{x}(0), \ldots, \mathbf{X}(T) = \mathbf{x}(T)\} \\ &\quad = P\{\mathbf{X}(0) = \mathbf{x}(0)\} \prod_{t=1}^{T} \prod_{i=1}^{n} A(\mathbf{x}(t-1), (\mathbf{x}(t))_i). \end{aligned} \tag{2.45}$$

2.6.2 Independent PBNs and DBNs

In order to be able to establish the relationship between PBNs and DBNs, we will add an extra feature to the definition of a PBN. We will assume that the initial state $\mathbf{X}(0)$ of a PBN can have any joint probability distribution, as in Eqs. (2.43) and (2.45). In particular, we assume $\mathbf{X}(0)$ to have the same distribution as defined by B_0 for the first slice of a DBN. For instance, in the context of genetic regulatory network modeling this initial distribution can be set equal to the stationary distribution of the corresponding Markov chain, as described in the previous section. Also note that this discussion pertains only to the class of binary-valued DBNs. This requirement can be relaxed, as demonstrated by Lähdesmäki *et al.* (2006).

We first illustrate a way of expressing an independent PBN $G(V,F)$ as a DBN (B_0, B_1). Let an independent PBN $G(V,F)$ be fixed. The nodes in the graphs G_0 and G_1 must clearly correspond to the nodes in the PBN. In order to distinguish between the nodes from different models, nodes in the PBN are denoted by X_i, $1 \leq i \leq n$, as above, and the corresponding nodes in the DBN are denoted by $\hat{X}_i$ (similarly for the vector-valued random variables $\mathbf{X}$ and $\hat{\mathbf{X}}$). We refer to nodes X_i and $\mathbf{X}$ as the PBN counterparts of $\hat{X}_i$ and $\hat{\mathbf{X}}$, respectively, and vice versa. Note that in order for us to establish these relationships, we must treat the nodes (genes) in the PBN as random variables, denoted by capital letters.

Because an initial Bayesian network B_0 is capable of representing any joint probability distribution over its nodes, the distribution of the initial state of the PBN, $P\{\mathbf{X}(0)\}$, can be represented by B_0. It is easy to see that both PBNs and DBNs obey the first-order Markovian property. Thus, we need only to consider one-step transition probabilities $P\{\mathbf{X}(t)|\mathbf{X}(t-1)\}$ when expressing the PBN as a DBN. Further, we can see that for PBNs and DBNs, the joint probability can also be decomposed over their nodes in the same way. Thus, when constructing a transition Bayesian network B_1, we can further concentrate only on a single node, say X_i, with other nodes being handled similarly. As above, let $\mathbf{X}_j^{(i)}(t-1) \subseteq \mathbf{X}(t-1)$ denote the set of essential variables (nodes) used by predictor function $f_j^{(i)}$ for gene X_i at time t. The set

$$\mathrm{Pa}(X_i(t)) = \bigcup_{j=1}^{l(i)} \mathbf{X}_j^{(i)}(t-1) \tag{2.46}$$

denotes the set of all variables used to predict the value of the gene X_i. Let us expand the domain of all the predictor functions in F_i by adding fictitious variables so that they are functions of variables in $\mathrm{Pa}(X_i(t))$. Thus, we can define $B_1 = (G_1, \Theta_1)$ as follows. The graph G_1 has directed edges from $\hat{\mathbf{X}}(t-1)$ to $\hat{\mathbf{X}}(t)$ such that the parents of $\hat{X}_i(t)$ are equal to the DBN counterparts of the nodes shown in Eq. (2.46). Given the distribution D_i over the predictor variables of node X_i (denoted as $\mathrm{Pa}(X_i) \sim D_i$), Eq. (2.41) can be interpreted as $P\{X_i(t) = 1 | \mathrm{Pa}(X_i(t)) \sim D_i\}$. Thus, by specifying D_i to be "deterministic" such that it corresponds to a particular predictor node configuration $\mathrm{Pa}(X_i(t)) = \mathbf{z}$, i.e., $D_i(\mathbf{x}) = 1$ if $\mathbf{x} = \mathbf{z}$ and $D_i(\mathbf{x}) = 0$ otherwise (a delta function), we have that

$$P\{X_i(t) = 1 | \mathrm{Pa}(X_i(t)) = \mathbf{z}\} = \sum_{j=1}^{l(i)} f_j^{(i)}(\mathbf{z}) c_j^{(i)}. \tag{2.47}$$

Then, the set Θ_1 of local conditional probability distributions (or CPTs in the discrete case) induced by the graph structure G_1 in the DBN has exactly the same entries as shown in Eq. (2.47) for each node. Thus, any PBN can be expressed as a binary DBN.

To establish the converse of the above result, let us see how a binary-valued DBN (B_0, B_1) can be represented as an independent PBN $G(V,F)$. Let a DBN (B_0, B_1) be given. The set of nodes V must clearly correspond to the nodes in the graphs G_0 and G_1, and the distribution of the first state $\mathbf{X}(0)$ in the PBN must be the same as for the DBN.

Following the same reasoning as above, we can again conclude that when constructing a PBN, one needs only to consider the predictor functions for a single node X_i between consecutive time instants $t-1$ and t. Let us assume that the local conditional probability distributions in Θ_1 are given in the form of CPTs, and the number of parents of the ith node is denoted as $q = |\mathrm{Pa}(\hat{X}_i(t))|$. Let us enumerate the entries in the CPTs that are assigned to the ith node as triplets $(z_l, \mathbf{y}_l, p_l)$, where $z_l \in \{0,1\}$, $\mathbf{y}_l \in \{0,1\}^q$,

$$p_l = P\{\hat{X}_i(t) = z_l | \mathrm{Pa}\left(\hat{X}_i(t)\right) = \mathbf{y}_l\},$$

and $1 \le l \le 2^{q+1}$. Let us also suppose that the triplets are enumerated such that the first $r = 2^q$ triplets have $z_l = 1$ and they are sorted in increasing order, i.e., $1 \le k < l \le r \Rightarrow p_k \le p_l$. Let $\mathbf{y}_l = y_{l_1} y_{l_2} \cdots y_{l_q}$ and interpret a sequence of symbols $\mathbf{x}_l = x_{l_1}^{y_{l_1}} x_{l_2}^{y_{l_2}} \cdots x_{l_q}^{y_{l_q}}$ as a conjunction, where $x_{l_i}^{y_{l_i}} = x_{l_i}$ if $y_{l_i} = 1$ and $x_{l_i}^{y_{l_i}} = \overline{x}_{l_i}$ if $y_{l_i} = 0$. The variables in the conjunction correspond to a set of specific variables, $\{X_{l_1}, \ldots, X_{l_q}\}$, in the PBN; i.e., they are the PBN counterparts of $\mathrm{Pa}(\hat{X}_i(t))$.

Then, for the ith node in the PBN, the set of functions F_i can be generated as follows: $F_i = \{f_1^{(i)}, \cdots, f_r^{(i)}, f_{r+1}^{(i)}\}$, where

$$f_l^{(i)} = \mathbf{x}_l \vee \mathbf{x}_{l+1} \vee \cdots \vee \mathbf{x}_r, \text{ for } 1 \le l \le r, \tag{2.48}$$

is a disjunction of conjunctions, $f_{r+1}^{(i)}$ is a zero function, i.e., $f_{r+1}^{(i)}(\mathbf{x}) \equiv 0$ for all $\mathbf{x} \in \{0,1\}^q$, and $r = 2^q$. Note that $f_1^{(i)}$ is essentially the unity function. Also, some of the Boolean expressions in Eq. (2.48) can possibly be expressed in more compact form. The corresponding selection probabilities are set to $c_1^{(i)} = p_1$, $c_l^{(i)} = p_l - p_{l-1}$ for $2 \le l \le r$, and $c_{r+1}^{(i)} = 1 - p_r$. If some of the $c_j^{(i)}$'s happen to be zero, the corresponding functions $f_j^{(i)}$ can be removed from F_i. By applying Eq. (2.47), we can verify that the above construction indeed gives an equivalent PBN to the given DBN. Let us first compute the one-step prediction probabilities for the cases where $X_i(t) = 1$ and the parent nodes have values $\mathbf{y}_l$. One gets

$$\begin{aligned} P\{X_i(t) = 1 | \mathrm{Pa}(X_i(t)) = \mathbf{y}_l\} &= \sum_{j=1}^{r+1} f_j^{(i)}(\mathbf{y}_l) c_j^{(i)} \\ &= \sum_{j=1}^{l} f_j^{(i)}(\mathbf{y}_l) c_j^{(i)} = p_1 + \sum_{j=2}^{l} (p_j - p_{j-1}) = p_l, \end{aligned}$$

where the second equality follows from the fact that only the first l functions $f_1^{(i)}, \ldots, f_l^{(i)}$ contain the conjunction $\mathbf{x}_l$ (see Eq. (2.48)), i.e., $f_j^{(i)}(\mathbf{y}_l) = 1$ only for $1 \le j \le l$. The probability of the corresponding complement event $P\{X_i(t) = 0 | \mathrm{Pa}(X_i(t)) = \mathbf{y}_l\}$ is clearly $1 - p_l$. So, a binary-valued DBN can be represented as an independent PBN. Thus, we can state the following theorem.

Theorem 2.7 (Lähdesmäki *et al.*, 2006). *Independent PBNs $G(V,F)$ and binary-valued DBNs (B_0, B_1) whose initial and transition Bayesian networks B_0 and B_1 are assumed to have only within and between consecutive slice connections, respectively, can represent the same joint distribution over their common variables.*

It is important to note that the mapping from a binary DBN to an independent PBN is not unique. Instead, there are many PBNs that have the same probabilistic structure. This is best illustrated by a toy example. Assume that gene X_1 is regulated by gene X_2 and all the values in the CPT of the gene X_1 are equal to 0.5. Then the following two function sets have the same conditional probabilities: $F_1 = \{f_1^{(1)}, f_2^{(1)}\}$, with $c_1^{(1)} = c_2^{(1)} = 0.5$ and where the functions $f_1^{(1)} = 0$ and $f_2^{(1)} = 1$ are constant zero and unity functions, respectively, or $F_{1'} = \{f_{1'}^{(1)}, f_{2'}^{(1)}\}$, with $c_{1'}^{(1)} = c_{2'}^{(1)} = 0.5$ and where the functions $f_{1'}^{(1)} = x_2$ and $f_{2'}^{(1)} = \overline{x}_2$ are the identity and the negation functions, respectively. In other words, two fundamentally different function sets can have the same probabilistic structure. In the case of many parent variables, this issue becomes more complicated. In practice, one may want to construct the predictor functions for each node in the PBN such that the predictor functions have as few variables as possible, or such that the number of predictor functions is minimized.

Let us first consider minimizing the number of variables in the predictor functions. The above construction method produces predictor functions with the maximal number of variables. In some cases, when the CPTs are "separable," one can construct predictor functions having fewer variables but at the same time being consistent with the original conditional probabilities. For example, consider a case where node X_1 is regulated by a set of nodes $X_2, \ldots, X_n$. Assume that the first parent node X_2 has a canalizing effect on the target node X_1 such that all the values in the CPT for which $X_1 = 1$ and $X_2 = 1$ are equal to 0.9. Assume further, for simplicity, that all other values in the CPT for which $X_1 = 1$ (i.e., $X_1 = 1$ and $X_2 = 0$) are smaller than 0.9, the largest of those being, say, 0.8. Instead of blindly using the aforementioned algorithm for generating the predictor functions for X_1, it can be useful to take the special form of the CPT into account. Following the above algorithm the first functions can be constructed as already explained. However, the effect of the canalizing variable X_2 can be accounted for using only a single one-variable predictor function $f^{(1)} = x_2$ with $c^{(1)} = 0.9 - 0.8 = 0.1$. Alternatively, if no single-variable predictor functions can be constructed as explained above, then combinations of variables, starting with two variables, three variables, etc., can be considered. For example, if all the values in the CPT for which $X_1 = 1$, $X_2 = 1$, and $X_3 = 1$ are equal, then a single two-variable function can be defined as $f^{(1)} = x_2 x_3$.

The above search for predictor variables is explained in terms of the original conditional probabilities (i.e., CPTs). Alternatively, one can also try to optimize the obtained predictor functions. That is, each function in F_i should be expressed in some optimal form. The first thing to be considered is the removal of fictitious variables from the functions. This step can potentially result in functions having far fewer input variables. One can further optimize the actual expressions of the Boolean functions using methods such as the well-known Quine–McCluskey algorithm, which minimizes the number of terms in the disjunctive normal form. The above discussion, however, does not provide any optimal method for the predictor function construction (apart from the optimal representation of the functions). An optimal method can be described in terms of the number of predictor functions which we describe next.

In the worst case, the minimum number of predictor functions is determined by the number of different values in the CPT. The above construction method automatically selects that number of predictors (plus a possible constant zero function). In some special cases, the number of predictor functions can be reduced. Again consider the same triplets as above $(z_l, \mathbf{y}_l, p_l)$ for a single node X_i and again focus on only those triplets for which $z_l = 1$. This leaves us with a set of probability values $p_1, \ldots, p_r$, $r = 2^q$. The general optimality criterion can be stated as follows. Find the smallest set of selection probabilities

$\{c_1^{(i)},\dots,c_\mu^{(i)}\}$, where each $c_j^{(i)} \in [0,1]$ and $\sum_{j=1}^{\mu} c_j^{(i)} = 1$, such that each p_l, $1 \le l \le r$, can be written as

$$p_l = \sum_{j \in I_l} c_j^{(i)}, \tag{2.49}$$

where $I_l \subseteq \{1,\dots,\mu\}$. Let $J_j = \{l | j \in I_l\} \subseteq \{1,\dots,2^q\}$ be the set of indices of those p_l for which $c_j^{(i)}$ is used in the sum in Eq. (2.49). Then the function set $F_i = \{f_1^{(i)},\dots,f_\mu^{(i)}\}$ is defined as

$$f_j^{(i)} = \bigvee_{l \in J_j} \mathbf{x}_l,$$

where $\mathbf{x}_l$ is as above. We can verify the correctness of the predictor functions and the corresponding selection probabilities

$$\begin{aligned} P\{X_i(t) = 1 | \mathrm{Pa}(X_i(t)) = \mathbf{y}_l\} &= \sum_{j=1}^{\mu} f_j^{(i)}(\mathbf{y}_l) c_j^{(i)} \\ &= \sum_{j\,:\, f_j^{(i)}(\mathbf{y}_l)=1} c_j^{(i)} = \sum_{j\,:\, l \in J_j} c_j^{(i)} = \sum_{j \in I_l} c_j^{(i)} = p_l, \end{aligned}$$

because (on the second line) $f_j^{(i)}(\mathbf{y}_l) = 1$ only if the function $f_j^{(i)}$ contains the conjunction $\mathbf{x}_l$, i.e., $l \in J_j$. The second to last equality follows from the fact that $l \in J_j \Leftrightarrow j \in I_l$, and the last equality follows from Eq. (2.49). Note that the optimal function set can still be nonunique.

The obtained functions in F_i can be modified further as explained above by removing the possible fictitious variables and by applying the Quine–McCluskey optimization algorithm. Although a computationally efficient algorithm for the search of the optimal function sets remains to be developed, fortunately, the search problem is usually limited in the sense that each gene contains only a few parent variables.

Theorem 2.7 says that the two model classes can represent the same probabilistic behavior. However, there are many statistically equivalent PBNs that correspond to a DBN. This means that the PBN formalism is redundant from the probabilistic point of view. On the other hand, the PBN formalism is richer from the functional point of view because it can explain the regulatory roles of different gene sets in more detail than the conditional probabilities in DBNs. A detailed discussion of these issues, including dependent PBNs and discrete-valued DBNs, is contained in Lähdesmäki *et al.* (2006).

The two model classes, PBNs and DBNs, were compared by Li *et al.* (2007) using a biological time series data set from the Drosophila Interaction Database in order to evaluate the performance of these two approaches. For construction of the PBN, the coefficient of determination was used (see Sections 1.5.1 and 3.2) and the performance was compared using recall and precision measures. It was found that both approaches have good performance, while the DBN identified more gene interactions and gave better recall than the PBN. However, the accuracy of inferring the network also depended on the particular DBN inference algorithms that were selected for the implementation (the DBN algorithms were also computationally more time-consuming). It should be noted, following the discussion in this section, that since PBNs are more complex than DBNs, the former being represented by Markov chains (with a loss of the rules), it follows that it is easier to infer DBNs (which are Markov chains). Furthermore, the coefficient of determination is not a directional measure, which complicates its use for the inference from time series data; this issue is discussed in Section 3.2.

2.7 Mappings between PBNs

One obstacle in applying PBNs to large-scale networks is the computational complexity of the model. Owing to the large number of states often present in full networks, it is sometimes necessary to construct computationally tractable subnetworks while still carrying sufficient structure for the application at hand. Construction of mappings to alter the PBN structure while at the same time maintaining consistency with the original probability structure has been studied by Dougherty and Shmulevich (2003). An important mapping among these is a *projection* onto subnetworks. Unfortunately, while projections maintain the probabilistic structure by reducing the number of genes, they also increase the complexity of the Boolean-function structure, by exponentially increasing the number of constituent Boolean networks. A different kind of size-reducing mapping, called the *reduction mapping*, was introduced by Ivanov and Dougherty (2004). In contrast to the projection mapping, the reduction mapping does not increase the number of predictors for the genes that remain in the new network, but this mapping may not preserve the probability structure of the original PBN. However, the structure is altered in such a way so as to maintain maximum consistency with the original probability structure and, once this notion of maximum consistency has been defined, the problem reduces to one of optimization. For both of these mappings, there is still no control over changes in the dynamics (i.e., state transition diagrams) of the constituent Boolean networks comprising the original PBN and, furthermore, both types of mappings rely on knowledge about the steady-state distribution of the original PBN. To address the need for reductions that preserve the dynamical structure of the network, Ivanov *et al.* (2007) developed a strategy to reduce the size of the network while maintaining the same number of constituent Boolean networks and preserving, in an optimal sense, the attractor structure and the relative sizes of the basins of attraction. The preservation of the attractor structure is important as the attractors of a PBN determine its steady-state behavior. Preservation of the relative sizes of the basins of attraction ensures that the basin entropy (Eq. (1.10)) remains the same in the reduced network, therefore maintaining the same amount of uncertainty as regards the future behavior of the network starting from a random state.

Let S denote the state space of a PBN (in the binary case, it is simply the n-dimensional hypercube). For the kth constituent Boolean network (BN) in a PBN, governed by the kth network function $\mathbf{f}_k$, the state transition diagram is a directed graph $\Gamma_k = (S, E_k)$ with the states in S as the set of its vertices, and with edges $(\mathbf{s}_i, \mathbf{s}_j) \in E_k$ connecting the state $\mathbf{s}_i$ with the state $\mathbf{s}_j$ if $\mathbf{f}_k(\mathbf{s}_i) = \mathbf{s}_j$. The state transition diagram Γ_k is partitioned into *level sets*, where the level set l_j^k is composed of all of the states of the kth network that transition to one of the attractor states in exactly j transitions. One can think of the set of attractor states as the level set l_0^k. Given any transient state, it belongs to a unique basin and unique level. The dynamics of the state space S in the entire PBN are represented by the collection of the state transition diagrams Γ_k where the decision about switching from one state transition diagram to another one is based on the model parameters q, p, and $c_1, c_2, \ldots, c_m$. One should also notice that the deterministic state transition diagrams Γ_k can be combined into a single probabilistic state transition diagram Γ which represents the Markov chain associated with the PBN, as we have discussed previously.

Projection mappings of independent PBNs were defined by Dougherty and Shmulevich (2003). They were introduced as an attempt to reduce the complexity of a PBN A while maintaining consistency with the original probability structure of the PBN. The basic projection Π_i is a mapping that transforms the given PBN A into a new one with the same parameters q and p, and such that the number of genes is reduced by one; i.e., the gene x_i in the original network is deleted. Without loss of generality we may assume that the deleted gene is x_n. Thus, for a PBN $A = A^{q,p}(V,F,C)$, where $V = \{x_1,\dots,x_n\}$, $x_i \in \{0,1\}$, $i = 1,\dots,n$, $F = \{\mathbf{f}_1,\mathbf{f}_2,\dots,\mathbf{f}_m\}$, $\mathbf{f}_j : \{0,1\}^n \to \{0,1\}^n$, $j = 1,\dots,m$, and $C = \{c_1,\dots,c_m\}$ is the list of selection probabilities,

$$\Pi_n : A \to \hat{A}_n,$$
$$\hat{A}_n^{q,p}(\hat{V},\hat{F},\hat{C}), \quad \hat{V} = \{x_1,\dots,x_{n-1}\},$$
$$\hat{F} = \{\hat{\mathbf{f}}_1,\hat{\mathbf{f}}_2,\dots,\hat{\mathbf{f}}_s\}, \quad \hat{C} = \{\hat{c}_1,\dots,\hat{c}_s\}.$$

Every predictor $\mathbf{f}_j^{(i)}$, $i = 1,\dots,n-1$, generates two predictors $\hat{\mathbf{f}}_{0j}^{(i)}$ and $\hat{\mathbf{f}}_{1j}^{(i)}$ according to the rule

$$\hat{\mathbf{f}}_{kj}^{(i)}(x_1,\dots,x_{n-1}) = \mathbf{f}_j^{(i)}(x_1,\dots,x_{n-1},k), \tag{2.50}$$
$$k \in \{0,1\} \quad \forall (x_1,\dots,x_{n-1}) \quad \text{in} \quad \hat{A}_n.$$

Thus, every network function $\mathbf{f}_j$ for A determines 2^{n-1} new network functions for $\hat{A}_n$ by combining $\hat{\mathbf{f}}_{0j}^{(i)}$'s and $\hat{\mathbf{f}}_{1j}^{(i)}$'s in all of the possible ways for every fixed j and $i = 1,\dots,n-1$. The new network functions have their corresponding selection probabilities given by the formula

$$c_j(P\{x_n = 1\})^l (P\{x_n = 0\})^{n-l-1}, \quad j = 1,\dots,s, \tag{2.51}$$

where l is the number of the components of the new network function that are coming as $\hat{\mathbf{f}}_{1j}^{(i)}$, and $P\{x_n = k\}$, $k \in \{0,1\}$, is the marginal probability for the gene x_n to have values 0 or 1, computed using the steady-state probability distribution of the original PBN A. For example, the new network function $(\hat{\mathbf{f}}_{11}^{(1)},\hat{\mathbf{f}}_{01}^{(2)},\dots,\hat{\mathbf{f}}_{01}^{(n-1)})$ has its selection probability equal to $c_1(P\{x_n = 1\})(P\{x_n = 0\})^{n-2}$. When two or more of the network functions for $\hat{A}_n$ happen to be identical their selection probabilities combine in a natural way. One should notice that while the projection mapping preserves the probability structure of A, it exponentially increases the number of the BNs that form $\hat{A}_n$. In this sense the projection mapping does not achieve the goal of reducing the complexity of the original PBN.

A different kind of size-reducing mapping (which also preserves the parameters q and p of the original PBN) is the *reduction mapping* of a PBN (Ivanov and Dougherty, 2004). As we mentioned above, this mapping might not preserve the probability structure of the original PBN. To better understand the motivation and the definition of the reduction mapping we consider the following portion of the probabilistic state transition diagram Γ for the original PBN containing the states $i_1 = (x_1,\dots,x_{n-1},1)$, $i_0 = (x_1,\dots,x_{n-1},0)$, $j_1 = (x'_1,\dots,x'_{n-1},1)$, and $j_0 = (x'_1,\dots,x'_{n-1},0)$, where $p_{i_1j_1}$, $p_{i_0j_0}$, $p_{i_1j_0}$, $p_{i_0j_1}$ are the corresponding transition probabilities (Fig. 2.11). If one deletes the node x_n, this diagram collapses to where $i = (x_1,\dots,x_{n-1})$ and $j = (x'_1,\dots,x'_{n-1})$ are the corresponding states (Fig. 2.12), and

$$p^*_{ij} = P\{x_n = 1\}(p_{i_1j_1} + p_{i_1j_0})$$
$$+ P\{x_n = 0\}(p_{i_0j_1} + p_{i_0j_0}).$$

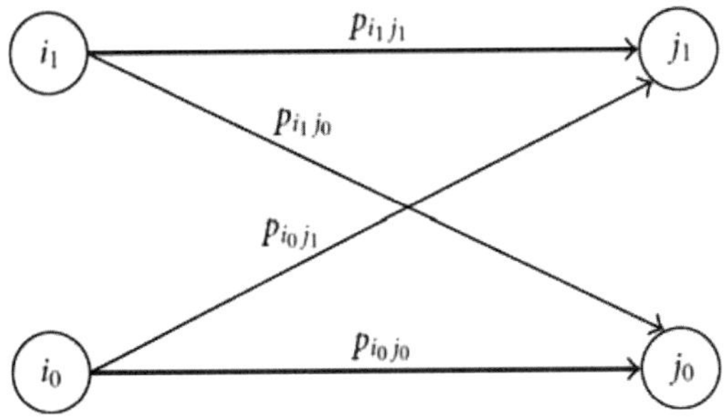

Figure 2.11. *Original state transition diagram (Ivanov and Dougherty,* 2004*).*

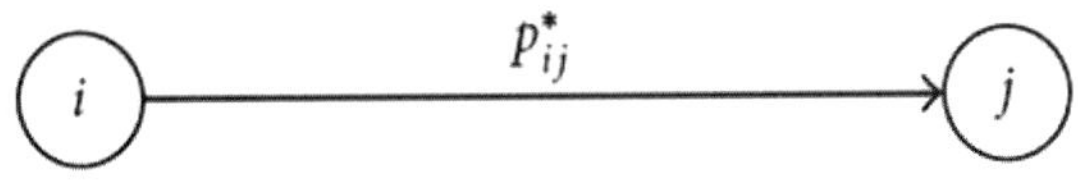

Figure 2.12. *Collapsed state transition diagram (Ivanov and Dougherty,* 2004*).*

If we formally define the reduction mapping acting on a PBN $A^{q,p}(V,F,C)$ by "deleting" one gene

$$\pi_n : A \rightarrow \tilde{A}_n,$$

$$\tilde{A}_n^{q,p}(\tilde{V},\tilde{F},\tilde{C}), \quad \tilde{V} = \{x_1,\dots,x_{n-1}\},$$

$$\tilde{F} = \{\tilde{\mathbf{f}}_1,\tilde{\mathbf{f}}_2,\dots,\tilde{\mathbf{f}}_s\}, \quad \tilde{C} = \{\tilde{c}_1,\dots,\tilde{c}_s\}$$

and require that $s \leq m$ for the new PBN $\tilde{A}_n$ with network functions having the same selection probabilities as their counterparts in A, $\tilde{c}_j = c_j$, $j = 1,\dots,s$, we can see that in order to maximally preserve the probability structure of the original PBN, the probabilistic state transition diagram for $\tilde{A}_n$ must have transition probabilities closely matching the transition probabilities p^*_{ij} of the "collapsed" state transition diagram described above. This goal is achieved by an optimization procedure which for every fixed network function $\mathbf{f}_j$, $j = 1,\dots,s$, from A combines the predictors $\hat{\mathbf{f}}^{(i)}_{0j}$ and $\hat{\mathbf{f}}^{(i)}_{1j}$ to form the new predictor $\tilde{\mathbf{f}}^{(i)}_j$, $i = 1,\dots,n-1$. A detailed discussion about the construction of π_n in the special case of an independent PBN is given in Ivanov and Dougherty (2004), and the construction carries on with no changes for the case of a general PBN.

One can immediately notice that the reduction mapping reduces the complexity of the original PBN not only by deleting one gene but also by not increasing the number of BNs that comprise the reduced PBN. Here, we want to point out the difference in defining the reduction and the projection mappings. While the projection is based on the probability distribution of a single gene, the reduction mapping is defined using the probability distribution of the entire collection of states of the given PBN which allows for the optimization procedure given in Ivanov and Dougherty (2004). In both cases, though, there is no control over the changes in the dynamics/state transition diagrams of the BNs comprising the original PBN. In addition, both mappings rely on knowledge of the steady-state distribution of the original PBN.

Another reduction mapping, referred to as *dynamics induced reduction* (DIRE), deletes genes from the PBN while preserving the dynamics of the constituent BNs and keeping their number unchanged (Ivanov *et al.*, 2007). This approach follows a procedure that collapses the state transition diagram for each individual BN in a manner similar to the reduction mapping. States in a given BN are aggregated together if they differ only in the expression value of the gene that has been deleted from the network. It is important to notice from the onset that the design of DIRE ensures control over the damage to the state transition diagram of every BN. In this way not only is the stationary/steady-state distribution of the reduced PBN similar to the one produced by the above described reduction mapping but, in addition, the structure of the state transition diagram is taken into account, which is the main advantage of the newly proposed algorithm. Several important issues must be addressed before describing the algorithm in greater detail. Assuming that the real genetic regulatory system being modeled by a PBN switches between different wirings for the genes participating in the regulation, depending on factors either external or internal to the cell, then deletion of a gene has to be properly interpreted. First, it is desirable that deleting a gene from a given regulatory network does not increase the number of constituent BNs. Second, it is important that the dynamics of a real genetic regulatory system do not strongly depend on how many of the genes participating in the regulation are observable or not. Thus, preserving properties of a PBN, such as its attractor structure, the relative sizes of the basins of attraction, as well as the level structure of the state transition diagrams of the BNs comprising the PBN, is a major goal of size reduction.

To appreciate the second goal, consider an example BN consisting of 3 genes, whose state transition diagram is given in Fig. 2.13. Suppose that the gene corresponding to the rightmost digit in the binary representation of the states is to be deleted. If we try to collapse the state transition diagram, we notice that, with respect to the attractor structure of the original BN, merging the leaf node (001) and the attractor node (000) can be done in two very different ways: either the merging happens towards the attractor state or it happens towards the leaf state. In the first case, the attractor state is preserved in the reduced BN as (00), and the basin of attraction of the attractor (111) in the original network loses one leaf. In the second case, the attractor structure of the reduced network differs significantly from that of the original BN, the only remaining attractor being the reduced state (11). Thus, if we consider the attractor structure of a BN as a representation of important biological characteristics of the real genomic regulatory system, then the merging of the states (000) and (001) should be done towards the attractor state. At the same time, we point out that those two states are the only states in the original state transition diagram that create a possibility of essentially altering the attractor structure of the original BN. The rest of the states will merge within the basin of attraction of the attractor state (111). Keeping all of this in mind we give the following definition and then state the DIRE algorithm.

Definition 2.8. *A state s is called an* ***inconsistency point*** *with respect to gene g if and only the state s' in the state transition diagram that differs from s only in the value of g belongs to a different basin of attraction than the basin of attraction to which s belongs.*

DIRE: Input - a PBN. Output - a reduced PBN.

for {*every BN in the PBN*} **do**
  {*label each attractor and its corresponding basin,*
    and label each leaf node in the state transition
    diagram} ;
  {*label each inconsistency point*} ;

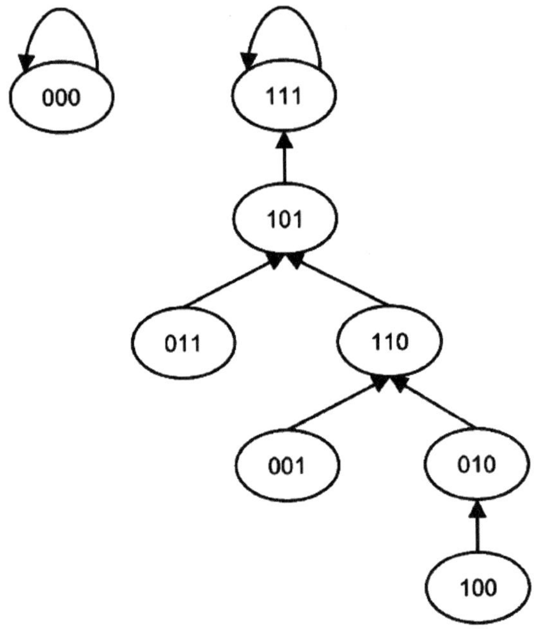

Figure 2.13. *State transition diagram of an example BN (Ivanov et al., 2007).*

merge *{states with respect to the gene being deleted,*
filling in at the same time the truth table
of the reduced BN} ;
end;

Notice that the complexity of the algorithm is determined by the number n of genes and by the number m of BNs comprising the PBN. Because the individual BNs model the different possible wirings of the real genomic regulatory network, one can assume that $m \leq m_{max} = \text{const}$. Biological considerations suggest that the constant m_{max} is not larger than 100, and in all practical applications $m_{max} \leq 20$. Therefore, the complexity of the algorithm depends on the number $M = 2^n$ of states, on the size of the truth table for the reduced BNs that are built up in the **merge** part of the algorithm, and on the number of visits to the states of the original BNs which is also part of the same procedure. The size of the truth table of the reduced BNs is $(n-1) \times 2^{n-1}$, which leads to a complexity estimate of $O((n-1)M/2)$. To analyze the average number of visits to the states of a BN we assume that its state transition diagram is a single-rooted well-balanced tree on M vertices. Such an assumption is motivated by the understanding that if a PBN were to model a real genomic regulatory system, then that PBN should have relatively short attractor cycles with many of them being singleton ones (i.e., fixed points). With this assumption in mind, one can observe that the number of state visits is approximately $2\times$ #leaves $\times$ #levels in the state transition diagram. The factor 2 comes from the observation that each state on the path from a leaf to the corresponding attractor could be visited twice within the **merge** procedure. The number of levels gives the average length of such a path, and for the well-balanced single-rooted tree we have #levels $\sim \log_2 M = n$ and #leaves $\sim M/2$. Thus, we

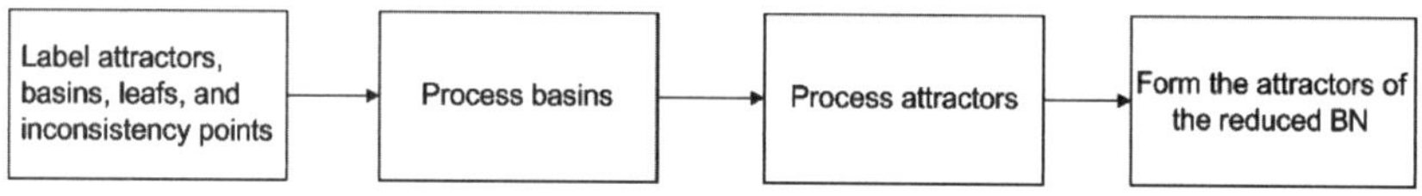

Figure 2.14. *General flowchart.*

obtain a complexity estimate of $O(nM)$ for the visits to the states of the state transition diagram of one of the original BNs. Next we discuss the **merge** part of the algorithm. When merging/collapsing the states that differ only by the value of the gene being deleted, one needs to take special care of two issues:

(1) controlling the damage to the attractor structure while properly handling the points of inconsistency;

(2) merging states without creating cycles within the basins of attraction.

In order to describe the merging part of the algorithm, we need a few more definitions.

Definition 2.9. *The states s and s' in a BN state transition diagram are called* ***dual states*** *with respect to a given gene g in the BN if and only if s and s' differ only in the value of that gene g.*

Definition 2.10. *The state produced after deleting the variable representing the gene g in a state s from the state transition diagram of the original BN is called the* ***reduced state*** *with respect to g and is denoted by s_g.*

Definition 2.11. *Assume that the attractor cycles and the singleton attractors of a BN are indexed. If S_j denotes the jth attractor cycle/singleton attractor, then $S_{jg} := \{$ pairwise distinct $s_g : s \in S_j\}$ or $S_{jg} := \{s_g\}$ if s is a singleton.*

Definition 2.12. *If the states s and s' are dual states with respect to the gene g in a BN, then the one that determines the position of the reduced state s_g in the state transition diagram of the reduced BN is called* ***absorbing*** *and is denoted by $\tilde{s}$.*

The flowchart of the algorithm is given in Figs. 2.14–2.18. Notice that the flowchart in Fig. 2.18 combines with the one in Fig. 2.15 to form the **process basins** procedure, and also with the flowchart in Fig. 2.16 to form the **process attractors** procedure. The last part of the algorithm, the **form attractors** procedure, forms an attractor cycle which is a random permutation of the reduced versions of the states that are not points of inconsistency in the corresponding attractor cycle of the original BN.

Before formally defining the algorithm, let us provide an example to show how the proposed algorithm works in the simple case of the BN on 3 genes shown on the left-hand side of Fig. 2.19. Assume that the gene corresponding to the rightmost digit in the states of the BN is to be deleted. The labeling of the states in the state transition diagram is produced by the first part of the algorithm given in the general flowchart, Fig. 2.14. For example the leaf node (001) has a label (1, I, L) indicating that, first, it belongs to the basin of the attractor 1, second that it is an inconsistency point, and last that it is a leaf node.

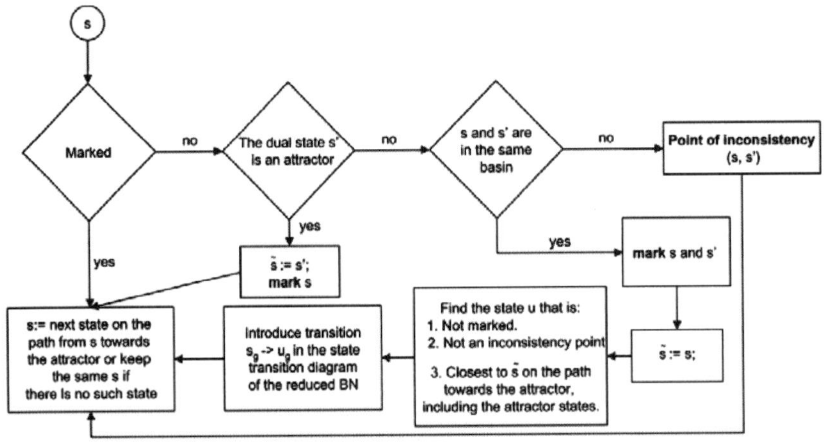

Figure 2.15. *Process state procedure.*

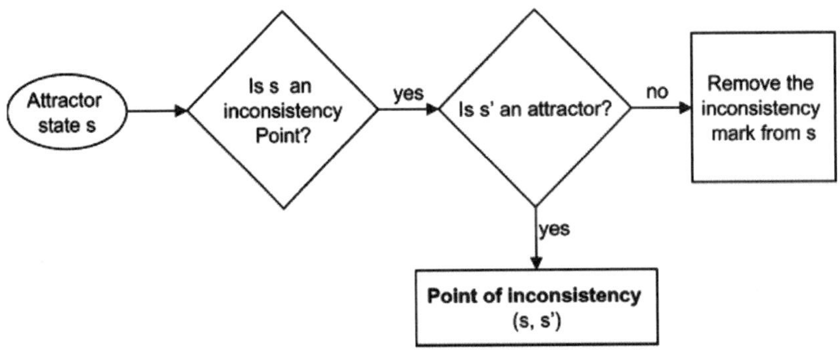

Figure 2.16. *Process attractors procedure (one pass of the* **for** *loop).*

One can notice that the states (000) and (001) are the only inconsistency points. When processing basins we assume that the leaves of the BN are ordered as (011), (001), (100). We point out that different orderings of the leaves could produce different state transition diagrams of the reduced BN. With the given leaf ordering, the first states processed and marked are (011) and (010), and $\tilde{s} := (011)$. Because there is a transition $(011) \to (101)$ the transition $(01) \to (10)$ is introduced in the state transition diagram of the reduced BN. The next available nonattractor state on the path from (011) towards the attractor, labeled as 1, is (101), and its dual state is the leaf (100). Thus, $\tilde{s} := (101)$, and the algorithm marks those states and introduces a transition $(10) \to (11)$ in the state transition diagram of the reduced BN. The reason for introducing such a transition is that the state $\tilde{s} \to (111)$. Because all of the nonattractor states on the path from the first leaf to the attractor 1 have been visited, the next input for the **process state** procedure is the second leaf (001) that happens to be an inconsistency point. Therefore the **point of inconsistency** procedure is invoked, which results in marking the state (001), because its dual state (000) is an attractor state. The next available state on the path from (001) to the attractor 1 is (110). Since its

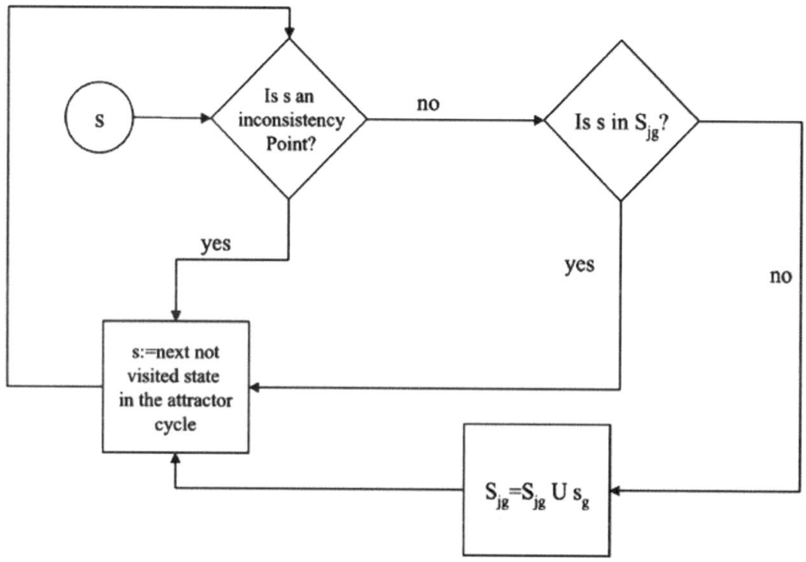

Figure 2.17. *Form the set S_{jg} (one pass of the* **for** *loop in the* **form attractors** *procedure).*

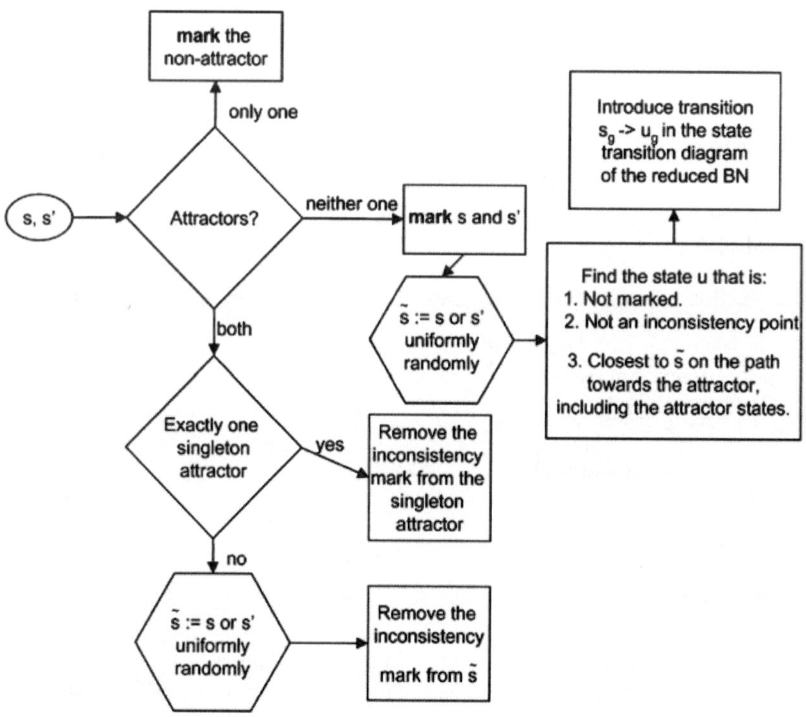

Figure 2.18. *Point of inconsistency procedure.*

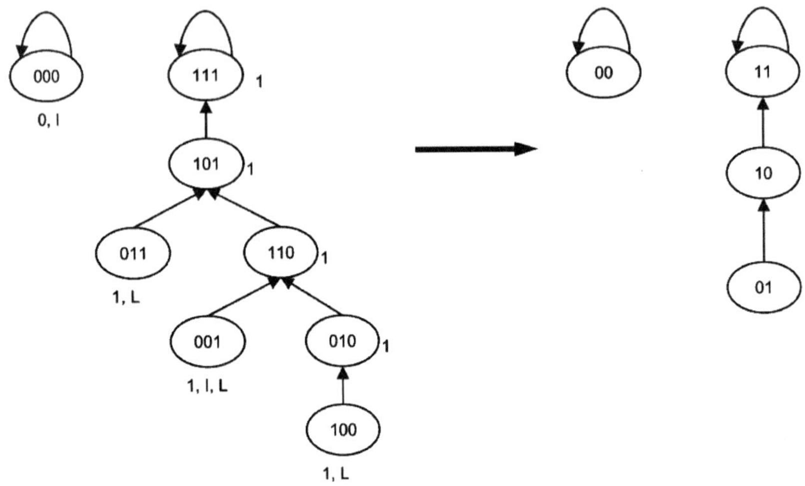

Figure 2.19. *Example BN for illustrating the reduction algorithm.*

dual is the attractor state (111) the state (110) is marked by the algorithm, and with that all of the basins of attraction for the BN have been processed. The attractor structure of the BN in consideration is relatively simple, and the **process attractors** procedure removes the inconsistency label from (000) because its dual state (001) is not an attractor state. Finally, the **form attractors** procedure uses the singleton attractors (000) and (111) from the original BN to create the singleton attractors (00) and (11) in the reduced BN. Thus, we arrive at the state transition diagram of the reduced BN given on the right-hand side of Fig. 2.19.

The following is the formal description of the proposed algorithm. Recall that $\tilde{s}$ refers to the absorbing state, and S_{jg} stands for the set of reduced, with respect to the gene g, attractor states from the j attractor cycle/singleton attractor of a BN that participates in the original PBN.

merge while deleting the gene g procedure: Input - a BN. Output - a reduced BN.

for {*every leaf node s in the BN*} **do**
    $s_1 :=$ **process state** (s);
    while $\{s_1 \neq s\}$ **do**
        $s := s_1$
        $s_1 :=$ **process state** (s);
    end;
end;
process attractors;
form attractors;

Remark: The **for** loop in the **merge** procedure ensures that all of the nonattractor states of a BN have been marked and the nonattractor transitions of the reduced BN have been established. The actual marking and the establishment of the new transitions are accomplished by successive calls to the **process state** procedure. The procedure returns a state s_1 such that an attractor state has been reached if and only if $s = s_1$.

process state (s) procedure: Input - a state s. Output - a state s.

if {s *is not marked*}
  if {*the dual* s' *of the state* s *is not an attractor state*}
    if {s *and* s' *belong to the same basin of attraction*}
    then
      mark {*the states* s *and* s'}; {$\tilde{s} := s$};
      {*fill in the truth table of the reduced BN by introducing a transition from* s_g *into* u_g, *where* $\tilde{s} \to u$ *is a path from* $\tilde{s}$ *to* u *in the state transition diagram of the original BN, and* u *is the closest, on the path towards the attractor, state to* $\tilde{s}$ *that is neither a point of inconsistency nor a marked state, or if all of the states on that path are either marked or points of inconsistency* u *is selected to be the first attractor state on that path*};
      else point of inconsistency (s, s');
    else{$\tilde{s} := s'$ *and mark* s};
if {*the state* w *from the transition* $s \to w$ *is a nonattractor state*} $s := w$;
return s;

Remark: Notice that the assignment $\tilde{s} := s$ is necessary in order not to increase the number of level sets of the state transition diagram of the BN being reduced. It should be noted that the other possible assignment for the absorbing state $\tilde{s} := s'$ can lead to a reduced BN with more level sets than the original BN.

process attractors procedure: Input - the attractors of a BN. Output - a labeling of the attractors.

for {*every attractor node* s *in the BN*} **do**
  if {s *is an inconsistency point and* s' *is an attractor node*}
  then
    point of inconsistency{s, s'};
  if {s *is an inconsistency point and* s' *is not an attractor node*}
  then {*remove the inconsistency mark from* s}
end;
return;

form attractors procedure: Input - the attractors of a BN together with the labeling. Output - the attractors of the reduced BN.

for{*every attractor cycle* j *in the BN*} **do**
  {**form** *the set* S_{jg}};
  {*fill in the truth table of the reduced BN by introducing a cycle connecting the reduced states in* S_{jg} *in the order of any one of their possible permutations*};
end;
return;

The next procedure takes care of the points of inconsistency in such a way that

(1) if s and s' are both nonattractor states, then the relative weight of a basin of attraction in the original BN is changed by one state only and with probability of $1/2$;

(2) if one of s and s' is an attractor state, then that state becomes the absorbing state;

(3) if both of the states are attractor states, the singleton attractors are favored and are labeled so that the **form attractors** procedure preserves them in the state transition diagram of the reduced BN.

point of inconsistency (s,s') procedure: Input - states s and s'. Output - a labeling for the states of a BN and transitions for the states in the basins of the state transition diagram of the reduced BN.

if{*s and s' are both nonattractor states*}
  mark{*the states s and s'*};
  {*uniformly randomly decide if $\tilde{s} := s$ or $\tilde{s} := s'$*};
  {*fill in the truth table of the reduced BN by introducing a transition from s_g into u_g, as described in the* process state *procedure*};
if{*either s or s' is an attractor state*}
  mark{*the nonattractor state*};
if{*s and s' are both attractor states*}
case 1 {*s and s' are either both singleton or nonsingleton attractors*}:
    {*uniformly randomly decide if $\tilde{s} := s$ or $\tilde{s} := s'$*};
    {*remove the inconsistency mark from $\tilde{s}$*};
case 2 {*exactly one of the dual states is a singleton attractor*}:
    {*$\tilde{s} :=$ singleton attractor*};
    {*remove the inconsistency mark from $\tilde{s}$*};

By the way it is constructed the proposed algorithm achieves the goals stated at the beginning of this section while taking care of the points of inconsistency. First, the algorithm does not increase the number of BNs comprising the original PBN, and in the special cases when some of the reduced BNs are identical, DIRE produces a reduced PBN with a lesser number of participating BNs compared to that of the original PBN. That is achieved by combining the identical BNs and assigning to the only remaining copy a selection probability that equals the sum of the individual selection probabilities of those BNs. Second, one can immediately notice that the merging and marking of states in the basins of attraction does not create cycles in the basins of attraction of the reduced BNs, the algorithm never increases the number of level sets present in the state transition diagram of the original PBN, and the points of inconsistency are treated in a way that controls the damage to the state transition diagram of the respective BN. The algorithm changes the relative weight of any basin of attraction, e.g., the **point of inconsistency** procedure, only in the case when a point of inconsistency leads to an absorbing state in a different basin of attraction. Moreover, the algorithm favors the original structure of the attractors of the BN being reduced. The only case when the number of attractors or the structure of an attractor could be changed is when a state s in that attractor is a point of inconsistency. If its dual state $s^{'}$ is not part of an attractor, then $s^{'}$ is merged into s. Otherwise, the two attractor states s and $s^{'}$ are merged so that the singleton attractor is always favored or merging is done uniformly randomly if none of the two states represents a singleton attractor. There is one exceptional case when all of the states in an attractor cycle are points of inconsistency, and then the attractor states in that cycle could be merged towards other attractor cycles with a probability of $1/2$. The intuition behind this treatment of the attractors and their basins is that the presence of latent variables should not affect in a significant way the observed dynamical behavior of the regulatory system modeled by the PBN. One should notice that every transition introduced in the state transition diagram of a reduced BN is equivalent to filling in the row in its truth table that is labeled by the reduced state s_g. Here we assume that the truth table of the BN has its rows labeled by the states (ordered as binary numbers) and its columns labeled by the individual genes. Marking the pairs of dual states ensures that every row in the truth table is visited only once. Finally, the basins and the attractors are handled in separate consecutive steps of the algorithm, and forming of the attractor cycles of the reduced BN is performed as the last step. All of this ensures that an attractor cycle of the reduced BN can be formed in any one of the $m!$ possible ways if there are m reduced states participating in that cycle, and in doing so none of the previously introduced state transitions between the states in the basins or from a state in the basin to an attractor state will change.

In summary, the proposed algorithm attempts to preserve the flow of information between states lying on the same path of the state transition diagram and handles the exceptions created by the inconsistencies arising from deletion of a gene by considering it to be a latent variable. Additionally, the algorithm does not require knowledge of the steady-state distribution of the original PBN. The design of the algorithm ensures the relative preservation of the structure of the attractor sets and their corresponding basins, which in turn helps in obtaining a reduced PBN whose steady-state distribution is close to the desired one.

Chapter 3

Inference of Model Structure

In previous chapters, we have discussed how genetic regulation can be modeled using PBN models. Various network properties and behaviors, particularly system dynamics, can be examined with this model class and related to biological function and phenotype. For example, the steady-state distribution of a network represents the tendency of the system, resulting from the dynamical interplay of the interactions among the system variables, to execute a specific limited set of coordinated gene activities out of a vast number of possible states. Such steady states correspond to functionally important cellular states. Furthermore, as we shall see in the next chapter, such models can be used to develop tools for rational therapeutic intervention in diseases such as cancer.

As we have seen in several examples, models (primarily deterministic as opposed to probabilistic) can be constructed "by hand" on the basis of known biological facts, such as mechanistic rules of regulation. However, a paramount problem is to identify such networks from experimental measurement data, particularly as in most cases, such networks are unknown or insufficiently characterized. Indeed, modern molecular biology technologies are now being used to generate increasingly large amounts of high-throughput systems-level measurement data for many organisms under a variety of experimental conditions. Such data can be used to learn the model structure, which in the context of PBNs entails inferring the Boolean functions for each gene along with their inputs (i.e., wiring).

The task of PBN inference generally requires the application of techniques and methods from the fields of statistical inference and computational learning theory. Some of the earliest attempts to infer genetic network structure from measurement data, such as gene expression microarray data, employed the BN formalism (Liang *et al.*, 1998; Akutsu *et al.*, 1999; Ideker *et al.*, 2000; Maki *et al.*, 2001; Shmulevich *et al.*, 2002e, Lähdesmäki *et al.*, 2003). Let us first focus on the inference of Boolean functions from data.

3.1 Consistent and Best-Fit Extensions

Perhaps the most straightforward inferential approach is the well-known *consistency problem*, which entails a search for a rule from examples (Pitt and Valiant, 1988; Boros *et al.*, 1998; Shmulevich *et al.*, 2001). The problem is formulated as follows. Given some sets T and F of "true" and "false" vectors, respectively, the aim is to discover a Boolean function

f that takes on the value 1 for all vectors in T and the value 0 for all vectors in F. It is also commonly assumed that the target function f is chosen from some class of possible target functions. In the context of BNs, such a class could be the class of canalizing functions (Harris *et al.*, 2002) or simply functions with a fixed number, K, of essential (nonfictitious) variables.

More formally, let

$$T(f) = \left\{v \in \{0,1\}^n : f(v) = 1\right\}$$

be the *on-set* of function f and let

$$F(f) = \left\{v \in \{0,1\}^n : f(v) = 0\right\}$$

be the *off-set* of f. The sets $T, F \subseteq \{0,1\}^n$, $T \cap F = \emptyset$, define a *partially defined* Boolean function $g_{T,F}$ as

$$g_{T,F}(v) = \begin{cases} 1, & v \in T, \\ 0, & v \in F, \\ * & \text{otherwise.} \end{cases}$$

A function f is called an *extension* of $g_{T,F}$ if $T \subseteq T(f)$ and $F \subseteq F(f)$. The consistency problem (also called the extension problem) can be posed as follows: given a class C of functions and two sets T and F, is there an extension $f \in C$ of $g_{T,F}$?

It is easy to see that if C is the class of all Boolean functions, an extension $f \in C$ exists if and only if $T \cap F = \varnothing$. This condition can be checked in time $O(|T| \cdot |F| \cdot \text{poly}(n))$ by comparing all examples pairwise, where poly(n) is the time needed to examine one pair of examples (Boros *et al.*, 1998). Sorting algorithms can be used to provide a more efficient solution to the consistency problem (Lähdesmäki *et al.*, 2003). Suppose that there are $|T| + |F| = m$ examples. By encoding all vectors $x \in T \cup F$ by positive integers, sorting them, and comparing consecutive elements in the sorted list to determine whether or not one belongs to T and the other to F, the problem can be solved in time $O(m \cdot \log m \cdot \text{poly}(n))$.

In reality, a consistent extension may not exist either due to measurement noise[5] or, more likely, due to a number of underlying latent factors, such as regulatory proteins or other external influences that may be outside the model. In this case, we may have to give up our goal of establishing a consistent extension and settle for a Boolean function that minimizes the number of misclassifications. This problem is known as the *best-fit extension problem* (Boros *et al.*, 1998) and is formulated as follows. Suppose we are given positive weights $w(x)$ for all vectors $x \in T \cup F$ and define $w(S) = \sum_{x \in S} w(x)$ for a subset $S \subseteq T \cup F$. Then, the *error size* of function f is defined as

$$\varepsilon(f) = w(T \cap F(f)) + w(F \cap T(f)). \tag{3.1}$$

If $w(x) = 1$ for all $x \in T \cup F$, then the error size is just the number of misclassifications. Other weights can be assigned on the basis of some other measure, such as a quality control score.[6] The goal is then to output subsets T^* and F^* such that $T^* \cap F^* = \emptyset$ and $T^* \cup F^* =$

[5]By measurement noise, we mean any random phenomena exhibited by a complex measurement process, such as gene expression microarrays, which may result from hybridization conditions, image processing techniques, and other sources of variability (see Zhang *et al.* (2004) for a detailed discussion of these issues).

[6]The weights can also be defined so as to take into account multiplicities of the measurements, where the same vector can occur multiple times, each associated possibly with a different weight, in both T and F (Lähdesmäki *et al.*, 2003).

$T \cup F$ for which the partially defined Boolean function g_{T^*,F^*} has an extension in some class of functions C and so that $w(T^* \cap F) + w(F^* \cap T)$ is minimum. Consequently, any extension $f \in C$ of g_{T^*,F^*} has minimum error size.

It is clear that the best-fit extension problem is computationally more difficult than the consistency problem, since the latter is a special case of the former, that is, when $\varepsilon(f) = 0$. Nonetheless, it has been shown that for many function classes, including the class of all Boolean functions, the best-fit extension problem for BNs is polynomial-time solvable (Shmulevich *et al.*, 2002e). Further, Lähdesmäki *et al.* (2003) have introduced a method that can be used to significantly accelerate the search for consistent gene regulatory networks (not just decide whether one exists) under the BN model. The same method, with minor changes, also applies to learning gene regulatory networks under the best-fit extension problem. We turn to this next.

Define a 4-ary column vector $\mathbf{f} \in \{0, 1, ?, *\}^{2^n}$ that is used to encode the truth table of a Boolean function f having n variables (we will discuss the case of $k \leq n$ variables next). $\mathbf{f}$ is indexed from 1 to 2^n (indices are shown as subscripts $\mathbf{f}_i$). So, the ith element of $\mathbf{f}$ defines the binary output value for the input vector $s^{-1}(i)$, where $s : \{0,1\}^n \to \{1,\dots,2^n\}$ can be used to encode all binary vectors of length n to positive integers. If $\mathbf{f}_i$ equals ?, the output corresponding to the input $s^{-1}(i)$ is undetermined, or can be either 0 or 1. Further, $\mathbf{f}$ is assumed to be initialized to $\mathbf{f} = (?,\dots,?)$. Let us call the nonbinary truth table, possibly containing ?'s, as the "generalized" truth table and for iteratively updating $\mathbf{f}$ define the jth version of $\mathbf{f}$ to be $\mathbf{f}^j$. Then, the following procedure answers the consistency problem and finds all consistent functions relative to the given examples.

This procedure runs through all examples $x \in T \cup F$ and during every cycle updates $\mathbf{f}^j$ as

$$\mathbf{f}^j_{s(x)} = \begin{cases} 0 & \text{if } x \in F \wedge \mathbf{f}^{j-1}_{s(x)} = ?, \\ 1 & \text{if } x \in T \wedge \mathbf{f}^{j-1}_{s(x)} = ?, \\ * & \text{otherwise,} \end{cases} \tag{3.2}$$

where $j = 1,\dots,m$ is the index of the current cycle (measurement) and $\mathbf{f}^0$ refers to the initial setting of vector $\mathbf{f}$, i.e., $\mathbf{f}^0 = (?,\dots,?)$. It can be easily seen that if the last branch of Eq. (3.2) is used, no consistent function can be found for the given examples. Otherwise there exists at least one consistent function and all of them are defined by $\mathbf{f}$ in the "generalized" form. Thus, when the time spent on memory initialization is ignored, the consistency problem can be answered and the "generalized" truth table is found in time

$$O((|T| + |F|) \cdot \text{poly}(n)) = O(m \cdot \text{poly}(n)).$$

The found generalized truth table $\mathbf{f}$ may now contain question marks. As mentioned above, each ? specifies an undetermined element of the truth table $\mathbf{f}$. If we prefer standard truth tables, each question mark must be changed to a single bit, either 0 or 1. In view of the consistency problem, each ? can be set to either 0 or 1 and the resulting truth tables, regardless of the chosen bits, are all consistent. Conversion to a standard truth table requires a run through the truth table $\mathbf{f}$ and each ? has to be set to either 0 or 1. Thus, $\mathbf{f}$ can be converted to a standard truth table in time $O(2^n)$.

It is easy to see that the total number of consistent functions scales exponentially with the number of question marks. Let $|f|_?$ denote the number of question marks in vector $\mathbf{f}$. Thus, conversion of all consistent functions into standard truth tables requires time $O(2^{|f|_?} \cdot 2^n)$ during which one needs to keep track of the previously set bits such that all possible combinations are found.

Let us now consider the consistency problem for Boolean functions having no more than $k \leq n$ variables. Such a constraint is frequently employed to mitigate model complexity, particularly in light of small sample sizes. The answer to the consistency problem and all consistent Boolean functions in the generalized form can be found simply by applying the above algorithm to all $\binom{n}{k}$ different variable combinations. The time needed for this is $O\left(\binom{n}{k} \cdot m \cdot \text{poly}(k)\right)$. The question marks can be removed in time

$$O\left(\binom{n}{k} \cdot 2^{|f|_{?,\max}} \cdot 2^k\right), \tag{3.3}$$

where $|f|_{?,\max}$ denotes the maximum number of question marks in any of the found generalized truth tables over all $\binom{n}{k}$ different variable combinations.

To generalize this for BNs, we must repeat the above process for all n genes, and thus the resulting time complexity is

$$O\left(\binom{n}{k} \cdot n \cdot m \cdot \text{poly}(k)\right). \tag{3.4}$$

Similarly to Eq. (3.3), the time complexity to convert the generalized truth tables to standard truth tables for the entire network is

$$O\left(\binom{n}{k} \cdot n \cdot 2^{|f|_{?,\max}} \cdot 2^k\right),$$

where now $|f|_{?,\max}$ denotes the maximum number of question marks in any of the found generalized truth tables over all n genes and $\binom{n}{k}$ different variable combinations.

A similar approach applies to the best-fit extension problem as well. As above, let us first consider the best-fit function for a single gene and n variables. In principle, we could solve the problem by computing the error size for all functions in some class of functions, requiring time $O\left(2^{2^n} \cdot m \cdot \text{poly}(n)\right)$, where 2^{2^n} is the total number of functions and $m \cdot \text{poly}(n)$ is the time needed to compute the error size for one function (see Eq. (3.1)). A faster solution can be found using a method similar to the one above.

Assume that we have two vectors $\mathbf{c}^{(0)}, \mathbf{c}^{(1)} \in \mathbb{R}^{2^n}$, where $\mathbb{R}$ denotes the set of real numbers, and $\mathbf{c}^{(0)}, \mathbf{c}^{(1)}$ are indexed from 1 to 2^n and are initially zero vectors. Then, during one pass over the given examples, $\mathbf{c}^{(0)}$ and $\mathbf{c}^{(1)}$ can be updated to

$$\begin{aligned} \mathbf{c}_i^{(0)} &= w(x) \quad \text{if } x \in F \wedge s(x) = i, \\ \mathbf{c}_i^{(1)} &= w(x) \quad \text{if } x \in T \wedge s(x) = i. \end{aligned} \tag{3.5}$$

Elements of $\mathbf{c}^{(0)}$ and $\mathbf{c}^{(1)}$ that are not set in Eq. (3.5) remain zero-valued due to initialization. Let a fully determined candidate function be f with the corresponding truth table $\mathbf{f}$ having only binary elements and let $f_{s(x)}$ define the output value for the input vector x as before. Let $\bar{\mathbf{f}}$ denote the complement of $\mathbf{f}$ (i.e., all zeros and ones are flipped). Then, the error size of function f can be written as

$$\varepsilon(f) = \sum_{i=1}^{2^n} \mathbf{c}_i^{(\bar{\mathbf{f}}_i)}. \tag{3.6}$$

It can be seen that the error size is minimized when $\mathbf{c}_i^{(\bar{\mathbf{f}}_i)}$ is minimum or, conversely, $\mathbf{c}_i^{(\mathbf{f}_i)}$ is maximum for all i. Thus, the truth table of the optimal Boolean function is $\mathbf{f}_i = \arg\max_j \mathbf{c}_i^{(j)}$ and the function minimizing the error size in Eq. (3.6) can be found within a single pass of the set elements of the vectors $\mathbf{c}^{(0)}$ and $\mathbf{c}^{(1)}$, requiring $O(m \cdot \text{poly}(n))$ time. The generalization for n genes and k variables is straightforward and similar to the consistency problem. The time complexity of the best-fit extension problem for the entire BN is then

$$O\left(\binom{n}{k} \cdot n \cdot m \cdot \text{poly}(k)\right). \tag{3.7}$$

It should be mentioned that the optimal function is often not unique after applying the above inference scheme. Particularly in the context of PBNs, we may be interested in finding all functions f having error size smaller than some specified threshold, i.e., $\varepsilon(f) \leq \varepsilon_{\max}$. The vectors $\mathbf{c}^{(0)}$ and $\mathbf{c}^{(1)}$ can also be used for this purpose, where the second-best function (in terms of error size) is found by flipping only one element of the truth table of the best function; this element, say the ith, has to be selected such that the corresponding absolute difference between elements $\mathbf{c}_i^{(0)}$ and $\mathbf{c}_i^{(1)}$ is minimum. Lähdesmäki *et al.* (2003) proposed a greedy algorithm to solve this problem and applied it to the Spellman *et al.* (1998) yeast cell-cycle data, finding the best-fit functions with limited error size for a selected set of genes. The results were found to be in good agreement with the known regulators of these genes. For instance, the complex Swi4/Swi6 is known to be a transcriptional regulator of the cyclin gene Cln1 (Simon *et al.*, 2001). The two-variable predictor {Swi4,Swi6} is among the top 0.5% of all two-variable predictors of the gene Cln1; similarly, the three-variable predictors {Swi4,Swi6,Mbp1} and {Swi4,Swi6,Ndd1} are both among the top 0.08% of all three-variable predictors of Cln1. Similar results were found for other genes, and the best-fit functions for the known transcriptional regulators were among the best ones in almost every case.

Since the search for a hidden Boolean function is essentially a mathematical programming problem, a variety of optimization approaches have been considered in the literature. For example, Kamath *et al.* (1992, 1993) cast the problem as a particular type of satisfiability problem which is then translated into an integer programming feasibility problem that is solved with an interior point algorithm for integer programming. One interesting aspect of several approaches in this genre is that they employ *continuous* mathematics to solve a discrete optimization problem.

3.1.1 Sensitivity regularization

Let us now turn to an example illustrating how the best-fit extension problem can be used to infer a BN while at the same time taking into account a global (i.e., system-wide) network dynamical property as an assumption in the inference process. The purpose of incorporating such an assumption is to reduce the inference error of the designed network. Indeed, any prior knowledge about the network structure, architecture, or dynamical rules is likely to improve the accuracy of the inference, especially in a small sample size scenario. Recall our discussion in Chapter 1 pertaining to the evidence that living systems, as manifestations of the underlying network of molecular interactions, are poised at the critical boundary between an organized and a disorganized state, affording optimality of cellular information processing and the ability of the cell to exhibit complex coordinated macroscopic behavior. If biological networks are indeed critical, a key question is whether this knowledge can be used to improve the inference of network structure and dynamics from measurements.

The average sensitivity, given in Eq. (1.7), is an order parameter for BNs, which can be computed directly from the Boolean functions specifying the update rules of the network and which measures the average response of the system to a minimum transient perturbation that is essentially equivalent to the Lyapunov exponent. The expectation of the average sensitivity under the random BN model is equivalent to the well-known phase transition curve given in Eq. (1.8). Liu *et al.* (2008) have studied whether the assumption of criticality, embedded in the inference objective function as a penalty term along with the best-fit error, improves the inference of BN models.

There are several ways to measure the performance of an inference method by comparing the state transitions, wiring, or (average) sensitivities with the original network. The state transition error captures the fraction of outputs that are incorrectly predicted, and can be defined as

$$\varepsilon = \frac{1}{n}\sum_{i=1}^{n}\frac{1}{2^n}\sum_{x\in\{0,1\}^n}\left(f_i(x)\oplus f_i'(x)\right) = \frac{1}{n}\sum_{i=1}^{n}\varepsilon_i, \tag{3.8}$$

where ε_i denotes the normalized error of the predicted function f'. Additionally, f_i and f' are extended such that they are functions of all the variables by adding fictitious (i.e., dummy) variables. The receiver operating characteristic (ROC) measure has been widely used in classification problems and is defined by the false positive ratio (FPR) and the true positive ratio (TPR) plotted on the x-axes and y-axes, respectively, depicting the relative trade-off between true positives and false positives. The FPR and TPR are defined as $FPR = FP/N$ and $TPR = TP/P$, where TP and FP represent true positive and false positive instances, respectively, while P and N represent the total positive and negative instances, respectively. The ROC distributions can be used to evaluate the accuracy of "wiring" for the inferred network. Finally, the sensitivity error measures the deviation in the average sensitivity of a predicted network and is defined as

$$\varepsilon_s = \frac{S'-S}{S} = \frac{\Delta S}{S}, \tag{3.9}$$

where S' is the sensitivity of the predicted network.

Consider two networks, each comprised of n genes, one being the original network and another being an inferred network, with average sensitivities S and S', respectively. Let $P = \{p_1,\ldots,p_n\}$ and $P' = \{p_1',\ldots,p_n'\}$ be the biases (see Section 1.3) with which functions f_i and f_i' are generated. Let $p_i' = p_i + \Delta p_i$. The expectation of the state transition error between the two networks can be written as

$$\begin{aligned} E[\varepsilon] &= E\left[\frac{1}{n}\sum_{i=1}^{n}\varepsilon_i\right] = \frac{1}{n}\sum_{i=1}^{n}E[\varepsilon_i] \\ &= \frac{1}{n}\sum_{i=1}^{n}\left[p_i(1-p_i') + p_i'(1-p_i)\right] \\ &= \frac{1}{n}\sum_{i=1}^{n}(p_i + p_i' - 2p_ip_i') \\ &= \frac{1}{n}\left[\sum_{i=1}^{n}2p_i(1-p_i) + \sum_{i=1}^{n}(1-2p_i)\Delta p_i\right]. \end{aligned} \tag{3.10}$$

Using the relationship between sensitivity and bias (see Eq. (1.8)), which is essentially the critical phase transition curve, we have

$$\begin{aligned} E[\Delta s_i]/2k_i &= \frac{E\left[s^{f_i'} - s^{f_i}\right]}{2k_i} \\ &= p_i' - p_i - \left((p_i')^2 - p_i^2\right) \\ &= (1-2p_i)\Delta p_i - (\Delta p_i)^2 \end{aligned} \tag{3.11}$$

and thus

$$E[\varepsilon] = \frac{1}{n}\left[\sum_{i=1}^{n}E\left[s^{f_i}/k_i\right] + \sum_{i=1}^{n}E\left[\Delta s_i/(2k_i)\right] + \sum_{i=1}^{n}(\Delta p_i)^2\right]. \tag{3.12}$$

If we further assume that both networks' connectivity is constant, $K = k_i$, $i = 1,\ldots,n$, then

$$\begin{aligned} E[\varepsilon] &= \frac{1}{nK}\sum_{i=1}^{n}E\left[s^{f_i}\right] + \frac{1}{2nK}\sum_{i=1}^{n}E[\Delta s_i] + \frac{1}{n}\sum_{i=1}^{n}(\Delta p_i)^2 \\ &= \frac{S}{K} + \frac{\Delta S}{2K} + \frac{1}{n}\sum_{i=1}^{n}(\Delta p_i)^2. \end{aligned} \tag{3.13}$$

This means that the expectation of the state transition error $E[\varepsilon]$ generally depends on the original network's connectivity K, its sensitivity S, the sensitivity deviation ΔS, and the mean quadratic terms of the bias deviation $\frac{1}{n}\sum_{i=1}^{n}(\Delta p_i)^2$. The simulations performed by Liu *et al.* (2008) indicated that the best-fit inference method typically yields networks with $\Delta S > 0$. This analysis suggests that the expected state transition error $E[\varepsilon]$ will have a linear relationship with ΔS, implying that the sensitivity deviation can be incorporated as a penalty term. Thus, if the true network is critical (i.e., its average sensitivity is equal to one), then any deviation from criticality in the inferred network will introduce additional error. The inference objective function $\hat{F}$ can then be constructed from two terms, the best-fit error ε' (actually estimated using a bootstrap procedure to avoid overly optimistic error estimates) and the sensitivity error ε_s, that is, $\hat{F} = \varepsilon' + |\varepsilon_s|$. The underlying hypothesis is

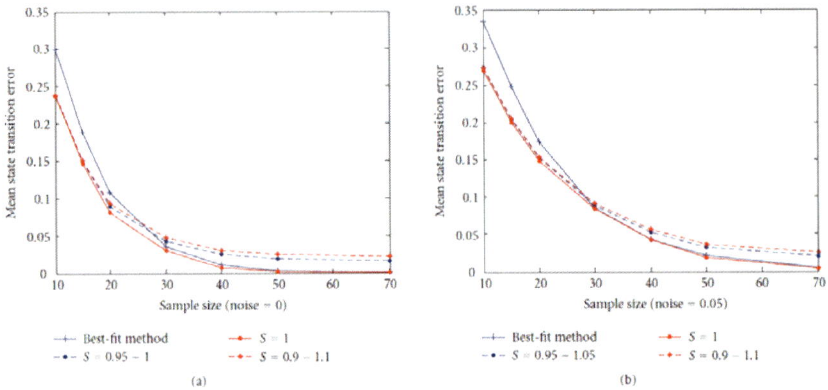

Figure 3.1. *Mean state transition error of different sensitivity deviation for sample sizes ranging from* 10 *to* 70, *computed with zero noise* (a) *and* 5% *noise* (b) *(Liu et al.,* 2008*).*

that the best inference should have a small error in both state transitions and sensitivity and, consequently, the minimal value of the objective function.

Since in practice we do not know the network sensitivity, the assumed value in the inference procedure may not agree with the actual value. Hence, the inference procedure must be robust to the difference. Under the assumption of criticality for living systems, it is natural to set $S = 1$ in the inference procedure. Moreover, assuming that the system remains near the border between order and disorder, the true sensitivity of the gene regulatory networks will remain close to 1 under the Boolean formalism. Thus, to investigate robustness, Liu *et al.* (2008) generated 1000 networks with sensitivities $S = 0.9, 0.95, 1.0, 1.05, 1.1$, and then inferred them using $S = 1$ with the method described above, which incorporates the deviation of the sensitivity as a penalty term. The mean state transition errors of both the best-fit method as well as the method that incorporates the penalty term are shown in Fig. 3.1, under different sample sizes ranging from 10 to 70 samples, for zero noise and 5% noise.[7]

When the actual sensitivity is equal to one, the method helps for small samples and the performances become close for large samples. When the true network deviates from the modeling assumption (i.e., $S = 1$), the proposed method helps for small samples and results in some loss of performance for large samples. This kind of behavior is what one would expect with an objective function that augments the error. In effect, the sensitivity is a penalty term in the objective function that imposes a constraint on the optimization. When the true sensitivity is not equal to 1, the sensitivity constraint $S = 1$ yields smaller sensitivity error than the best-fit methods in small sample situations, while the sensitivity error of the best-fit method is smaller for large samples. In sum, the constraint is beneficial for small samples. In terms of the ROC measure, the method that incorporates the sensitivity deviation penalty can achieve the same TPR with a lower FPR under a small sensitivity deviation in small sample situations (Liu *et al.*, 2008). The above discussion demonstrates that the best-fit inference approach can be combined in a straightforward way with global

[7]The noise is added to the output states of each gene by flipping its value with probability 0.05.

structural parameters of a network, such as the network sensitivity, which captures the operating dynamical behavior, so as to improve the performance of the inference.

3.2 Coefficient of Determination as an Inferential Tool

We have already introduced the coefficient of determination (CoD) in Section 1.5.1 when we discussed discretization of gene expression data. Recall that the CoD, given in Eq. (1.14), measures the relative decrease in prediction error when using the predictor variables in comparison to using the best estimate of the target in the absence of knowledge concerning other variables.

Although many inference methods generally assume time-course data, there are situations, such as microarray studies involving human samples (e.g., tumors), that have no notion of time. In this context, it is generally assumed that the data come from the steady state of the network. Although the construction of dynamical models generally requires temporal data, it is nonetheless possible to infer the structure of such models from steady-state data, as has been demonstrated for the case of DBNs (Lähdesmäki and Shmulevich, 2008).

Recall from Section 1 that the long-run behavior of a BN is characterized by its attractor cycles: given an initial state, the (deterministic) network will transition into a cycle of states and remain cycling forever. The attractor cycles are associated with functional states on physiological time scales and cell types on developmental time scales. In the absence of time-course data, it can be assumed that one is sampling from the steady state. Furthermore, let us assume that the sampled data states represent singleton attractors (i.e., fixed points).

The CoD has been used for inference of PBNs from steady-state data (Shmulevich *et al.*, 2002a). However, a fundamental issue is that, without time-course data, the CoD cannot provide information on the direction of prediction. This problem manifests itself in the situation where if gene **a** is a high CoD predictor of gene **b**, then gene **b** is typically a high CoD predictor of gene **a**. We refer to this situation as a *bidirectional relationship* between genes **a** and **b**. Not only do bidirectional relationships impact the inference process by producing spurious connections in the regulatory graph of the network, thereby deleteriously affecting the graph topology of the inferred network, they also affect the attractor structure to the extent that the inferred network possesses spurious attractor cycles (which likely do not correspond to any biological state). Put more succinctly, given steady-state data, there should not be a significant amount of steady-state probability mass in the inferred network lying outside the mass of the data distribution. The problem is sufficiently troublesome that it has suppressed use of the CoD inference method for steady-state data.

The inference methods that have taken its place are primarily based on the attractor structure (Pal *et al.*, 2005b; Dougherty and Xiao, 2006) or graph topology (Zhao *et al.*, 2008), both of which will be discussed subsequently. In the methods that emphasize attractor structure, the key concern is that the inference algorithm produces an attractor structure close to that of the true network, and in the methods that emphasize graph topology, the key concern is the agreement between the graph connections, for instance, as measured by the Hamming distance between the regulatory graphs. While these methods tend to achieve their goals, they pay only secondary concern for the functional relations among genes in the network, that is, the ability to predict the value of a specific target gene based on the values of a set of regulatory genes for that gene.

To address these issues, Vahedi *et al.* (2009a) proposed an approach to accomplish both goals in network design: preservation of attractor structure and connectivity based on strong gene prediction. It should be pointed out that two genes with a bidirectional relationship are a special case of relevant nodes (Section 1.2). Let the set of genes $W_i = \{x_{i_1}, \dots, x_{i_{k_i}}\}$ be the predictor gene set of the target gene x_i, with f_i being the predictor function of x_i. Of particular interest is the manner in which genes that are predictors of each other affect the attractor structure of the network. This motivates the following definition.

Definition 3.1. *The genes x_i and x_j in a BN are said to have a bidirectional relationship if $x_i \in W_j$ and $x_j \in W_i$. The relationship is said to be of connectivity k if $|W_i| = |W_j| = k$.*

To say that x_i and x_j have a bidirectional relationship of connectivity k is to say that they have a bidirectional relationship and each has connectivity k. In order to investigate the effect of bidirectionality on the attractor structure, let us first consider connectivity 1 and show that there is at least one nonsingleton attractor cycle in the BN.

Proposition 3.2 (Vahedi *et al.*, 2009a). *If there are two genes in a BN having a bidirectional relationship of connectivity* 1, *then the BN has at least one nonsingleton attractor cycle.*

Proof. Let us refer to the two genes as x_1 and x_2. There are four possible transition pairs of predictor functions for these genes: (1) $f_1 \equiv x_2$ and $f_2 \equiv x_1$; (2) $f_1 \equiv x_2$ and $f_2 \equiv \overline{x_1}$; (3) $f_1 \equiv \overline{x_2}$ and $f_2 \equiv x_1$; and (4) $f_1 \equiv \overline{x_2}$ and $f_2 \equiv \overline{x_1}$, where the overbar denotes negation.

Consider the first possible pair: $f_1 \equiv x_2$ and $f_2 \equiv x_1$. If the transitions start from the point $01\mathbf{y}$, then after finitely many transitions, the BN will enter an attractor $01\mathbf{x_0}$ or $10\mathbf{y_0}$, where $\mathbf{y}$, $\mathbf{y_0}$, and $\mathbf{x_0}$ denote vectors of the remaining gene values. Assume that the first visited attractor state is $01\mathbf{x_0}$ (the other possibility $10\mathbf{y_0}$ can be considered in the same way). Because x_1 and x_2 depend only on each other and $01\mathbf{x_0}$ is an attractor state, from this point on, the network must follow a transition sequence of the form $01\mathbf{x_0}, 10\mathbf{x_1}, 01\mathbf{x_2}, \dots, 01\mathbf{x}_k$, where $\mathbf{x}_k = \mathbf{x_0}$ and $\mathbf{x}_r \neq \mathbf{x_0}$ for $1 \le r < k$. Thus, the sequence forms an attractor cycle of length $k = 2m > 1$, where m is a natural number. It is straightforward to show that similar cycles are formed when $f_1 \equiv \overline{x_2}$ and $f_2 \equiv \overline{x_1}$.

Next, we consider the predictor functions pair: $f_1 \equiv x_2$ and $f_2 \equiv \overline{x_1}$. If the transitions start from any point of the form $x_1 x_2 \mathbf{y}$, then after finitely many transitions the BN will enter an attractor state that is of one of the following forms: $00\mathbf{x_0}$, $01\mathbf{y_0}$, $10\mathbf{z_0}$, or $11\mathbf{u_0}$. Here we consider the case when the first visited attractor state is $00\mathbf{x_0}$ (the other possibilities can be considered similarly). Because x_1 and x_2 depend only on each other and $00\mathbf{x_0}$ is an attractor state, from this point on the network must follow a transition sequence of the form $00\mathbf{x_0}, 01\mathbf{x_1}, 11\mathbf{x_2}, 10\mathbf{x_3}, \dots, 00\mathbf{x}_k$, where $\mathbf{x}_k = \mathbf{x_0}$ and $\mathbf{x}_r \neq \mathbf{x_0}$ for $1 \le r < k$. Thus, the sequence forms an attractor cycle of length $k = 4m > 1$. It is straightforward to show that similar cycles are formed when $f_1 \equiv \overline{x_2}$ and $f_2 \equiv x_1$. In short, one of the following four possible cases happens and guarantees the formation of nonsingleton attractors:

$01x_0$	$10x_1$	$\dots$	$01x_k$		
$00x_0$	$11x_1$	$\dots$	$00x_k$		
$00x_0$	$01x_1$	$11x_1$	$10x_1$	$\dots$	$00x_k$
$10x_0$	$11x_1$	$01x_1$	$00x_1$	$\dots$	$10x_k$

$\square$

A similar result holds for connectivity 2, although even such a minimal increase of the cardinality of the predictor sets complicates the analysis of the attractor structure.

Proposition 3.3 (Vahedi *et al.*, 2009a). *If a BN possesses a pair of genes that have a bidirectional relationship of connectivity* 2, *then at least* $1/8$ *of the states in its state space cannot be singleton attractors of the network.*

Proof. Suppose x_1 and x_2 have a bidirectional relationship of order 2 with $W_1 = \{x_2, x_4\}$ and $W_2 = \{x_1, x_3\}$. Because all predictor variables are assumed to be essential, the following conditions cannot occur (see Table 3.1):

1. $(a_1 = c_1$ and $b_1 = d_1)$ or $(a_2 = c_2$ and $b_2 = d_2)$.
2. $(a_1 = b_1$ and $c_1 = d_1)$ or $(a_2 = b_2$ and $c_2 = d_2)$.

These conditions exclude cases where at least one gene is nonessential. The first condition, $(a_1 = c_1$ and $b_1 = d_1)$ excludes three cases: (1) f_1 being constant, (2) f_1 being equal to x_4, and (3) f_1 being equal to $\overline{x_4}$. Under (1), both genes are nonessential. Under (2) and (3), x_2 is nonessential. Furthermore, $(a_2 = c_2$ and $b_2 = d_2)$ excludes three cases: (1) f_2 being constant, (2) f_2 being equal to x_3, and (3) f_2 being equal to $\overline{x_3}$. Under (1), and both genes are nonessential. Under (2) and (3), x_1 is nonessential. Likewise, condition 2 excludes cases where x_3, x_4, or both are nonessential.

Moreover, any combination of f_1 and f_2 belongs to at least one of the following (not mutually exclusive) classes:

F_1: $a_1 = \overline{c_1}$ and $a_2 = \overline{c_2}$
F_2: $a_1 = \overline{c_1}$ and $b_2 = \overline{d_2}$
F_3: $b_1 = \overline{d_1}$ and $a_2 = \overline{c_2}$
F_4: $b_1 = \overline{d_1}$ and $b_2 = \overline{d_2}$

To prove the proposition, consider the following four cases: (a) $(f_1, f_2) \in F_1$ and $x_3x_4 = 00$; (b) $(f_1, f_2) \in F_2$ and $x_3x_4 = 01$; (c) $(f_1, f_2) \in F_3$ and $x_3x_4 = 10$; and (d) $(f_1, f_2) \in F_4$ and $x_3x_4 = 11$.

For case (a), first consider $(f_1, f_2) \in F_1$ such that $a_1 = \overline{c_1} = 0$ and $a_2 = \overline{c_2} = 0$. Examination of the truth tables of f_1 and f_2, Table 3.1, where $a_1 = \overline{c_1} = 0$ and $a_2 = \overline{c_2} = 0$, together with the assumed constant values of x_3 and x_4, shows that any state with $x_3x_4 = 00$ and $x_1 = \overline{x_2}$ cannot be a singleton attractor. A simple counting argument shows that the states where $x_3x_4 = 00$ and $x_1 = \overline{x_2}$ account for exactly $1/8$ of all of the states in the state space. Reasoning in the same way, one can check that when $(f_1, f_2) \in F_1$ with $a_1 = \overline{c_1} = 1$ and $a_2 = \overline{c_2} = 1$, the states with $x_3x_4 = 00$ and $x_1 = x_2$ cannot be singleton attractors, and that there are exactly $1/8$ such states in the state space. To complete the analysis of case (a), consider the situation where $(f_1, f_2) \in F_1$ with $a_1 = \overline{c_1}$, $a_2 = \overline{c_2}$, and $a_1 = \overline{a_2}$. In this case, examination of the truth tables of f_1 and f_2 shows that all of the states where $x_3x_4 = 00$ cannot be singleton attractors. It is straightforward to count that exactly $1/4$ of the states in

Table 3.1. *Truth tables for f_1 and f_2.*

x_2	x_4	f_1
0	0	a_1
0	1	b_1
1	0	c_1
1	1	d_1

x_1	x_3	f_2
0	0	a_2
0	1	b_2
1	0	c_2
1	1	d_2

the state space are of this type. Using similar arguments and symmetry considerations, one can show that the proposition holds for cases (b), (c), and (d). □

While the preceding two propositions explain why BNs with bidirectional relationships possess significant numbers of states that cannot be singleton attractors, one should not jump to the conclusion that the presence of bidirectional relationships is equivalent to the absence of singleton attractors in a BN. Indeed, it can be shown that among BNs possessing a pair of genes with a bidirectional relationship of connectivity 2, there exists a BN with at least one singleton attractor cycle (Vahedi *et al.*, 2009a). Thus, the bidirectional relationship between pairs of genes in a BN is not a sufficient condition for the presence of nonsingleton attractors.

The previous propositions explain why very often the CoD-inferred BNs possess spurious nonsingleton attractors. An algorithm proposed by Vahedi *et al.* (2009a), described below, overcomes the aforementioned shortcoming by making the typical assumption that the data come from the steady state and that each data point is a singleton attractor. The algorithm limits the number of bidirectional relationships when predictor sets are chosen using the CoD method.

Since the predictor function of each target gene is estimated from the steady-state data, not time series data, each gene is a perfect estimator of itself (CoD equal to 1). To eliminate this trivial case, no gene can be a member of its own predictor set.[8] Therefore, for each target gene, there are $\sum_{m=1}^{k_i} \binom{n-1}{m}$ possible combinations for W_i, where k_i is the maximum cardinality of W_i, and n is the number of genes in the network.

The algorithm's input is the binary gene expression data. The output of the algorithm is a BN with no nonsingleton attractors. The following parameters are set in advance: (1) a threshold, T_{CoD}, for the CoD; (2) M_{BR}, the maximum number of permitted bidirectional relationships; and (3) m_A, the minimum number of points in the sample that appear as singleton attractors in the inferred BN. Any predictor function that exceeds T_{CoD} is referred to as a *high CoD predictor function*. We now describe the Singleton Attractor CoD (SA-CoD) algorithm (Vahedi *et al.*, 2009a).

Singleton Attractor CoD (SA-CoD) Inference Algorithm

1. Estimate the CoD and f_i for all the combinations of predictor sets W_i, for $i = 1, 2, \ldots, n$.

2. Save all W_i and f_i with CoD exceeding T_{CoD}. For each target gene, save the high CoD predictor sets and their associated predictor functions into two columns. The length, C_i, $i = 1, \ldots, n$, of both columns depends on T_{CoD}.

3. For each gene, randomly select W_i from the high CoD predictor sets obtained in step 2. Form a BN from W_i and the corresponding f_i. If the total number of bidirectional relationships in the BN does not exceed M_{BR}, go to step 4. Otherwise, repeat step 3.

4. If there is a nonsingleton attractor in the BN, then go to step 3; otherwise, continue.

5. If the number of data points appearing as singleton attractors in the BN is less than m_A, then go to step 3; otherwise, STOP.

[8]Although dynamically such self-loops are encountered frequently in biomolecular networks, self-regulation cannot be captured by steady-state data.

The steps of the algorithm accomplish certain goals. Step 3 constrains bidirectional relationships, thereby limiting spurious attractor cycles resulting from bidirectional relationships. The algorithm never allows bidirectional relationships for connectivity 1 since this case guarantees the formation of nonsingleton attractor cycles. It should be noted that for each gene, there could be more than one possibility for the selection of W_i and f_i among high CoD predictor sets. Step 3 selects W_i for each target gene randomly among high CoD predictor sets such that the total number of bidirectional relationships in the generated BN does not exceed M_{BR}. Step 4 checks to see if any nonsingleton attractor cycles have "slipped through." If the generated BN possesses at least one nonsingleton attractor, the algorithm goes back to step 3 and randomly selects another set of W_i and f_i among high CoD predictor sets for all genes. The goal of this iteration is to generate a BN with only singleton attractors with at least m_A data points as singleton attractors. Step 5 ensures that some minimal number of data points appear as singleton attractors in the inferred BN. The algorithm does not guarantee that the inferred BN will not contain singleton attractors that are not data points, but it does guarantee that there will be no nonsingleton attractors. It is spurious nonsingleton attractors that are ubiquitous in unconstrained CoD design. The algorithm does not guarantee that all data points will be singleton attractors, although it guarantees a minimum number, m_A, of these. The selection of the parameters in the SA-CoD algorithm is discussed in Vahedi *et al.* (2009a). The algorithm can be run a number of times to produce a number of BNs, with each data point appearing in one or more BNs as a singleton attractor.

Using a set of gene expression profiles from a study of malignant melanoma samples (Bittner *et al.*, 2000) taken directly from biopsies (hence, steady-state data), Vahedi *et al.* (2009a) applied the SA-CoD algorithm 500 times to generate 500 BNs. These BNs possessed no nonsingleton attractors and there were at least three data points as singleton attractors in each of them. Then, 10 BNs were randomly chosen from the pool of 500 and a PBN was generated from them, with small perturbation and switching probabilities (the BNs were selected randomly with equal probability). The PBN was run for a sufficiently long time so that its steady-state distribution could be estimated. The steady-state probability corresponding to each attractor represents the probability of being in that attractor in the long run. The proportion of the steady-state probability mass of the generated PBN lying outside the data states was computed. This procedure was repeated 1000 times to generate 1000 PBNs, in each case the proportion of the steady-state probability mass outside the data states being computed. It was found that the proportion of the steady-state probability mass outside the data states was concentrated very close to zero, whereas when the same procedure was repeated for the standard unconstrained CoD method, the proportion of steady-state probability mass outside the data states was much higher (up to 35%). In summary, the above algorithm can be used to infer BNs from nontemporal data using the coefficient of determination. This algorithm avoids nonsingleton attractors and yields few attractors that are not data points while at the same time capturing data points as attractors in the designed networks.

As discussed above, the coefficient of determination has also been used directly for the construction of PBNs (Shmulevich et al., 2002a), as it is a natural way to select a set of predictors for a given gene as well as the set of selection probabilities for the predictors. Let x_i be the target gene; $W_1^{(i)}, W_2^{(i)}, \ldots, W_{r(i)}^{(i)}$ be sets of genes; and $f_1^{(i)}, f_2^{(i)}, \ldots, f_{r(i)}^{(i)}$ be function rules such that $f_1^{(i)}(W_1^{(i)}), \ldots, f_{r(i)}^{(i)}(W_{r(i)}^{(i)})$ are optimal predictors of x_i relative to some probabilistic error measure $\varepsilon(x_i, f_k^{(i)}(W_k^{(i)}))$. Thus, according to Eq. (1.14), for each k, the coefficient of determination for x_i relative to the conditioning set $W_k^{(i)}$ is defined by

$$\theta_k^i = \frac{\varepsilon_i - \varepsilon(x_i, f_k^{(i)}(W_k^{(i)}))}{\varepsilon_i}, \tag{3.14}$$

where ε_i is the error of the best (constant) estimate of x_i in the absence of any conditional variables. For instance, in the case of minimum mean-square error estimation, ε_i is the error of the mean of x_i, which is the best constant estimate, and $f_k^{(i)}(W_k^{(i)})$ is the conditional expectation of x_i given $W_k^{(i)}$, that is, $f_k^{(i)}(W_k^{(i)}) = E[x|W_k^{(i)}]$.

In practice, the CoD must be estimated from training data with designed approximations being used in place of $f_1^{(i)}, f_2^{(i)}, \ldots, f_{r(i)}^{(i)}$. Consequently, the complexity of the functions $f_1^{(i)}, f_2^{(i)}, \ldots, f_{r(i)}^{(i)}$ and the amount of training data become an issue. As we have seen in Section 2.1, for the microarray-based analysis of Kim *et al.* (2002), the number of genes in each predictor was kept to a maximum of three. The framework afforded by the PBN model is well suited for dealing with design imprecision due to limited sample size, where the domain of each predictor may need to be constrained due to lack of training data, but several predictors, with possibly different domains (i.e., $W_k^{(i)}$), are collectively employed (this was one of the original motivations of PBNs).

Let us now assume that a class of gene sets $W_1^{(i)}, W_2^{(i)}, \ldots, W_{r(i)}^{(i)}$ possessing high CoDs has been selected. We can take the designed approximations of the optimal function rules $f_1^{(i)}, f_2^{(i)}, \ldots, f_{r(i)}^{(i)}$ as the rule set for gene x_i with the probability of $f_k^{(i)}$ being selected (see Eq. (2.5)) given by

$$c_k^{(i)} = \frac{\theta_k^i}{\sum_{j=1}^{r(i)} \theta_j^i}, \tag{3.15}$$

where the CoDs are the estimates formed from the training data. According to the above expression, those functions corresponding to the highest CoDs will be selected more often in the probabilistic network (here we are assuming an independent PBN). The number of chosen predictors, $r(i)$, can be a user-selectable parameter and determines the amount of uncertainty that the model can handle. Another approach for selecting the probabilities $c_k^{(i)}$ is to use the best-fit error size instead of the CoD, followed by normalization such that the probabilities sum to 1.

In an alternative approach that makes use of the CoD, proposed by Zhou *et al.* (2003b), gene predictors in PBNs (and their multivalued generalizations) are constructed using a perceptron model consisting of a linear and a nonlinear component. Then, a reversible jump Markov chain Monte Carlo (MCMC) technique is used to calculate the model order and parameters. The CoD is employed to compute the probability of selecting different predictors for each gene, as in Eq. (3.15). The number of possible predictors, $r(i)$, is chosen by means of a clustering technique based on mutual information minimization, which is solved by simulated annealing.

We note that Zhou *et al.* (2004) have proposed a related fully Bayesian approach (that does not use the CoD) for constructing PBNs that emphasizes network topology. The method computes the possible parent (predictor) sets of each gene, the corresponding predictors, and the associated probabilities based on the same nonlinear perceptron model, using reversible jump MCMC. An MCMC method is then employed to search the network configurations to find those with the highest Bayesian scores to construct the network. This method can be naturally applied for context-sensitive PBNs. This approach produces models whose steady-state distribution contains attractors that are either identical or very similar to the states observed in the data, with many of the attractors being singleton, mimicking the biological propensity to stably occupy a given state. We now discuss approaches that construct networks with such requirements, namely, so that the designed network possesses certain global properties, such as a steady-state distribution or attractor structure, that agree with the empirical distribution of the data.

3.3 Design of Networks under Data Consistency Requirements

Let us first consider the design of BNs possessing a desirable stationary distribution or attractor structure. This attractor inverse problem is generally ill-posed, meaning that there will be many networks, or perhaps none, having the desired properties. This problem is important for network inference from steady-state data, as already discussed. Indeed, in many microarray-based gene-expression studies, it is assumed that the data result from sampling from the steady state. If one considers a more general Boolean-type model, such as a BN with random perturbations or a PBN with perturbations, then the dynamical system represented by the network is an ergodic Markov chain and there exists a steady-state distribution; nevertheless, under mild stability assumptions that reflect biological state stability, most of the steady-state probability mass is concentrated in the attractors and it is expected that most data correspondingly come from the attractors (see the discussion in Section 2.5.2).

As before, let W_i be the predictor set for gene x_i and $W = (W_1, \ldots, W_n)$ be the predictor set for the BN. For a deterministic (noise-free) BN, the directed graph $\tilde{\Gamma} = (\tilde{V}, \tilde{E})$ represents the state transition diagram (see Fig. 2.3 for an example), where the states of the BN are the set $\tilde{V}$ of vertices and the state transitions correspond to the edges in $\tilde{E}$ (also see Section 2.7). We say that $\tilde{\Gamma}$ is *compatible* with W if the truth table induced by $\tilde{\Gamma}$ has W as the predictor set for the BN associated with that truth table. Recall (Section 2.7) that the state transition diagram is partitioned into *level sets*, where level set l_j is composed of all states that transition to one of the attractor states in exactly j transitions, with attractors composing level set l_0.

Pal *et al.* (2005b) proposed two algorithms for solving the attractor inverse problem. The problem can be formulated in the following manner. Given a set V consisting of n nodes (genes), a family of n subsets $W_1, W_2, \ldots, W_n$ of V with cardinalities not less than m and not larger than M, $0 < m \leq M$, a set A containing k states, and two positive integers $l \leq L$, accordingly construct a BN with node set V, having predictor set $W = (W_1, W_2, \ldots, W_n)$, possessing attractor cycles $A_1, A_2, \ldots, A_r$, where $A = \cup_{j=1}^{r} A_j$, and containing between l and L level sets. The requirement on the predictor set means that the state transition diagram of the designed network must be compatible with W. There may exist none or many compatible networks. The algorithms are typically initiated by specifying a minimum and maximum number of predictors for each gene, randomly selecting $W_1, W_2, \ldots, W_n$ subject to the specified maximum and minimum, and randomly selecting attractor cycles of given sizes. In this way, one can utilize the algorithms to search for BNs constrained by the connectivity of the network. For instance, if $|W_i| \leq M$ for $i = 1, 2, \ldots, n$, then the algorithms find networks with connectivity bounded by M – if any exist with the required attractor structure.

A state transition diagram constitutes a single-rooted tree if it possesses exactly one singleton attractor (a single-state attractor cycle): the network reaches its attractor via the tree. If it possesses k singleton attractors, then it is composed of k single-rooted trees and is called a *k-forest*: the network reaches an attractor state via one of the single-rooted trees. Let us examine the size of the search space under the assumption of singleton attractors. To simplify the formulas, it is assumed that $m = 1$ and $l = 1$. One can easily make the necessary changes in the general case.

There are $A = A_{n,M} = \sum_{i=1}^{M} \binom{n}{i}$ possible predictor sets W_i for each gene x_i, $i = 1,\ldots,n$. Thus, there are A^n possible predictor sets W for the BN to select from when searching for a compatible state transition diagram $\tilde{\Gamma}$. The different choices for $\tilde{\Gamma}$ depend on the number of attractor states and on the level-set structure of the state transition diagram. There are $\binom{N}{k}$ possible ways of selecting the k singleton attractors for $\tilde{\Gamma}$, where $N = 2^n$. The remaining $N_k = N - k$ nonattractor states will form the level sets of $\tilde{\Gamma}$. There are $n(l_{i+1})^{n(l_i)}$ ways for connecting any two successive level sets l_i and l_{i+1} with $n(l_i)$ and $n(l_{i+1})$ states in them, respectively. Therefore, the number of possible ways to structure the level sets for $\tilde{\Gamma}$ with no more than L level sets is

$$\Lambda = \Lambda_{L,k,n} = \sum_{i=1}^{L} \sum_{i} \frac{N_k!}{n(l_1)!\ldots n(l_i)!} k^{n(l_1)} n(l_1)^{n(l_2)} \ldots n(l_{i-1})^{n(l_i)}, \tag{3.16}$$

where $\sum_i$ is a summation over all of the different choices of the positive integers $n(l_1),\ldots,$ $n(l_i)$ such that $\sum_{j=1}^{i} n(l_j) = N_k$. Combining this with the choices for selecting attractor states, and with the choices for selecting predictor sets, yields the size of the search space,

$$S = S_{n,m,k} = A^n \binom{N}{k} \Lambda. \tag{3.17}$$

To appreciate the size of the search space, consider an example of a very small BN, where $n = 4$, $m = 4$, $k = 4$, and the state transition diagram has exactly 4 levels. The computations show that $S \geq 10^{17}$. The following theorem (see Pal *et al.* (2005b) for a proof) extends a well-known result (Deo, 1974) about 1-trees, or single-rooted trees.

Theorem 3.4 (Pal *et al.*, 2005b). *The cardinality of the collection of all forests on N vertices is $(N+1)^{N-1}$.*

No restrictions are imposed on the structure of the level sets. Consequently the theorem provides us with an upper bound for the term $\binom{N}{k}\Lambda$ appearing in Eq. (3.17). While this upper bound is by no means tight, it is much tighter in comparison to the number of all possible directed graphs on N vertices, N^N. One can easily see that the ratio $\frac{(N+1)^{N-1}}{N^N}$ is asymptotically equal to $\frac{e}{N}$. Since $N = 2^n$, the probability mass decreases exponentially relative to the number of genes n. This shows that a brute force search for an acceptable BN by randomly filling in a BN truth table has very little chance of success. Therefore, if one wants to efficiently generate a BN with the desired characteristics, one has to incorporate information from the state transition diagram, as well as the information about the predictor set of the network, into the algorithm.

The following two algorithms, developed by Pal *et al.* (2005b), are able to generate BNs given attractor and connectivity information. The first algorithm works directly with the truth table, incorporating at the same time the information about the attractor set, as well as the information about the predictor set of the BN. There is no control over the level-set structure, and the transition diagram generated by the algorithm has to be checked for the presence of cycles. The second algorithm employs a state transition diagram $\tilde{\Gamma}$ that satisfies the design goals about attractor structure and level-set structure and checks if the truth table associated with $\tilde{\Gamma}$ has a predictor set W satisfying the design goals. The algorithms are presented for the case of singleton attractors, though they can be adapted to multiple-state cyclic attractors.

ALGORITHM 3.1.
(Pal *et al.*, 2005b)

Step 1: Randomly generate a set of k attractor states. If Step 1 has been repeated more than a prespecified number of times, terminate the algorithm.

Step 2: Randomly pick up a predictor set W, where each W_i contains not less than m and not more than M elements. If Step 2 has been repeated more than a prespecified number of times, go back to Step 1.

Step 3: Check if the selected attractor set is compatible with W; i.e., only, the attractor set of the state transition diagram is checked for compatibility against W. If the attractor set is not compatible with W, go back to Step 2; otherwise continue to Step 4.

Step 4: Fill in the entries of the truth table that correspond to the attractors generated in Step 1. Using the predictor set W randomly fill in the remaining entries of the truth table. If Step 4 has been repeated more than a prespecified number of times, go back to Step 2.

Step 5: Search for cycles of any length in the state transition diagram $\tilde{\Gamma}$ that is, associated with the truth table generated in Step 4. If a cycle is found, go back to Step 4; otherwise continue to Step 6.

Step 6: If $\tilde{\Gamma}$ has less than l or more than L level sets, go back to Step 4; otherwise continue to Step 7.

Step 7: Save the generated BN and terminate the algorithm.

ALGORITHM 3.2.
(Pal *et al.*, 2005b)

Step 1: Randomly generate a state transition diagram $\tilde{\Gamma}$ that satisfies the design goals about the attractor structure and level-set structure. If Step 1 has been repeated more than a prespecified number of times, terminate the algorithm.

Step 2: Fill in the truth table using $\tilde{\Gamma}$.

Step 3: If there is at least one W_i in the predictor set W given by the truth table that has less than m or more than M elements, go back to Step 1; otherwise continue to Step 4.

Step 4: Save the generated BN and terminate the algorithm.

The following example provides a walk-through illustration to show how Algorithm 3.1 works in the particular case of 3 genes. Suppose that $k = 2$, $m = 1$, $M = 2$, $l = 1$, and $L = 5$. Next, suppose that the states 000 and 011 are generated by Step 1.

In Step 2, suppose W is generated, where $W_1 = \{x_2, x_3\}$, $W_2 = \{x_1, x_3\}$, $W_3 = \{x_1, x_2\}$. For Step 3, Table 3.2 shows that the attractors generated in Step 1 are compatible with W. The remaining entries $a_1, \ldots, a_6$ in the truth table are filled in randomly in the next step of the algorithm. One can notice certain patterns in the entries in each one of the three columns of the table. These reflect the structure of the predictor set W and reduce the number of the possible ways to randomly fill in the missing entries during the next step of the algorithm. On the other hand, if the attractors generated in Step 1 were 000 and 100, then for the predictor function of the first gene x_1, we must have $f_1(0,0) = 0$, while from the second attractor, we get $f_1(0,0) = 1$, which is a contradiction. Therefore, that attractor set is not compatible with the set W generated in Step 2.

In Step 4, we randomly fill in the remaining entries of the truth table. Suppose that this produces $a_1 = 0, a_2 = 0, a_3 = 1, a_4 = 0, a_5 = 0, a_6 = 1$. This selection produces the following transitions in the state transition diagram: $0 \to 0; 1 \to 2; 2 \to 1; 3 \to 3; 4 \to 2; 5 \to 0; 6 \to 3; 7 \to 1$, where we have used the decimal representation of the states. It is clear that during Step 5 of the algorithm the cycle $1 \to 2; 2 \to 1$ will be discovered, which will cause the BN generated by the present truth table to be discarded, and we will be returned to Step 4. On the other hand, if we had $a_1 = 0, a_2 = 1, a_3 = 1, a_4 = 0, a_5 = 0, a_6 = 1$ produced in Step 4, then the transitions in the state transition diagram would be $0 \to 0; 1 \to 2; 2 \to 5; 3 \to 3; 4 \to 2; 5 \to 0; 6 \to 7; 7 \to 1$. Since the only cycles here are those within the attractor set, Step 5 of the algorithm will take us to Step 6. Step 6 will detect that there are 5 level sets, and this will take us to Step 7.

The above algorithms produce many distinct networks satisfying a given set of constraints. The presence of multiple solutions allows for optimization procedures when designing PBNs from microarray data. Let us now describe a procedure for designing a PBN

Table 3.2. *Truth table for example of Algorithm* 3.1.

$x_1\ x_2\ x_3$	f_1	f_2	f_3
0 0 0	0	0	0
0 0 1	a_1	1	0
0 1 0	a_2	0	1
0 1 1	0	1	1
1 0 0	0	a_3	a_4
1 0 1	a_1	a_5	a_4
1 1 0	a_2	a_3	a_6
1 1 1	0	a_5	a_6

from data (Pal *et al.*, 2005b). The sizes of the basins are used to select BNs from a group of generated networks and to combine them in a PBN whose steady-state distribution closely matches the observed frequency distribution of the data. The assumption that these data correspond to the steady-state of the underlying gene regulatory system provides a reference point of how closely the dynamics of a generated PBN approximate the data. The designed PBN should have these data points as attractors—and no other attractors because there is no reason in the data for having other attractors. We focus on singleton attractors.

The design procedure begins by selecting a random number N_1 between 2 and 5 and then randomly selecting N_1 distinct states as singleton attractors from the original data according to the data frequency distribution. Repeating this procedure 10 times yields 10 sets, $A_1, A_2, \dots, A_{10}$, of singleton attractors, with A_i possessing N_i attractors, $2 \leq N_i \leq 5$, and $i = 1,2,\dots,10$. After that, Algorithm 3.1 is employed to generate 100 BNs, $\mathcal{B}_{i1}, \mathcal{B}_{i2}, \dots, \mathcal{B}_{i,100}$, for each of the 10 attractor sets. The generated networks have state transition diagrams satisfying two additional constraints. First, the number of their level sets ranges from 2 to 10. This constraint manifests the understanding that in the underlying gene regulatory network the steady-state/fixed points of the system are not achieved with too few or too many consecutive transitions. The second constraint is that all gene predictor sets have between 1 and 3 genes, the number being randomly set.

The BNs generated for each one of the 10 attractor sets are then used as a sample space for the selection of a PBN whose steady-state distribution matches closely the frequency distribution of the data in the mean-square error (MSE) sense. One BN from each group of 100 BNs is randomly selected, and the basin size of each singleton attractor is calculated. These numbers are used as estimates of the steady-state probabilities of the corresponding attractors (keeping in mind that very little time is spent in transient states and that random perturbations and switching randomly put the network in different basins). For example, if the BN $\mathcal{B}_{ij}$ has attractor states $\mathbf{a}_1^{ij}, \mathbf{a}_2^{ij}, \dots, \mathbf{a}_{N_{ij}}^{ij}$ with corresponding basin sizes S_1^{ij}, S_2^{ij}, ..., $S_{N_{ij}}^{ij}$, respectively, then our estimate of the steady-state probability for the attractor $\mathbf{a}_l^{ij}$ is given by $\pi_{ij}(\mathbf{a}_l^{ij}) = S_l^{ij} / \sum_{k=1}^{N_{ij}} S_k^{ij}$. One can take the average of these estimates of the steady-state probabilities of the singleton attractor $\mathbf{a}_l^{ij}$ over the 10 BNs comprising the PBN as an estimate of the steady-state probability of that attractor in the PBN. If a particular singleton attractor is not present in a constituent BN, then its contribution to the steady-state probability is set to 0.

Continuing in this fashion, one obtains an estimate of the steady-state probabilities of each one of the data states used in the generation of a PBN. Let us denote these states by $\mathbf{b}_1, \mathbf{b}_2, \dots, \mathbf{b}_m$ and their corresponding estimated steady-state probabilities by $\pi_1, \pi_2, \dots, \pi_m$. The procedure calculates the MSE between $\pi_1, \pi_2, \dots, \pi_m$ and $\phi_1, \phi_2, \dots, \phi_m$, where ϕ_i is the relative frequency of $\mathbf{b}_i$ in the sample. The designed PBN is selected as the one that minimizes the MSE among a randomly selected subset of 10000 PBNs from the set of all possible PBNs that can be generated using the BNs produced by Algorithm 3.1 for the selected attractor sets. One must settle on 10000 PBNs because an exhaustive search is prohibitive, there being a total of 100^{10} possible PBNs.

In summary, these procedures apply a pattern recognition principle of constraining the solution space when making inferences from limited data, in this case by making assumptions on the dynamical structure of the network. Since the algorithms generate networks in the constrained solution space, they can be used to provide synthetic networks to test the proposed inference algorithms, both for BNs and PBNs.

3.3.1 Contextual data consistency

Except in rare circumstances, the optimal function for a gene will not be a perfect predictor because there will be inconsistencies in the data. This means that a specific vector of values for a set of regulatory genes will not necessarily correspond to a single value of the target gene. Thus, network design is inherently probabilistic. One approach, taken by Dougherty and Xiao (2006), is to model these inconsistencies in a way that reflects context changes in genomic regulation. The network can be in any of a number of contexts, as discussed in Section 2.2. Within a context, the network behaves deterministically and the generated data is consistent. If a regulatory set takes on a specific vector of values, then the target gene associated with the regulatory set must take on a single value, and this is reflected in the data. Dougherty and Xiao (2006) proposed an inference procedure for PBNs whose contexts model the data in such a way that they are consistent for each context, the intent being to view data inconsistencies as being due to latent variables.

As we have seen in this chapter, for the most part, PBN inference has been based on classical binary optimization, where the predictor variables for each target gene have been selected using the coefficient of determination. If we fix ahead the number of predictor genes (i.e., the connectivity) that can compose a regulatory set for a target gene, then the design method is to choose the regulatory set with the largest CoD and then define a binary regulatory function based on the genes in the regulatory set. For instance, suppose genes g_1, g_2, and g_3 have the highest collective CoD among all triples for predicting gene g. Let $wxyz$ denote a binary vector of values for (g_1, g_2, g_3, g). If 0001 appears more often in the data than 0000, then for $xyz = 000$ the predictor function is defined by $\psi(000) = 1$; otherwise, it is defined by $\psi(000) = 0$ (ties being broken either by convention or randomly).[9] If both 0000 and 0001 appear in the data, then the data is inconsistent relative to predicting g via g_1, g_2, and g_3.

Consider a network with two contexts, C_0 and C_1. If the regulatory genes g_1, g_2, and g_3 form the vector 001 in context C_0, then their target gene g must take on a specific value, say 0, in C_0. This uniqueness condition holds for all vectors of values for g_1, g_2, and g_3. It may well be that in context C_1 the regulatory genes take the vector 001 while gene g has value 1, but the data is consistent so long as a single context is maintained. Unless the contexts are known when data from the network is sampled, it would appear that the network is not operating consistently. Since the context is generally not known, an experiment to predict g when $g_1g_2g_3 = 001$ is likely to yield n_0 and n_1 observations of 0 and 1, respectively, meaning that 0010 and 0011 for genes g_1, g_2, g_3 and g have been observed n_0 and n_1 times in contexts C_0 and C_1, respectively. The regulatory function ψ for g would then be defined for 001 by $\psi(001) = 0$ if $n_0 > n_1$ and $\psi(001) = 1$ if $n_1 > n_0$, with some convention determining $\psi(001)$ if $n_1 = n_0$.

Dougherty and Xiao (2006) approached this problem as follows. If the data reveal two values for a target gene for a single vector for the regulatory set, then we will construct the network in such a way that there are two distinct functions, ψ_0 and ψ_1, such that $\psi_0(001) = 0$ and $\psi_1(001) = 1$. The two functions represent two different network contexts. The probabilities of the two functions being selected for regulation will be in agreement with the context probabilities (in a manner to be defined shortly).

Since the context is selected by external variables, we cannot know deterministically when the system is in a certain context, but we can infer the probability of the system being

[9]Here, we use the letter ψ instead of f to denote the predictor function to emphasize that it need not be a Boolean function.

in a particular context from the data. The basic criterion for network design is that the distribution of expected state observations for the system, if it is observed over a long period of time, agrees with the observed distribution of states for the data. As for consistency, that holds *ipso facto* because the system behaves deterministically so long as it remains in a fixed context (is determined by the unique set of functions defining that context). Whereas with the previously discussed design methods the number of constituent networks depends on the number of high-CoD predictor sets (Shmulevich *et al.*, 2002a) or high Bayes-score predictor sets (Zhou *et al.*, 2004), and these depend on the designer's choice of a threshold, using the approach of Dougherty and Xiao (2006), the number of constituent networks is determined by the data.

Owing to their importance, it is key that attractors are properly modeled in an inferred PBN. If the switching and perturbation probabilities are very small, which is typical if the network is sufficiently self-contained not to be subject to frequent latent-variable effects, then it behaves as a single BN for long periods of time. As a result, it spends the vast majority of its time in attractors.

The assumption that we are sampling from the steady state has two implications for inference. First, and most importantly, since data states are, with probability near one, attractor states, we would like them to be attractors in the model. For small samples, it is very possible that sampling misses biological attractor states in the data; however, with large samples the likelihood grows for observing biological attractor states in the data and therefore incorporating them in the model. This is precisely what one would expect in a learning paradigm. As for the converse of the first implication, while network design can result in nondata states being attractor states in the model, we will show that in a number of cases a nondata state will not be an attractor. In addition, as might be expected in a learning environment, avoiding nondata states as attractors depends on design generalization beyond that immediately implied by the data.

Since the key to network design is designing the functions, we begin by treating data consistency in the more general framework of designing a single Boolean operator on random inputs. Let $S = \{\mathbf{x}^1, \mathbf{x}^2, \ldots, \mathbf{x}^m\}$ be the set of $m = 2^n$ vectors associated with the binary-valued observation variables $X_1, X_2, \ldots, X_n$ and let Y be a target binary random variable to be predicted via $X_1, X_2, \ldots, X_n$. A data set D composed of observations of the form $(\mathbf{x}^k, y)$ is said to be *consistent* if $(\mathbf{x}^k, 0)$ and $(\mathbf{x}^k, 1)$ are not both in D. Going the other way, a random predictor-target pair $(\mathbf{X}, Y)$ is said to be consistent with the data set D if D is consistent relative to the observation pairs resulting from $(\mathbf{X}, Y)$. In such a case, there exists a predictor ψ for Y via $\mathbf{X}$, defined on S, possessing zero error on the data. ψ is said to be consistent relative to D. ψ may not be unique, since for any vector $\mathbf{x}^k$ for which neither $(\mathbf{x}^k, 0)$ nor $(\mathbf{x}^k, 1)$ appears in the data, ψ can be defined arbitrarily.

Consider a random operator Ψ on S. Every realization ψ of Ψ defines a function on the random vector $\mathbf{X}$ or, equivalently, on the state space S endowed with the probability measure corresponding to $\mathbf{X}$. If D is any data set generated by ψ, then obviously ψ is consistent relative to D. The number of observations in the data corresponding to any vector $\mathbf{x}^k$ is related to the probability of $\mathbf{x}^k$ in S, not ψ. Specifically, letting $\nu(\mathbf{x}^k)$ denote the number of observations of $\mathbf{x}^k$ in an arbitrary data set of size N, then $E[\nu(\mathbf{x}^k)] = N\pi(\mathbf{x}^k)$, where $\pi(\mathbf{x}^k)$ is the probability of $\mathbf{x}^k$ in S. In accordance with the empirical distribution of $\mathbf{x}^1, \mathbf{x}^2, \ldots, \mathbf{x}^m$ for D, define the probability measure π_D on S by $\pi_D(\mathbf{x}^k) = \nu_D(\mathbf{x}^k)/N$, where $\nu_D(\mathbf{x}^k)$ is the number of observations of $\mathbf{x}^k$ in D. Then $\pi_D(\mathbf{x}^k)$ is an estimate of $\pi(\mathbf{x}^k)$. A key to operator design is the following observation: if ψ_0 and ψ_1 are two realizations of

Ψ and they agree on all vectors except $\mathbf{x}^i$, for which $\psi_0(\mathbf{x}^i) = 0$ and $\psi_1(\mathbf{x}^i) = 1$, then both pairs $(\mathbf{x}^i, 0)$ and $(\mathbf{x}^i, 1)$ may lie in a data set generated by ψ_0 and ψ_1, but the data will be consistent for all $\mathbf{x}^k \neq \mathbf{x}^i$.

Operator construction

Case 1 : Suppose the data set D has the property that there is a single vector, $\mathbf{x}^i$, possessing different Y values, and all other vectors possess a single Y value in D. Suppose there are $\nu_D(\mathbf{x}^i, 0)$ and $\nu_D(\mathbf{x}^i, 1)$ pairs $(\mathbf{x}^i, 0)$ and $(\mathbf{x}^i, 1)$, respectively. Define two functions, ψ_0 and ψ_1, that agree on all vectors except $\mathbf{x}^i$ and are thereon defined by $\psi_0(\mathbf{x}^i) = 0$ and $\psi_1(\mathbf{x}^i) = 1$. Let Ψ_D be a random function possessing two realizations ψ_0 and ψ_1. We define the probability structure for Ψ_D by

$$P(\Psi_D = \psi_a) = \frac{\nu_D(\mathbf{x}^i, a)}{\nu_D(\mathbf{x}^i)}$$

with $a = 0, 1$. ψ_0 is consistent relative to the data set $D_i(0)$ consisting of the original data set D with all pairs $(\mathbf{x}^i, 1)$ removed, and ψ_1 is consistent relative to the data set $D_i(1)$ consisting of D with all pairs $(\mathbf{x}^i, 0)$ removed.

Case 2 : Suppose the data set D has the property that there exist two vectors, $\mathbf{x}^i$ and $\mathbf{x}^j$, possessing different Y values, and all other vectors possess a single Y value in D. Let there be $\nu_D(\mathbf{x}^i, 0)$, $\nu_D(\mathbf{x}^i, 1)$, $\nu_D(\mathbf{x}^j, 0)$, and $\nu_D(\mathbf{x}^j, 1)$ pairs of $(\mathbf{x}^i, 0)$, $(\mathbf{x}^i, 1)$, $(\mathbf{x}^j, 0)$, and $(\mathbf{x}^j, 1)$, respectively. Define four functions ψ_{00}, ψ_{01}, ψ_{10}, and ψ_{11} that agree on all vectors except $\mathbf{x}^i$ and $\mathbf{x}^j$ and are thereon defined by $\psi_{00}(\mathbf{x}^i) = 0$, $\psi_{00}(\mathbf{x}^j) = 0$, $\psi_{01}(\mathbf{x}^i) = 0$, $\psi_{01}(\mathbf{x}^j) = 1$, $\psi_{10}(\mathbf{x}^i) = 1$, $\psi_{10}(\mathbf{x}^j) = 0$, $\psi_{11}(\mathbf{x}^i) = 1$, and $\psi_{11}(\mathbf{x}^j) = 1$. Define the following probability structure for Ψ:

$$P(\Psi_D = \psi_{ab}) = \frac{\nu_D(\mathbf{x}^i, a)\nu_D(\mathbf{x}^j, b)}{\nu_D(\mathbf{x}^i)\nu_D(\mathbf{x}^j)} \tag{3.18}$$

for $a, b = 0, 1$. For instance, $P(\Psi_D = \psi_{00}) = \nu_D(\mathbf{x}^i, 0)\nu_D(\mathbf{x}^j, 0)/\nu_D(\mathbf{x}^i)\nu_D(\mathbf{x}^j)$. ψ_{ab} is consistent relative to the data set $D_{ij}(ab)$ consisting of the original data set D with all pairs $(\mathbf{x}^i, 1-a)$ and $(\mathbf{x}^j, 1-b)$ removed.

Case *k* : The preceding definition and probability structure can be inductively defined for any k vectors possessing different Y values, with all the other vectors possessing a single Y value. We say that the resulting random function is *order-k consistent* relative to the data set D.

We now state the basic theorem for consistent-data operator design.

Theorem 3.5 (Dougherty and Xiao, 2006). *If the random function Ψ_D is order-k consistent relative to the set D, then* (1) *when restricted to any of its realizations, Ψ_D produces consistent data,* (2) *the estimate of the expected distribution of the data generated by Ψ_D using π_D in place of π agrees with the distribution of the data in D, and* (3) *the latter condition cannot be accomplished with less than 2^k functions, the number of realizations of Ψ_D.*

Proof. We first prove Case 1. For a random data set $\mathcal{D}$ of size N generated by the random function Ψ_D, let $\eta(\mathbf{x}^i,0)$ and $\eta(\mathbf{x}^i,1)$ be the random variables giving the number of times $\mathbf{x}^i$ is 0 and 1, respectively, in $\mathcal{D}$. Since Ψ_D is designed from the given data set D and thereafter applied to random data sets, the probability $P(\Psi_D = \psi_0)$ is fixed upon the design of Ψ_D and is independent of the probability of observing any particular state vector in $\mathcal{D}$. Thus,

$$\begin{aligned} E[\eta(\mathbf{x}^i,0)] &= NP(\Psi_D(\mathbf{x}^i)=0)\pi(\mathbf{x}^i) \\ &= N\pi(\mathbf{x}^i)[P(\Psi_D=\psi_0)P(\psi_0(\mathbf{x}^i)=0)+P(\Psi_D=\psi_1)P(\psi_1(\mathbf{x}^i)=0)] \\ &= NP(\Psi_D=\psi_0)\pi(\mathbf{x}^i) \\ &= N\frac{\nu_D(\mathbf{x}^i,0)}{\nu_D(\mathbf{x}^i)}\pi(\mathbf{x}^i). \end{aligned} \tag{3.19}$$

If we replace $\pi(\mathbf{x}^i)$ by its estimate $\pi_D(\mathbf{x}^i)$ based upon the data set D, then we obtain the estimate

$$\hat{E}[\eta(\mathbf{x}^i,0)] = \nu_D(\mathbf{x}^i,0) \tag{3.20}$$

of the expectation $E[\eta(\mathbf{x}^i,0)]$. Equation (3.20) states that the estimate of the expectation of the number of times that $\mathbf{x}^i$ has the label 0, based on the estimate π_D equals the number of times $\mathbf{x}^i$ has the label 0 in the data. Similarly,

$$\hat{E}[\eta(\mathbf{x}^i,1)] = \nu_D(\mathbf{x}^i,1). \tag{3.21}$$

For $j \neq i$, $\hat{E}[\eta(\mathbf{x}^j,0)]$ is either 0 or $\nu_D(\mathbf{x}^j)$, depending on the common value of $\psi_0(\mathbf{x}^j)$ and $\psi_1(\mathbf{x}^j)$. In sum, when restricted to either ψ_0 or ψ_1, the random function produces consistent data, and the expected distribution of the data agrees perfectly with the empirical distribution π_D. Clearly, this could not have been accomplished by a single realization.

For Case 2, for an arbitrary data set $\mathcal{D}$ of size N generated by Ψ_D, let $\eta(\mathbf{x}^i,0)$, $\eta(\mathbf{x}^i,1)$, $\eta(\mathbf{x}^j,0)$, and $\eta(\mathbf{x}^j,1)$ be random variables giving the number of times $\mathbf{x}^i$ is 0, $\mathbf{x}^i$ is 1, $\mathbf{x}^j$ is 0, and $\mathbf{x}^j$ is 1, respectively, in $\mathcal{D}$. Then

$$\begin{aligned} E[\eta(\mathbf{x}^i,0)] &= NP(\Psi_D(\mathbf{x}^i)=0)\pi(\mathbf{x}^i) \\ &= N\pi(\mathbf{x}^i)[P(\Psi_D=\psi_{00})P(\psi_{00}(\mathbf{x}^i)=0)+P(\Psi_D=\psi_{01})P(\psi_{01}(\mathbf{x}^i)=0) \\ &\quad +P(\Psi_D=\psi_{10})P(\psi_{10}(\mathbf{x}^i)=0)+P(\Psi_D=\psi_{11})P(\psi_{11}(\mathbf{x}^i)=0)] \\ &= N\pi(\mathbf{x}^i)[P(\Psi_D=\psi_{00})+P(\Psi_D=\psi_{01})] \\ &= N\pi(\mathbf{x}^i)\left(\frac{\nu_D(\mathbf{x}^i,0)\nu_D(\mathbf{x}^j,0)}{\nu_D(\mathbf{x}^i)\nu_D(\mathbf{x}^j)}+\frac{\nu_D(\mathbf{x}^i,0)\nu_D(\mathbf{x}^j,1)}{\nu_D(\mathbf{x}^i)\nu_D(\mathbf{x}^j)}\right). \end{aligned} \tag{3.22}$$

If we replace $\pi(\mathbf{x}^i)$ by its estimate $\pi_D(\mathbf{x}^i)$ based on the data set D, then we obtain the estimate of the expectation $E[\eta(\mathbf{x}^i,0)]$,

$$\hat{E}[\eta(\mathbf{x}^i,0)] = \nu_D(\mathbf{x}^i,0)\left(\frac{\nu_D(\mathbf{x}^j,0)}{\nu_D(\mathbf{x}^j)}+\frac{\nu_D(\mathbf{x}^j,1)}{\nu_D(\mathbf{x}^j)}\right) = \nu_D(\mathbf{x}^i,0). \tag{3.23}$$

Similarly, $\hat{E}[\eta(\mathbf{x}^i,1)] = \nu_D(\mathbf{x}^i,1)$, $\hat{E}[\eta(\mathbf{x}^j,0)] = \nu_D(\mathbf{x}^j,0)$, and $\hat{E}[\eta(\mathbf{x}^j,1)] = \nu_D(\mathbf{x}^j,1)$. For $l \notin \{i,j\}$, $\hat{E}[\eta(\mathbf{x}^l,0)]$ is either 0 or $\nu_D(\mathbf{x}^l)$, depending on the common value of $\psi_0(\mathbf{x}^l)$ and $\psi_1(\mathbf{x}^l)$ for l. In sum, when restricted to either ψ_{00}, ψ_{01}, ψ_{10}, or ψ_{11}, the estimate of the expected distribution of the data, using the estimate π_D, agrees with the data distribution. This cannot be accomplished with less than four functions. Indeed, since any function must agree with the single value for vectors other than $\mathbf{x}^i$ and $\mathbf{x}^j$, were there only three functions, these would be a subset of $\{\psi_{00},\psi_{01},\psi_{10},\psi_{11}\}$ and there would still be four equations of the kind in Eq. (3.22). These would require solution with only three variables of the kind $P(\Psi_D = \psi_{ab})$ instead of the four variables $P(\Psi = \psi_{00})$, $P(\Psi = \psi_{01})$, $P(\Psi_D = \psi_{10})$, and $P(\Psi_D = \psi_{11})$.

The proof for Case 2 extends directly to any order k, albeit, with increased notational complexity. □

The third part of the theorem is critical because it says that the constructed random function solves the problem with which we are concerned in an optimal way relative to minimizing the number of its realizations. By addressing data inconsistency under the assumption that inconsistencies result from the data arising from a random function of the state space, optimal operator design becomes one of finding the realizations of a random function and the probability mass on those realizations so that the resulting random operator best fits the data relative to the expectation of its output and does so using a minimal number of randomizations. In effect, we have presented an algorithm to solve this optimization problem.

It is important in understanding Theorem 3.5 to recognize that the third part of the theorem refers to the second part; that is, the number of realizations required to accomplish the distributional requirement is 2^k. If we were not concerned with the expected concordance between the expected distribution of data generated by the random function Ψ_D and the distribution of the data in D, then we would need only two realizations to achieve consistent design. To see this, suppose in D there exist m vectors $\mathbf{x}^{i_1},\mathbf{x}^{i_2},\ldots,\mathbf{x}^{i_m}$ possessing different Y values and for any other vector $\mathbf{x}$ there is a single observed Y value $a_{\mathbf{x}}$. Define $\psi_0(\mathbf{x}^{i_j}) = 0$ and $\psi_1(\mathbf{x}^{i_j}) = 1$ for $j = 1,2,\ldots,m$, and $\psi_0(\mathbf{x}) = \psi_1(\mathbf{x}) = a_{\mathbf{x}}$ for any other $\mathbf{x}$. These two realizations can account for all of the inconsistencies; however, the expected distribution of data generated by Ψ_D will not be concordant with the data distribution in D.

Data-consistent design of PBNs Adaptation of the consistent-data predictor design to PBNs is straightforward, though there are some issues regarding generalization and attractors that need to be addressed. Consider designing a PBN from a data set for the set $S = \{\mathbf{x}^1,\mathbf{x}^2,\ldots,\mathbf{x}^m\}$ of $m = 2^n$ binary vectors. For a PBN, each gene is taken in turn as the target to be predicted via the remaining genes by a predictor function. This means that the consistent-data design procedure is applied to each gene in turn to derive its predictor set. A network function for the PBN is defined by taking one predictor function for each gene. For a network with n genes, if there are m_k predictors for gene k, then there are $m_1 m_2 \cdots m_n$ network functions. Each network function defines a context of the network in which the data are consistent. This means that, so long as a network is in the context of a network function, it will generate consistent data. Each context defines a standard (constituent) BN (see Section 2.2). The selection probability of a network function is the product of the selection probabilities for the individual functions composing the network function (Eq. (2.6)). We now present a summary of the design algorithm.

ALGORITHM 3.3.
(Dougherty and Xiao, 2006)

Step 1: Let the data set be $D = \{\mathbf{x}^1, \mathbf{x}^2, \ldots, \mathbf{x}^{m_D}\} \subseteq S, m_D \leq m$.

Step 2: Let g_i be the target gene. If there exist l_i pairs of data,

$$\mathbf{x}^{k_{10}}, \mathbf{x}^{k_{11}}, \ldots, \mathbf{x}^{k_{l_i0}}, \mathbf{x}^{k_{l_i1}},$$

such that the vectors in each pair differ only on the value of target gene, namely,

$$\mathbf{x}^{k_{j0}} = \left[x_{k_j1}, x_{k_j2}, \ldots, x_{k_j(i-1)}, 0, x_{k_j(i+1)}, \ldots, x_{k_jn}\right]$$

and

$$\mathbf{x}^{k_{j1}} = \left[x_{k_j1}, x_{k_j2}, \ldots, x_{k_j(i-1)}, 1, x_{k_j(i+1)}, \ldots, x_{k_jn}\right],$$

$j = 1, 2, \ldots, l_i$, then $m_i = 2^{l_i}$ functions can be defined for g_i, namely

$$g_i = \varphi_j^i(g_1, \ldots, g_{i-1}, g_{i+1}, \ldots, g_n),$$

$j = 1, 2, \ldots, m_i$. Each function is assigned a probability

$$P\left\{\varphi_j^i\right\} = \prod_{h=1}^{l_i} p_{hb}^{(i)},$$

where $p_{hb}^{(i)} = \nu\left(\mathbf{x}^{k_{hb}}\right) / \left[\nu\left(\mathbf{x}^{k_{h0}}\right) + \nu\left(\mathbf{x}^{k_{h1}}\right)\right]$, $b = 0$ if φ_j^i is consistent with $\mathbf{x}^{k_{h0}}$, $b = 1$ if it is consistent with $\mathbf{x}^{k_{h1}}$, and $P\left\{\varphi_1^i\right\} + \cdots + P\left\{\varphi_{m_i}^i\right\} = 1$. Apart from the l_i pairs, φ_j^i is consistent with the rest of the data in D. If $l_i = 0$, then only one function is defined and it is consistent with all data in D.

Step 3: Apply Step 2 to each gene in turn. By choosing a function for each gene and making all possible combinations, we obtain $m_1 m_2 \cdots m_n$ network functions and associated selection probabilities; namely, there are $r = m_1 m_2 \cdots m_n$ contexts.

Attractors are important to understanding a PBN. Each context corresponds to a BN, and by definition the attractors of the PBN are the attractors of its constituent BNs. Relative to attractors, there is a fundamental difference between data states and nondata states. We say that a nondata state $\mathbf{x} = x_1 x_2 \cdots x_n$ is *partially mapped* by the data if there exists at least one subvector, $x_1 x_2 \cdots x_{k-1} x_{k+1} \cdots x_n$, which has been observed in the data, so that there exists a function ψ_k for x_k for which $\psi_k(x_1 x_2 \cdots x_{k-1} x_{k+1} \cdots x_n)$ has been determined by the data, not by generalization. A nondata state $\mathbf{x} = x_1 x_2 \cdots x_n$ is *fully unmapped* by the data if no subvector $x_1 x_2 \cdots x_{k-1} x_{k+1} \cdots x_n$ has been observed in the data. A nondata state is *fully mapped* if all subvectors have been observed in the data. We now provide some formal propositions regarding attractors.

Proposition 3.6. *A data state is a singleton attractor in at least one context.*

Proof. If $\mathbf{x} = x_1x_2\cdots x_n$ is a data state, then for each gene x_k, there is at least one function, ψ^k, inferred from the data for which $\psi^k(x_1x_2\ldots x_{k-1}x_{k+1}\cdots x_n) = x_k$. $\mathbf{x}$ is a singleton attractor for the context $\{\psi^1,\psi^2,\ldots,\psi^n\}$. □

Proposition 3.7. *A fully or partially mapped nondata state is not a singleton attractor in any context.*

Proof. If $\mathbf{x} = x_1x_2\cdots x_n$ is a fully or partially mapped nondata state, then there exists a gene x_k determined from the data relative to $x_1x_2\cdots x_n$. Suppose $x_1x_2\cdots x_n \rightarrow x_1x_2\cdots x_n$ in some context $\{\psi^1,\psi^2,\ldots,\psi^n\}$. Then $x_k = \psi^k(x_1x_2\cdots x_{k-1}x_{k+1}\cdots x_n)$. Since this relationship has been determined from the data, $x_1x_2\cdots x_n$ must be a data state, which is a contradiction. □

Proposition 3.8. *If a nondata state and its complement are both fully mapped, then they form a two-state attractor cycle in every context.*

Proof. If $\mathbf{x} = x_1x_2\cdots x_n$ is fully mapped, then in any context $\{\psi^1,\psi^2,\ldots,\psi^n\}$,

$$\mathbf{x} \rightarrow \psi^1(x_2x_3\cdots x_n)\psi^2(x_1x_3\cdots x_n)\cdots\psi^n(x_1x_2\cdots x_{n-1}).$$

It must be that $\psi^k(x_1x_2\cdots x_{k-1}x_{k+1}\cdots x_n) = x_k^c$; otherwise $x_1x_2\cdots x_{k-1}x_{k+1}\cdots x_n$ having been observed in the data would mean that $x_1x_2\cdots x_n$ having been observed in the data, which it has not. Hence $\mathbf{x} \rightarrow \mathbf{x}^c$. The same argument applied to $\mathbf{x}^c$ shows that $\{\mathbf{x},\mathbf{x}^c\}$ is a two-state attractor. □

Proposition 3.9. *Generalization can always make a given fully unmapped nondata state be or not be a singleton attractor.*

Proof. If $\mathbf{x} = x_1x_2\cdots x_n$ is a fully unmapped nondata state, then there are no data-determined functions $\psi^k(x_1x_2\cdots x_{k-1}x_{k+1}\cdots x_n)$. In order to make $\mathbf{x}$ an attractor, define $\psi^k(x_1x_2\cdots x_{k-1}x_{k+1}\cdots x_n) = x_k$ for all k; to make $\mathbf{x}$ not a singleton attractor, define ψ^k in any other manner. □

Let us call any attractor composed solely of nondata states an *artificial attractor*. As we have noted previously, it may not be possible to make two fully unmapped nondata states into singleton attractors. According to Proposition 3.7, artificial singleton attractors are fully unmapped. Every singleton attractor is either a data state or an artificial attractor. According to Proposition 3.8, if a nondata state and its complement are both fully mapped, then they form an artificial two-state attractor cycle in every context.

The state transitions for a PBN produce an ergodic Markov chain possessing a steady-state distribution (Section 2.5). When a PBN is designed from data, the implicit assumption is that the data have been obtained in the steady state. This means that the state transitions of the designed PBN do not correspond to transitions in biological time but to synthetic (mathematical) time. Hence, there is no direct correspondence between transient states of the PBN and data states. There should be, however, correspondence between steady-state behavior and the data states. Since we expect network switching to be infrequent in a real system, most of the steady-state mass should belong to the attractors, and since the data has been drawn from the steady state, we would expect it to be highly likely that the data states are attractors. In this sense, Proposition 3.6 provides strong support for the context-switching model. Proposition 3.7 is also encouraging relative to steady-state and data distribution correspondence. Proposition 3.8, while not encouraging, posits the strong requirement that a nondata state be fully mapped. Finally, Proposition 3.9 asserts only existence and says nothing about the consequences of a reasonable generalization.

Issues of standard and contextual design Whereas a BN is assured for a single observed data state, two data states may require a PBN. At the other extreme, only a BN is required for consistency if the data states are 001, 010, 100, and 111, and all four states would be singleton attractors. The issue of the number of contexts is related to a deeper issue of how we have chosen to use the data for inference. At its root, the matter concerns learning predictors for a dynamical system from steady-state data.

To explain, we consider a three-gene BN with vectors xyz and data set $\{000, 001\}$. If we observe 000 more often than 001, why define the prediction $\psi^z(00) = 0$? After all, in the real system, 000 might transition to another state, and therefore $xy = 00$ may predict z being 1. For instance, if in the actual system $000 \to 001$, then would it not be better to predict z by $\psi^z(00) = 1$? Perhaps it would be if we had dynamical data to indicate so, but we have only steady-state data. The original use of prediction for gene expression was to measure multivariate gene interaction (Kim *et al.*, 2000a): based on the data, if $xy = 00$ is observed in the steady state, then what is the best prediction for z? The prediction methodology is purely statistical and makes no inference regarding causality. Clearly, if we observe 000 in the data more often than 001, then the best prediction on observing $xy = 00$ in a future observation would be to predict $z = 0$. This approach has been adopted for network inference and represents a kind of generalization because a network involves dynamical behavior. Nonetheless, under the assumption that the data come from the steady state, and assuming that when in the steady state the network spends the great majority of its time in its attractors, when choosing between the singleton attractor 000 ($\psi^z(00) = 0$) and the singleton attractor 001 ($\psi^z(00) = 1$), a majority decision based on the data indicates the singleton attractor 000. Indeed, if 001 were observed only rarely in the data, one might conjecture it to be a noisy version of 000 or a transient state of the form $001 \to 000$.

The situation becomes more flexible with the use of PBNs. We reconsider the three-gene situation with data set $\{000, 001\}$. At first glance it may appear that we have three possibilities: (1) 000 and 001 compose an attractor cycle in the same BN; (2) they are singleton attractors in a single BN; or (3) they are singleton attractors in different contexts. But the first situation is not possible because $000 \to 001$ requires $\psi^z(00) = 1$, and $001 \to 000$ requires $\psi^z(00) = 0$. As for the second possibility, it involves the choice just discussed. If we choose $\psi^z(00) = 0$, then to have the network remain in an attractor, we must have $\psi^x(00) = 0$ and $\psi^y(00) = 0$, in which case 000 is an attractor and 001 is a transient state; if we choose $\psi^z(00) = 1$, then to have the network remain in an attractor, we must have $\psi^x(01) = 0$ and $\psi^y(01) = 0$, in which case 001 is an attractor and 000 is a transient state. Thus, we choose $\psi^z(00)$ based on the majority decision. The third possibility occurs by using context: 000 and 001 are singleton attractors in different contexts, in which case we have $\psi^z(00) = 0$ in one context and $\psi^z(00) = 1$ in the other, with all conflicts being resolved. Note that this same analysis applies whenever there are two data points and they differ only for a single gene.

For another situation, consider the data set $\{000, 111\}$. The same three apparent possibilities appear, but now they are all truly possible. We could have the cycle $000 \rightarrow 111$. This would not create a conflict in any predictor definitions: $\psi^x(00) = \psi^y(00) = \psi^z(00) = 1$ and $\psi^x(11) = \psi^y(11) = \psi^z(11) = 0$. They could also form two singleton attractors in the same BN, with $\psi^x(00) = \psi^y(00) = \psi^z(00) = 0$ and $\psi^x(11) = \psi^y(11) = \psi^z(11) = 1$. Last, they could be singleton attractors in different contexts of a PBN. Using either noncontextual or contextual design, they appear as singleton attractors in a single BN. Were the data actually reflective of a cycle in a real regulatory system, then the inference would be erroneous. Because the steady-state data is insufficient to infer dynamics, a learning assumption has been made that favors short cycles over long, in this case favoring singleton attractors. Moreover, the number of contexts is minimized by assuming them to be singleton attractors in a BN. Note that the same analysis applies whenever there are two data points and they differ by more than a single gene.

To help clarify the issue, we define two states to be neighbors if they differ by a single gene. A data state is said to be *isolated* if it has no neighbors in the data and *nonisolated* otherwise. If two data states are neighbors, as are 000 and 001, then they require two contexts to avoid data inconsistency. Since context selection depends on the data frequencies, the frequencies of 000 and 001 affect the resulting PBN probabilities. On the other hand, if a data state is isolated, as is the case of 000 for the data set $\{000, 110, 111\}$, then it does not generate contexts. This is what happened in the preceding illustration using the data states 000 and 111. Both are isolated in the data and therefore there is a single context. When a data state is isolated, its frequency in the data does not affect the PBN probabilities.

The number of constituent networks is determined by how inconsistencies appear in the data, not the number of states appearing in the data. For instance, a PBN with data set $\{000, 001, 010, 011\}$ has 4, 4, and 1 functions for x, y, and z, respectively, with a total of 16 contexts. A PBN with data set $\{010, 100, 101, 110\}$ has 2, 2, and 2 functions for x, y, and z, respectively, with a total of 8 contexts.

Let us now address two issues arising with consistency-based model design. First, based strictly on the data, typically there is more than one PBN complying with the same data under the consistency requirement. Thus, we must find a criterion that enables us to perform a model selection on the candidate PBNs, thereby giving rise to the *generalization* problem. Second, for both computational and biological reasons, we may wish to constrain the connectivity optimization; that is, we may wish to restrict the maximum number of variables allowed in a regulatory function. This may require loosening the consistency requirement. Thus, there is the question of how to adapt the network with minimal loss of fidelity.

3.3.2 Optimization of consistency-based design

Greater data inconsistency results in a greater number of contexts. Increased inconsistency can result from observation noise or sample heterogeneity. Indeed, the entire issue of contextual modeling relates to data heterogeneity: the data relating to a specific set of genes composing a network derive from heterogeneous sources because each source is conditioned by conditions external to the network. The desire to make PBN design specific to the subpopulations (contexts) arising from external latent variables is what has motivated data-consistent design. When there is excessive sample heterogeneity there can be a very large number of contexts. To reduce the number of contexts arising from excessive data heterogeneity or from observation noise, Dougherty and Xiao (2006) proposed to apply a Hamming-distance filter; namely, if two profiles are very close, they are joined, thereby identifying their individual contexts.

ALGORITHM 3.4. Profile Reduction.
(Xiao and Dougherty, 2006)

1: Given a batch of n-gene (binary) expression profiles, split them into a single-copied data group $\mathcal{G}_1$ and a multiple-copied data group $\mathcal{G}_2$. Set $i := 1, j := 1, k := 1$.

2: Select the ith profile ρ_{1i} from group $\mathcal{G}_1$ and compare it with the jth profile ρ_{2j} in $\mathcal{G}_2$. If their Hamming distance does not exceed k, merge ρ_{1i} into ρ_{2j}; i.e., delete ρ_{1i} and increase the number of copies of profile ρ_{2j} by 1. If ρ_{1i} does not merge with ρ_{2j}, let $j := j+1$, and repeat until either ρ_{1i} is deleted or the end of $\mathcal{G}_2$ is reached.

3: Let $i := i+1$, and repeat step 2 until the end of $\mathcal{G}_1$.

4: Let $k := k+1$; if k is less than or equal to a prescribed value K $(K \geq 1)$, repeat steps 2 and 3.

Selection of K in the algorithm is a heuristic decision and contingent on n. The reason for introducing the loop $k = 1, \ldots, K$ is that we always merge a profile to its nearer neighbor with higher priority.

Generalization via connectivity optimization When applying the preceding design procedure, the truth tables are usually incompletely specified by available data, and these ambiguous entries must be assigned values according to some protocol. Addressing this generalization issue is essential because it affects the structure and final expression of Boolean functions of the designed PBN.

Example 3.10 Suppose gene Z is potentially regulated by genes W, X, and Y, with the following truth table for Z inferred from the data: $[1,1,0,\times,0,0,\times,1]$, arranged in the usual lexicographic order, where $\times$ denotes "don't care"—entries that cannot be inferred from available information, thus requiring generalization. If we assign 0 to $\times$ in the fourth entry and 1 to $\times$ in the seventh entry, then the Boolean function will be $Z = f_2(W,X) = \bar{W}\bar{X} + WX$. If both $\times$'s are assigned 1, then we obtain the Boolean function $Z = f_1(W,X,Y) = \bar{W}\bar{X} + \bar{W}XY + WX$. It can be seen that the former generalization involves fewer variable and requires fewer logic gates. ∎

This example suggests that we can look for a generalization that achieves a minimally interconnected PBN, with the simplest logic rules. In other words, for each gene, we would like its predictors to be as few as possible with a simple prediction rule; hence the PBN is minimally connected, and the realization of Boolean functions is minimized. The significance of minimizing connectivity is not limited to its own sake but also reflects the biological propensity for low connectivity.

There are many ways to describe a Boolean function via logic gates. Here, we adopt the standard sum of products (SOP) form,[10] which can be implemented with two-level Boolean logic circuits consisting solely of AND and OR gates (besides NOT gates whenever necessary), where AND gates are used only in the first level, and OR gates are used only in the second level. Therefore, we have the following objectives in mind when generalizing a PBN: (1) achieve a minimally connected network; (2) seek the simplest SOP Boolean realizations.

Notice that a simplest SOP Boolean realization may not be one that has minimal variables, and vice versa. For instance, take $WX+Y$ versus $WX+\bar{W}\bar{X}$. Thus in achieving the objective, it is a matter of balancing goal (1) and goal (2). One can either give goal (2) a higher priority by seeking a simplest SOP realization with as few variables as possible or give goal (1) a higher priority by achieving minimal variables first and finding a simplest SOP realization based on the minimal variables.

Since we are often faced with a small sample of data, there can be a considerable number of "don't cares" in the truth tables. Taking account of only the number of genes and contexts, the number of possible ways to generalize is enormous. To address the generalization problem, we employ a Boolean function minimization technique tailored to the case of incompletely specified truth tables. Given any completely fixed truth table, an initial Boolean function in SOP form can be obtained by the following steps:

1. Pick out every combination of input variables that yields output 1.

2. Represent each combination (called a *minterm*) as a product (logic AND) of input variables or their complements. If an input variable X value is 1, use the variable itself; otherwise, use its inverse $\bar{X}$.

3. Write the Boolean function as a sum (logic OR) of all minterms.

A Boolean function obtained directly from a truth table (without any simplification) in SOP form is often written as a sum using binary string representations of the minterms. For instance, $f(W,X,Y) = WX\bar{Y} + WXY$ can be written as $f(W,X,Y) = 110+111$. It can also be expressed as a sum of row numbers of the minterms, i.e., each minterm is represented by a decimal number converted from its binary string, so that $f(W,X,Y) = \sum(6,7)$.

After finding an initial Boolean function, one can often simplify its expression by applying the following logic rules repeatedly: (1) $X\cdot 1 = 1\cdot X = X$; (2) $X+\bar{X} = 1$; and (3) $WX+WY = W(X+Y)$. During simplification, if a variable vanishes, it will be replaced by $-$ in binary string representation. The following definitions provide the nomenclature for Boolean function minimization.

Definition 3.11. *A Boolean expression φ is said to imply another Boolean expression ψ, designated as $\varphi \Rightarrow \psi$, if, when φ and ψ are considered as functions, ψ has the value* 1 *at least at every combination at which φ has the value* 1.

Definition 3.12. *Let ξ be a product of literals (a literal is a variable or its complement) and ψ be a Boolean function. If $\xi \Rightarrow \psi$, then ξ is said to be an implicant of ψ.*

[10] It is also known as the disjunctive normal form (DNF).

Definition 3.13. *A product of literals ξ is a prime implicant of ψ if $\xi \Rightarrow \psi$ and if deleting any literal from ξ results in a new product term that does not imply ψ.*

It is well known that, to find the minimal expression of a Boolean function, it suffices to look for its smallest set of prime implicants. That is, any minimal SOP expression of a Boolean function ψ is equivalent to a sum of prime implicants of ψ (see e.g., Langholz *et al.* (1998).

Simplification of a Boolean function can be achieved via the well-known Quine–McCluskey (Q–M) method, which is a tabular algorithm for Boolean reduction that can be adapted to deal with incompletely specified truth tables (McCluskey, 1956). The Q–M method lists all the minterms and applies the adjacency rule repeatedly to combine qualified minterm pairs. After finding every prime implicant (PI) of the Boolean function, it then tabulates the results to search for a smallest set of PIs that covers the Boolean function. Since the objective is to find a generalization resulting in the simplest Boolean function (in SOP form), we are not only concerned with minimization given a (fixed) truth table, but also optimization upon an incompletely specified truth table, with uncertain outputs ("don't cares"). This can be achieved by a slight variation of the original Q–M method, allowing don't cares along with minterms but excluding them when searching for the minimum set of prime implicants.

ALGORITHM 3.5.
(Xiao and Dougherty, 2006)

1: Construct truth tables of the PBN from sample data, leaving undetermined entries as $\times$'s. In each table, pick out all minterms and don't cares and group them according to the number of 1's (the index) in each term; e.g., the index of 01101 is 3. Put them in ascending order of index in table $\mathcal{T}_j$, initially setting $j := 1$.

2: In the group of index i, for each term, find all its partners in the adjacent group of index $i+1$ such that each partner differs from it by one digit only (i.e., 0 and 1, respectively). Combine them to form a new term in which the formerly different digit is replaced by $-$; e.g, 01001 (index = 2) combines with 01011 (index = 3) to form 010$-$1 (index = 2); while 01$-$10 (index = 2) combines with 01$-$11 (index = 3) to form 01$-$1$-$ (index = 2). Put the newly formed terms into table $\mathcal{T}_{j+1}$. If a term has no partner at all, put itself to $\mathcal{T}_{j+1}$.

3: Let $i := i+1$ and repeat step 2 until the end of table $\mathcal{T}_j$.

4: Let $j := j+1$ and repeat the above process until no further combination is possible. The final table $\mathcal{T}_f$ now contains all generalized prime implicants (GPIs). They are GPIs but not real PIs, because some of them actually imply don't cares in the truth table, not the original 1's.

5: Construct a two-dimensional PI table $\mathcal{P}$, where each column corresponds to a minterm (but not don't care), and each row corresponds to an initial GPI. Mark the (i, j) position of $\mathcal{P}$ by $*$ if the ith GPI implies the jth minterm. Delete rows (GPIs) without $*$. The remaining rows are PIs.

6: If a column in $\mathcal{P}$ has only one $*$, then the corresponding row is identified as an essential prime implicant (EPI). Find all EPIs and include them in the set of minimum prime implicants Ξ. Put the variables contained in the EPIs to set $\mathcal{V}$. Then remove all EPIs and corresponding minterms (i.e., columns with single $*$) from $\mathcal{P}$.

7: Find a row with the maximum number of $*$'s, denoting this number s. **Search all rows that contain s $*$'s and pick the PI that adds the least number of new variables to $\mathcal{V}$.** Include this PI in Ξ and update the variable set $\mathcal{V}$. Remove this row and its corresponding columns from $\mathcal{P}$.

8: Repeat the above step until there is no $*$ in $\mathcal{P}$. Now, Ξ is the minimum set of PIs and the variables contained in $\mathcal{V}$ form the minimum predictor set.

We point out that standard Boolean minimization focuses on a minimal realization rather than connectivity. The modification in step 7 (highlighted) favors a realization with lower connectivity when two or more realizations are equally simple. In spite of the improvement made, Algorithm 3.5 favors minimal SOP realization more than minimal connectivity. If a minimal connectivity is more preferable, we may use the following alternative "prune-and-minimize" algorithm. The idea is to find through exhaustive search a smallest predictor set and the corresponding pruned truth table and then perform Algorithm 3.5 to obtain a minimal realization with regard to the smallest predictor set.

ALGORITHM 3.6. Prune-and-Minimize.
(Xiao and Dougherty, 2006)

1: Construct the original truth tables of the PBN.

2: For each truth table, suppose the set of variables are $\mathcal{V}$, with cardinality $|\mathcal{V}| = k$.

3: Let $m = 1$. If there exists a subset $\mathcal{V}_m = \{v_1, \ldots, v_m\} \subset \mathcal{V}$ such that for all rows in the truth table which have the same values of $v_1, \ldots, v_m$, the function values do not contain both 0 and 1 (it is all right to have don't cares), then $\mathcal{V}_m$ is a minimal predictor set. Prune the truth table accordingly. In determining the function value of the pruned truth table, if in the original table, under the same $v_1, \ldots, v_m$, the function values are all don't cares, then the corresponding function value in the pruned table will also be don't care. Otherwise, if the original function values under the same $v_1, \ldots, v_m$ contain at least one 0 (or 1, but not both), then the pruned function value will be 0 (or 1, accordingly). If there does not exist such a subset $\mathcal{V}_m$, then let $m := m + 1$ and repeat the search.

4: After pruning all the truth tables, perform Algorithm 3.5 (with the following modification) to find minimal SOP realizations of Boolean functions, thus completing the generalization of the PBN.

5: Modification: change highlighted part in Algorithm 3.5 to the following: **Search all rows that contain s $*$'s and pick the PI with the least variables.**

Proposition 3.14 (Xiao and Dougherty, 2006). *A PBN designed by the procedure in Algorithm* 3.6 *has* (a) *the minimum connectivity, and* (b) *the minimal SOP realizations on the found predictor sets.*

Proof. The exhaustive search on the predictor set with lowest cardinality guarantees (a). As for (b), once a smallest predictor set is found and the truth table pruned, Algorithm 3.6 uses Algorithm 3.5, which always selects with priority the PI which covers as many minterms as possible. Thus, the realization found will have the least number of products (incurring the smallest number of "OR" gates). The modification in Algorithm 3.6 chooses the PIs with the least variables (thus shortest product), so that the number of "AND" gates is also minimal. Therefore, the SOP realization on the found predictor set must be minimal. □

From a general computational perspective, Algorithm 3.5 has the same level of computational complexity as the Q–M algorithm. However, for a fixed number of variables, the more sparsely specified the truth table, the more time it takes to perform the Q–M algorithm. For a large-scale gene network with n genes, since the number of samples $N \ll 2^n$, the resulting functions will be extremely sparsely specified. Therefore, it is recommended that it be applied on PBNs of no more than 15 genes. Algorithm 3.6 may handle a larger gene network because it reduces predictors of any gene from $n-1$ down to n^* before Boolean minimization, and in practical problems the genetic regulatory network usually has low connectivity, typically $n^* \leq 5$. The time used for searching n^* variables is well compensated by performing Boolean minimization on fewer variables.

Constrained connectivity optimization Let us now consider the constrained optimization problem in which an upper limit is set on the maximum number of predictors allowed in a PBN; i.e., we restrict the connectivity to be no more than a prescribed integer κ^* in the entire network. Under the constraint, the resultant PBN may no longer reflect the original data with complete information; instead, there is a trade-off between information capacity and network connectivity. From previous discussion, we are able to find the smallest predictor sets (called *full predictor sets*) for genes within the network such that consistency is satisfied. We make the following assumption: *Under the constraint on maximum allowed connectivity, the suboptimal predictor set is a subset of the full predictor set.*

Under this assumption, the constrained-connectivity optimization problem can be interpreted as a lossy coding problem: we are to transmit a random row of a truth table by a fixed code length. The truth table contains $k+1$ binary variables, where $\mathbf{x} = (x_1, \ldots, x_k)$ is the input vector, and y is the output variable. For the lth row, the values of the input and output variables are denoted by $\mathbf{x}^l = (x_1^l, \ldots, x_k^l)$ and y^l, respectively. Each row of truth table can be encoded into a string of $k+1$ bits, and this coding will be lossless. When code length is limited to $\kappa^* + 1$ with $\kappa^* < k$, however, we must select a subset of κ^* input variables from $x_1, \ldots, x_k$ so that the code will convey as much correct information as possible so that the original truth table can be recovered from received shortened codes with the least error.

Suppose the Boolean function of the original truth table is $f(x_1, \ldots, x_k)$, which in the case of reduced input variables has become $f'(x_{n_1}, \ldots, x_{n_{\kappa^*}})$, with $\{x_{n_1}, \ldots, x_{n_{\kappa^*}}\} \subset \{x_1, \ldots, x_k\}$. As a result, for the lth row of original truth table, $f'(x_{n_1}^l, \ldots, x_{n_{\kappa^*}}^l)$ may or may not equal y^l. More generally, if a weight w_l is assigned to the lth row, then we can define the cost function concerning chosen reduced inputs $\mathbf{x}_n = (x_{n_1}, \ldots, x_{n_{\kappa^*}})$ to be the expected (i.e., the weighted average of) probability of error, P_e, written as

$$J = E[P_e] = \sum_{l=0}^{2^k-1} I_{f'(\mathbf{x}_n^l) \neq y^l}(\mathbf{x}_n^l) w_l, \tag{3.24}$$

where w_l satisfies $\sum_{l=0}^{2^k-1} w_l = 1$.

This idea can naturally be extended to the constrained-connectivity-optimization problem for PBNs. Consider the ith context in a PBN, supposing gene g_j has κ_{ij} predictors. By forcing connectivity to be no more than κ^* ($\kappa^* < \kappa_{ij}$), $\kappa_{ij} - \kappa^*$ predictors have to be dropped, and doing so will create discrepancies when predicting g_i. We desire the best κ^* predictors out of the original κ_{ij} such that the target gene can be predicted with the least probability of error. If we assume that all combinations of values for the predictor genes are equally likely, then the following algorithm suffices.

ALGORITHM 3.7. Prune-and-Minimize.
(Xiao and Dougherty, 2006)

1: Apply Algorithm 3.6 to obtain a minimally connected n-gene PBN, which has r contexts. For the ith context and jth gene, the optimal predictor set is $\mathcal{V}_{ij}$, and its cardinality $|\mathcal{V}_{ij}| = \kappa_{ij}$. Set $i := 1, j := 1$.

2: If $\kappa_{ij} \leq \kappa^*$, there is nothing to be done. Otherwise, if $\kappa_{ij} > \kappa^*$, choose a subset $\mathcal{V}_{ij}^{(k)} \subset \mathcal{V}_{ij}$ with $|\mathcal{V}_{ij}^{(k)}| = \kappa^*$, $k = 1, \ldots, \binom{\kappa_{ij}}{\kappa^*}$. Variables in $\mathcal{V}_{ij}^{(k)}$ can take on 2^{κ^*} different combinations of values, ranging from $00\cdots0$ to $11\cdots1$. For the lth combination they take, those remaining variables (not included in $\mathcal{V}_{ij}^{(k)}$) can have $2^{\kappa_{ij}-\kappa^*}$ different combinations, from which one can count the times that the function value (i.e., target gene g_j) is 1 or 0 in the truth table, and denote them by $t_{1,ijl}^{(k)}$ and $t_{0,ijl}^{(k)}$, respectively (the sum of the two equaling $2^{\kappa_{ij}-\kappa^*}$). Let $t_{ijl}^{(k)} := \min(t_{1,ijl}^{(k)}, t_{0,ijl}^{(k)})$. Compute $t_{ij}^{(k)} = \sum_{l=1}^{2^{\kappa^*}} t_{ijl}^{(k)}$ and the probability of error for $\mathcal{V}_{ij}^{(k)}$,

$$\epsilon_{ij}^{(k)} = \frac{t_{ij}^{(k)}}{2^{\kappa_{ij}}}.$$

3: Compute $\epsilon_{ij}^{(k)}$ for all subsets $\mathcal{V}_{ij}^{(k)}$, $k : 1 \leq k \leq \binom{\kappa_{ij}}{\kappa^*}$, and let

$$\epsilon_{ij}^* := \min_k \epsilon_{ij}^{(k)}, k_{ij}^* := \arg\min_k \epsilon_{ij}^{(k)}.$$

Then $\mathcal{V}_{ij}^* := \mathcal{V}_{ij}^{(k_{ij}^*)}$ is the suboptimal predictor set for gene g_j in context i.

4: Repeat for all genes and all contexts. Redefine the Boolean functions according to suboptimal predictor sets.

The algorithms presented here address generalization in consistency-based PBN design by considering the connectivity and realization complexity in terms of an optimality measure. Algorithm 3.6 is guaranteed to produce a PBN with minimum connectivity and the minimal SOP realizations on the predictor sets. Algorithm 3.7 treats constrained connectivity as a rephrased lossy coding problem and attains the required connectivity by removing some of the predictors of each target gene in a way that minimizes the probability of error in every regulating Boolean function.

3.4 Information Theoretic Approaches

The discussion and results presented in the previous section motivate the idea that in order for predictors to generalize well, that is, to have good predictive accuracy, model complexity must be taken into account. If a model is overly complex, which is to say that it has many free parameters, it may describe the data very well, by virtue of "tuning" these parameters to fit the data, but may exhibit very poor predictive accuracy over the population from which the data are presumably generated (Shmulevich, 2003). In a sense, the selection of the best model consists of choosing the right trade-off between goodness of fit and model complexity. The field of information theory has provided powerful tools to address these issues. Perhaps the most widely studied and powerful approach in this genre is the minimum description length (MDL) principle (Rissanen, 1978).

The MDL principle states that given a set of data and class of models, one should choose the model providing the shortest encoding of the data. The coding amounts to storing both the network parameters and any deviations of the data from the model (i.e., residuals), thereby attempting to strike a balance between precision and complexity. From the perspective of inference, the MDL principle represents a form of complexity regularization, the intent being to capture the goodness of fit using some error measure as well as some measure of complexity so as not to overfit the data—a critical issue for inferring genetic networks from limited data sets. In a sense, the fundamental question is whether a model is describing the data generation mechanism "cheaply enough"; if so, then it will be cheaper to use the model along with the residuals to describe the data than using the data in its raw form.

Perhaps the first application of the MDL principle to gene prediction using Boolean and ternary models was proposed by Tabus and Astola (2001), who used ternary gene expression data from a melanoma study by Bittner *et al.* (2000). As there were 30 measurements used in the experiments, encoding the data without a model would require $\log_2(3^{30}) = 47.55$ bits. The residuals can be encoded in such a way as to penalize the errors. For each error that the predictive model makes, we can store the actual (correct) value using 2 bits (since the data are ternary, we require 2 bits to uniquely specify each of the three values, reserving one of the 2-bit codes, say, 00, to signal the end of the residual sequence). After the actual value, we also need to store the location of the error using $\log_2 31 = 5$ bits. Thus, the error will be transmitted using 7 bits.

Let us consider the specification and design of a Boolean predictor. Suppose that the ternary data, which typically takes values in $\{-1,0,1\}$, are transformed into the alphabet $\{0,1,2\}$. Then, define the thresholded vectors $x^{b_1} = [x_1^{b_1},\ldots,x_n^{b_1}]$ and $x^{b_2} = [x_1^{b_2},\ldots,x_n^{b_2}]$ as

$$x_i^{b_1} = \begin{cases} 0 & \text{if } x_i = 0, \\ 1 & \text{if } x_i \geq 1 \end{cases}$$

and

$$x_i^{b_2} = \begin{cases} 0 & \text{if } x_i \leq 1, \\ 1 & \text{if } x_i = 2. \end{cases}$$

The prediction is then defined as

$$\hat{y} = f\left(x^{b_1}\right) + f\left(x^{b_2}\right),$$

where $f(\cdot)$ is a Boolean function with n variables. Designing the predictor is straightforward using the methods already discussed; for example, we can quantize the conditional expectation to two intervals and assign $f^*(x^b) = 0$ if

$$\frac{E\left[y^{b_1}|x^{b_1} = x^b\right] + E\left[y^{b_2}|x^{b_2} = x^b\right]}{2} \leq 0.5$$

and $f^*(x^b) = 1$ otherwise. To compute the description length, consider the two-part code: first encode the model itself, which requires 2^n bits, and then, for N_e nonzero prediction errors, we require $7N_e$ bits (plus 2 bits to signify the end of the sequence), totaling

$$L(n) = 2^n + 7N_e + 2 \text{ bits}.$$

The particularly elegant aspect of this approach is that the best combination of gene predictors is automatically selected. In other words, by sequentially designing the predictors with $n = 2, 3, 4, \ldots$ inputs (i.e., predictor gene sets), the MDL principle selects as an optimal predictor size that n for which $L(n)$ is minimum. Furthermore, the MDL principle is able to decide on the best model class (e.g., Boolean, ternary, perceptron, etc.), if that is what is desired, simply by comparing their description lengths. Indeed, the two-part code, balancing the model complexity with its goodness of fit (residuals), is explicitly built into the code length $L(n)$. Therefore, whereas appending an additional predictor gene to an already good predictor would seemingly decrease the error even more (as it would, for example, in the case of the best-fit error estimated on the entire data set), it may not improve the descriptive power of the model according to the MDL criterion. For example, for the melanoma data set, all predictors of size 3 were actually worse than the best predictor of order 2, which had a shorter description length (Tabus and Astola, 2001). This is why other methods, such as best-fit extension or coefficient of determination, must be estimated with additional nonparametric approaches, such as cross-validation or bootstrap (as was done in Section 3.1.1), so as to avoid overfitting the data. MDL, on the other hand, implicitly penalizes models that are too large and has been shown to be a consistent order estimator, whereas several cross-validation methods are not.

3.4.1 Minimum description length-based network inference from time series data

The aforementioned work of Tabus and Astola (2001) focused on using the MDL principle in the prediction of gene expressions. Zhao *et al.* (2006) considered the more general task of learning the network structure, using a probabilistic network modeling framework that is compatible with the family of models represented by PBNs and DBNs (see Section 2.6.2). That is, all regulatory functions are assumed to be represented as conditional probability tables (Eq. (2.38) is an example of such a table) and can therefore be viewed as the transition network in DBNs (B_1 in Fig. 2.10). The strength of temporal relationships was evaluated using a cross-time mutual information metric,[11] and the MDL principle was utilized to determine the threshold that differentiates between strong and weak relationships. The method proposed by Zhao *et al.* (2006) used an *ad hoc* measure of description length that necessitated a tuning parameter for artificially balancing the model coding length and the data encoding length, which directly conflicts with the MDL principle's implied universality. To avoid this drawback, Dougherty *et al.* (2008) proposed a novel MDL-based method in which the description length is a theoretical measure derived from a universal normalized maximum likelihood model.

Let us begin by considering the standard BN formalism defined by a set $V = \{v_i\}_{i=1}^n$ of n binary-valued nodes representing genes, a collection of structure parameters $\Lambda = \{\lambda_i\}_{i=1}^n$ indicating the regulatory sets of genes (i.e., "wiring diagram"), and the Boolean functions $F = \{f_i\}_{i=1}^n$ used to update their values over time. We note a change of notation here in order to distinguish the genes from their actual values. Earlier (Section 3.2) we used $W_i = \{x_{i_1}, \ldots, x_{i_{k_i}}\}$ to be the predictor gene set of the target gene x_i, with f_i being the predictor function of x_i. Here, each structure parameter $\lambda_i = \left\{i_1, \ldots, i_{k_i}\right\}$ represents the collection of *indices* associated with v_i's regulatory nodes. We assume that the nodes are observed over $m+1$ time points and write $y_{i,t} \in \{0,1\}$ to denote the values of node v_i for $t = 0, 1, \ldots, m$, which progress according to

$$y_{i,t} = f_i\left(y_{i_1,t-1}, \ldots, y_{i_{k_i},t-1}\right). \tag{3.25}$$

We then define the inputs of f_i as the column vector $\mathbf{x}_{i,t} = [y_{i_1,t-1}, \ldots, y_{i_{k_i},t-1}]^T$, allowing us to rewrite Eq. (3.25) as

$$y_{i,t} = f_i\left(\mathbf{x}_{i,t}\right). \tag{3.26}$$

Thus, we reserve the letters x and y to cast the formulation as a multivariate Boolean regression problem, which can be converted into matrix notation, dropping the time index t:

$$\mathbf{Y}_i = f_i\left(\mathbf{X}_i\right) \oplus \varepsilon_i \in \{0,1\}^n, \tag{3.27}$$

where $\oplus$ denotes summation modulo 2, f_i acts independently on each column of $\mathbf{X}_i$, and ε_i is a vector of independent Bernoulli random variables with $P\left\{\varepsilon_{i,t} = 1\right\} = \theta_i \in [0,1]$. The fundamental problem to be addressed is the estimation of Λ and F. It is further assumed that the errors for different nodes are independent and we allow θ_i to depend on i, as it can be interpreted as the probability that node v_i disobeys the network rule, and it is natural to consider different nodes to have varying propensities for misbehaving.

We observe that λ_i and f_i can be estimated separately for each gene and the model stipulates independence between all random variables. Thus, we can restrict ourselves to estimating the parameters for one node and rewrite Eq. (3.27) as

$$\mathbf{Y} = f(\mathbf{X}) \oplus \varepsilon, \tag{3.28}$$

noting that θ_i and k_i now become θ and k, respectively. Finally, the model is specified by extending the parameter space for the error rates by replacing θ with $\Theta = \{\theta_l\}_{l=0}^{2^k-1}$,

[11] One of the first published approaches for learning BNs from gene expression data, called the REVEAL algorithm, made use of mutual information (Liang *et al.*, 1998).

where each θ_l corresponds to one of the 2^k possible values of $\mathbf{x}_t$. This allows the degree of reliability of the network function to vary based upon the state of a gene's regulators. Note that 2^k is only an upper bound on the number of error rates because we will not necessarily observe all 2^k possible regressor values. Adopting notation from Tabus *et al.* (2002), we can refer to the collection of all possible parameter settings as the model class $\mathcal{M}(\Theta, \lambda, f)$.

As already mentioned, no unique manner of codelength evaluation is specified by the MDL principle, meaning that no unique MDL solution exists. Several examples of such different encoding lengths are given in Dougherty *et al.* (2008), along with a demonstration, using a Monte Carlo simulation, of the risk of using the MDL principle with *ad hoc* encodings. The normalized maximum likelihood (NML) model is a universal model for measuring code length which, instead of a two-part description length, uses distributions over the class of all models. The parameters are chosen such that the probability of the data is maximized (Rissanen, 2007).

Following the exposition and notation of Dougherty *et al.* (2008), for $\mathcal{M}(\Theta, \lambda, f)$ with a fixed λ, the NML model is introduced by the standard likelihood optimization problem $\max_\Theta \log P(\mathbf{y}; \Theta, \lambda, f)$. The solution is obtained for $\Theta = \hat{\Theta}$, the maximum likelihood estimate. However, since $P(\mathbf{y}; \hat{\Theta}, \lambda, f)$ does not integrate to unity, it cannot be used as a model. Instead, a distribution $q(\mathbf{y})$ is used such that its ideal code length $-\log_2 q(\mathbf{y})$ is as close as possible to the code length $-\log_2 P(\mathbf{y}; \hat{\Theta}, \lambda, f)$. A minimax solution is chosen such that for the worst-case $\mathbf{y}$, the difference between $\log_2 q(\mathbf{y})$ and $\log_2 P(\mathbf{y}; \hat{\Theta}, \lambda, f)$ is minimized, as

$$\min_q \max_{\mathbf{y}} \log_2 \frac{P(\mathbf{y}; \hat{\Theta}, \lambda, f)}{q(\mathbf{y})}. \tag{3.29}$$

If the residuals $\mathbf{e} = f(\mathbf{X}) \oplus \mathbf{y}$ are modeled by the Bernoulli distribution, then

$$P(Y_t = y_t; \mathbf{x}_t = \mathbf{b}_l) = \theta_l^{y_t \oplus f(\mathbf{b}_l)} (1 - \theta_l)^{1 - y_t \oplus f(\mathbf{b}_l)}, \tag{3.30}$$

where $\mathbf{b}_l$ denotes the binary representation of integer l. For any $\mathbf{y} \in \{0,1\}^n$ and fixed λ, let m_l be the number of times each vector $\mathbf{b}_l$ occurs in $\mathbf{X}$, and let m_{l_1} be the number of times $f(\mathbf{b}_l) = 1$. Then, the NML model is given by

$$\hat{P}(\mathbf{y}) = \frac{P(\mathbf{y}; \lambda, \hat{f}, \mathbf{X}, \hat{\Theta})}{\prod_{l:\mathbf{b}_l \in \mathbf{X}} C_{m_l}}, \tag{3.31}$$

where

$$P(\mathbf{y}; \lambda, \hat{f}, \mathbf{X}, \hat{\Theta}) = \prod_{l:\mathbf{b}_l \in \mathbf{X}} \left(\frac{m_{l_1}}{m_l} \right)^{m_{l_1}} \left(1 - \frac{m_{l_1}}{m_l} \right)^{m_l - m_{l_1}} \tag{3.32}$$

and

$$C_{m_l} = \sum_{i=0}^{m_l} \binom{m_l}{i} \left(\frac{i}{m_l} \right)^i \left(1 - \frac{i}{m_l} \right)^{m_l - i}. \tag{3.33}$$

The total code length of the model is now the so-called stochastic complexity of the data, defined as

$$-\log \hat{P}(\mathbf{y}) = \sum_{l:\mathbf{b}_l \in \mathbf{X}} \left[m_l h \left(\frac{m_{l_1}}{m_l} \right) + \log C_{m_l} \right], \tag{3.34}$$

where $h(p) = -p\log p - (1-p)\log(1-p)$ is the binary entropy function. The total code length must also include the structure parameter λ_i that specifies the set of regulator genes for each gene. For a regulatory set of size k, $\log\binom{n}{k}$ bits would be required to select one of the regulatory sets. Using additional arguments concerning the in-degrees of genetic networks and code lengths favoring smaller in-degrees, Dougherty *et al.* (2008) suggest the total code length

$$L_T(\mathbf{y},\lambda) = -\log\hat{P}(\mathbf{y}) + \min\left\{n, \log\binom{n}{k} + \log(k+1) + \log(1+\ln n)\right\}. \quad (3.35)$$

If the quantity $L_T(\mathbf{y},\lambda)$ is computed for every possible λ, then the one with the shortest code length can be selected. This requires computing 2^n code lengths. To avoid such excessive computations, Kolmogorov's structure function can be used; we refer the reader to Dougherty *et al.* (2008) for the details.

Dougherty *et al.* (2008) evaluated the performance of the NML approach on simulated data generated from synthetic BNs, with different sample sizes, parameter values, and classes of functions, including random and canalizing networks, and compared the error rates with the REVEAL (Liang *et al.*, 1998) algorithm and the MDL approach in Zhao *et al.* (2006). The NML approach was superior to the other algorithms tested in terms of speed and accuracy.

3.5 Inference of PBNs from Time Series Data

Let us now consider the inference of a PBN from time series observations. While the inference is relatively straightforward for a BN, using any number of approaches already discussed, inferring the PBN structure also involves inference of the probabilities of perturbations, the probabilities of switching between constituent BNs, and the selection probabilities of the constituent networks. This essentially requires one to separate the observed temporal sequence of network states into subsequences that correspond to the constituent BNs, then given each such subsequence infer the corresponding constituent BN, and finally infer all of the necessary probabilities. Because of the large number of parameters, particularly for situations in which there are many inconsistencies in the data, which may be due to latent variables or measurement uncertainty, it should be expected that the sample size requirements are rather high for inference of PBNs.

We now present the inference procedure proposed by Marshall *et al.* (2007) for inferring PBNs from temporal data. Let us first consider the inference of a BN with perturbation. The characteristics of a BN, with or without perturbation, can be estimated by observing its pairwise state transitions, $\mathbf{x}(t) \to \mathbf{x}(t+1)$, $\mathbf{x}(t) \in \{0,1\}^n$. Given a temporal data sequence $\mathbf{x}(0),\ldots,\mathbf{x}(m)$, the transition counting matrix C contains entries c_{ij} equal to the number of state transitions from the ith state to the jth state. As we have seen earlier, for a BN without perturbations, each row of the matrix C contains at most one nonzero value.

For a BN with a small perturbation probability, C will have nonzero entries replacing some or all of the zero entries. Owing to perturbation and the consequent ergodicity, a sufficiently long data sequence will sufficiently populate the matrix to determine the entries caused by perturbation, as well as the functions and inputs underlying the model, thus determining the mapping $\mathbf{x}(t) \to \mathbf{x}(t+1)$ linking pairs of state vectors.

Given a gene, x_i, and keeping in mind that some observed state transitions arise from random perturbations rather than transition functions, we wish to find the k_i variables that control x_i, that is, the predictor set $W_i = \{x_{i_1}, \ldots, x_{i_{k_i}}\}$. To determine the predictors for a given variable, x_i, we will define a cost function. Assuming k variables are used to predict x_i, there are $n!/(n-k)!k!$ ways of choosing them. Each k with a choice of variables has a cost. By minimizing the cost function, we can identify k such that $k = k_i$, as well as the predictor set. In a BN without perturbation, if the value of x_i is fully determined by the predictor set W_i, then this set will not change for different combinations of the remaining variables, which are fictitious insofar as x_i is concerned (i.e., their activities are zero; see Eq. (1.6)). Hence, so long as $x_{i_1}, \ldots, x_{i_{k_i}}$ are fixed, the value of x_i should remain 0 or 1, regardless of the values of the remaining variables.

For any given realization $(x_{i_1}, \ldots, x_{i_k}) = (a_{i_1}, \ldots, a_{i_k})$, $a_{i_j} \in \{0,1\}$, let

$$u(a_{i_1}, \ldots, a_{i_k}) = \sum_{x_{i_1}=a_{i_1}, \ldots, x_{i_k}=a_{i_k}} f_i(x_{i_1}, \ldots, x_{i_k}) \tag{3.36}$$

be the sum of the next-state values assuming $x_{i_1}, \ldots, x_{i_k}$ are held fixed at $a_{i_1}, \ldots, a_{i_k}$, where f_i is the regulatory function corresponding to x_i. It can be seen that there will be 2^{n-k} terms in the above summation. Equation (3.36) attains its maximum (2^{n-k}) or minimum (0) if the value of $f_i(x_{i_1}, \ldots, x_{i_k})$ remains unchanged for all the terms in the summation. Thus, a good predictor set has $u(a_{i_1}, \ldots, a_{i_k})$ close to 0 or 2^{n-k}.

We define the cost function

$$R(a_{i_1}, \ldots, a_{i_k}) = \sum_{a_{i_1}, \ldots, a_{i_k}} r(a_{i_1}, \ldots, a_{i_k}), \tag{3.37}$$

where the quantity

$$\begin{aligned} r(a_{i_1}, \ldots, a_{i_k}) = {} & u(a_{i_1}, \ldots, a_{i_k}) 1[u(a_{i_1}, \ldots, a_{i_k}) \le 2^{n-k-1}] \\ & + \left(2^{n-k} - u(a_{i_1}, \ldots, a_{i_k})\right) 1[u(a_{i_1}, \ldots, a_{i_k}) > 2^{n-k-1}] \end{aligned} \tag{3.38}$$

is minimized when $u(a_{i_1}, \ldots, a_{i_k})$ is close to either 0 or 2^{n-k}. The predictors for variable x_i are chosen to be the k variables that minimize the cost in Eq. (3.37) and k is selected as the smallest integer to achieve the minimum. We emphasize on the smallest because if k ($k < n$) variables can perfectly predict x_i, then adding one more variable also achieves the minimum cost. For small numbers of variables, the k inputs may be chosen by a full search, with the cost function being evaluated for every combination. For larger numbers of variables, genetic algorithms can be used to minimize the cost function.

In many instances, we have only a partially defined Boolean function, as discussed in Section 3.1, meaning that there are "don't care" outputs. Marshall *et al.* (2007) report that the input variables may still be identified correctly even for 90% of missing data. As already mentioned, the inference of PBNs will first entail splitting the temporal sequence into "pure" subsequences and then inferring the perturbation, switching, and selection probabilities.

3.5.1 Splitting the temporal data sequence into pure subsequences

The underlying assumption of the approach proposed by Marshall *et al.* (2007) is that between any two successive switch points, there exists a pure temporal subsequence generated by a single constituent BN. Thus, assuming such a subsequence is sufficiently long, the empirically generated transition counting matrix C is expected to have one large value in each row, corresponding to the actual state transition of the constituent BN, with the remainder of the values being small, due to random perturbations. The idea is that a measure of "purity" should maximize the difference between the largest value in each row and any other value. Marshall *et al.* (2007) propose the measure

$$P = \frac{\sum_{i=1}^{2^n}(c_i^{(2^n)} - c_i^{(2^n-1)})}{\sum_{i,j} c_{ij}}, \tag{3.39}$$

where $c_i^{(k)}$ is the kth order statistic of the ith row of C. The measure can be applied by running through the temporal sequence and at each point, comparing the left-hand and right-hand sides of the sequence. The values of P on either side should be maximized at the true switch point. For example, the product of the left-hand side and right-hand side purities could be used as an overall measure that can be thresholded for estimating the position of the switch point.

There are certain situations in which such an approach does not work well. For instance, if the switch point occurs close to the end or beginning of a sequence, then the right-hand or left-hand subsequence, respectively, may be insufficiently long to populate the transition counting matrix. Other purity measures, based on the entire distribution of state transition counts, which may have better behavior under small sample sizes, could be envisioned. Another difficult situation occurs when the two constituent BNs are different, yet have many common state transitions, particularly if they have common attractors. Furthermore, the actual values of the purity function may vary greatly between different subsequences. Marshall *et al.* (2007) suggest a normalized purity value

$$P_{\text{norm}} = \frac{P - T}{T},$$

where T is either the mean or geometric mean of the values in the window in which the purity is calculated. Such a normalization removes differences in the ranges and average values of points in different subsequences, making it easier to identify genuine switches between constituent BNs.

3.5.2 Estimation of switching, selection, and perturbation probabilities

Typically, much longer data sequences are required to estimate the switching, selection, perturbation probabilities in PBNs. To estimate the switching probability, one can simply divide the number of detected switch points by the total sequence length. To estimate the perturbation probability, one can identify the state transitions not determined by constituent network functions. In other words, for each data point, the next state is predicted using the inferred model and each predicted state that does not match the actual state is recorded as being caused by a perturbation.

The perturbation probability is computed by dividing the total number of such recorded perturbations by the length of the sequence. For selection probabilities, it is assumed that a constituent network cannot switch into itself. Thus, the selection probabilities are conditional, depending on the current network, taking the form q_{AB}, which gives the probability of selecting network B during a switch, given that the current network is A. This quantity is estimated by dividing the number of times the data sequence switches from A to B by the number of times it switches out of A.

Marshall *et al.* (2007) performed a number of simulated experiments to assess the performance of the inference algorithm relative to switching probability, perturbation probability, connectivity (k), and length of the data sequence. For instance, for a low switching probability, the resulting pure subsequences will be long, making it possible to infer each constituent BN from such subsequences, but it may be necessary to observe a very long sequence in order to encounter all of the constituent BNs. As expected, it is easier to recover predictors for smaller values of k; interestingly, larger values of the perturbation probability and switching probability are also beneficial. In general, if one does not wish to infer the switching, perturbation, and selection probabilities, then the data requirements become significantly reduced. Additionally, if one were interested only in restricting the inference to state transitions in attractor cycles (as opposed to all state transitions, including transient ones), then the amount of data necessary for inference would be substantially reduced as well. The performance of the inference algorithms was assessed by computing the percentage of predictors identified correctly as well as the percentage of function outputs identified correctly (Marshall *et al.*, 2007). In general, a number of different validation procedures for network inference can be used.

3.6 Validation of network inference procedures

In order to meaningfully compare different inference algorithms, it is necessary to quantify their performance. The validity of an inference algorithm must be evaluated relative to its ability to infer a model network close to the ground truth network from which the data have been generated. Thus, validation of an inference procedure can be formulated in terms of the distance between two networks, or the distance between two structures deduced from the networks, such as steady-state distributions (Section 2.5).

It is required only that a measure μ quantifying the difference between two networks, M and N, be a semimetric (Dougherty, 2007), meaning that it satisfies

$$\begin{aligned}
&\mu(M,N) \geq 0, \\
&\mu(M,M) = 0, \\
&\mu(M,N) = \mu(N,M), \\
&\mu(M,N) \leq \mu(M,P) + \mu(P,N).
\end{aligned}$$

We do not require μ to be a distance, which must satisfy

$$\mu(M,N) = 0 \Longrightarrow M = N,$$

since we often compare networks in terms of some characteristics, such as the structure of their regulatory graph or steady-state distribution. Such a semimetric can be used to measure the goodness of an approximation to a network N. If M_1 and M_2 are approximations of N, then M_1 is better than M_2 if $\mu(M_1,N) < \mu(M_2,N)$.

A rule-based distance is essentially the normalized Hamming distance between all of the corresponding truth tables in the two networks N and M (see Eq. (3.8)). If one's focus is on the topology of the network, then a topology-based distance can be defined as a Hamming distance between the adjacency matrices corresponding to the two networks N and M. Obviously, such a measure ignores dynamics, since two different networks can have a zero topology-based distance. Nonetheless, measures such as sensitivity, specificity, precision, and recall, constructed in terms of true positive and false positive ratios, are frequently employed in studies aimed at inferring mechanistic regulatory rules in genetic networks (Stolovitzky, 2009). The area under the curve of the receiver operating characteristic (ROC) curve is one common summary measure.

Another distance between networks can be defined in terms of their respective transition probability matrices, using matrix norms. If P_M and P_N are the state transition matrices corresponding to networks M and N, respectively, then

$$\mu_{\text{prob}}(M,N) = \|P_M - P_N\|_r, \tag{3.40}$$

where $\|\cdot\|_r$ is the r-norm defined by

$$\|P\|_r = \left(\sum_{i,j} |p_{ij}|^r\right)^{1/r}.$$

It is easy to see that the measure in Eq. (3.40) is a network semimetric since two distinct networks may have the same transition probability matrix.

Another distance that is based on system dynamics is the long-run (steady-state) distance, which measures the difference between steady-state distributions. If π_M and π_N are steady-state distributions corresponding to networks M and N, respectively, then the long-run distance can be defined as

$$\mu_{\text{ss}}(M,N) = \|\pi_M - \pi_N\|_r. \tag{3.41}$$

For cases where the steady-state distribution does not exist, for example in a BN with no perturbations, an analogous measure that compares the attractor sets of the two networks can be defined by

$$\mu_{\text{att}}(M,N) = \frac{1}{2^n}|A_M \bigtriangleup A_N|, \tag{3.42}$$

where $|\cdot|$ denotes the cardinality of a set and $\bigtriangleup$ is the symmetric set difference defined by

$$A_M \bigtriangleup A_N = (A_M \backslash A_N) \cup (A_N \backslash A_M).$$

Network semimetrics, such as the ones stated above, can be used to define equivalence classes of networks, where two networks are said to be equivalent if the distance between their respective network characteristics, such as the regulatory graphs, transition probability matrices, or steady-state distributions, is zero.

An inference procedure operates on data generated by a network N and constructs an inferred network M to serve as an estimate of N. Alternatively, the procedure constructs a characteristic to serve as an estimate of the corresponding characteristic of N, such as the steady-state distribution. The inference procedure ψ is a mapping from a space of samples to a space of networks, so that for a data set S, $\psi(S) = M$ is the estimated network.

The goodness of an inference procedure ψ is measured relative to a semimetric $\mu(\psi(S), N)$, where N is the true network generating the data. It should be noted that the data S represent a realization of a random set process, Σ, governing data generated from N. Thus, $\mu(\psi(\Sigma), N)$ is a random variable and the performance of the inference procedure ψ is characterized by the distribution of $\mu(\psi(\Sigma), N)$, with mean $E_\Sigma[\mu(\psi(\Sigma), N)]$.

Rather than considering an inference procedure relative to a single network, we can consider a distribution, $\mathbf{N}$, over the space of networks from which realizations of networks are generated. Averaging over the class of random networks, the relevant performance measure becomes

$$\mu^*(\mathbf{N}, \Sigma, \psi) = E_{\mathbf{N}}[E_\Sigma[\mu(\psi(\Sigma), \mathbf{N})]], \tag{3.43}$$

making it possible to compare two inference procedures ψ_1 and ψ_2 using

$$\mu^*(\mathbf{N}, \Sigma, \psi_1) \gtrless \mu^*(\mathbf{N}, \Sigma, \psi_2).$$

In practice, expectations are estimated by averages. An important issue is that of consistency, meaning that for large data sets, the inferred network will be close to the true network, i.e., $\mu^*(\mathbf{N}, \Sigma, \psi) \to 0$.

Suppose we do not know the random network, $\mathbf{N}$, generating the data for which we want to evaluate the inference procedure, ψ, but know a network $\mathcal{N}$ that we believe to be a good approximation to the networks in $\mathbf{N}$. Thus, the inferred network can be compared to $\mathcal{N}$, which essentially approximates $\mu^*(\mathbf{N}, \Sigma, \psi)$ by $E_\Sigma[\mu(\psi(\Sigma), \mathcal{N})]$. It can be shown (Dougherty, 2007) that this occurs when $E_{\mathbf{N}}[\mu(\mathcal{N}, \mathbf{N})]$ is small. On the other hand, when $E_{\mathbf{N}}[\mu(\mathcal{N}, \mathbf{N})]$ is not small, the entire validation procedure is flawed because the approximation of $\mathbf{N}$ by $\mathcal{N}$ is confounding the procedure.

Such a validation procedure is often carried out, whereby the inferred network is compared to some "ground truth" network derived from prior knowledge (e.g., biological literature or interaction databases). However, this is valid only to the extent that a ground truth network is a sufficiently good approximation of the system generating the data on which the inference procedure is applied. If not, meaning $E_{\mathbf{N}}[\mu(\mathcal{N}, \mathbf{N})]$ is not small, then the entire validation procedure is flawed due to the confounding effects of approximating $\mathbf{N}$ by $\mathcal{N}$.

Another form of approximation results from using experimental data for validation rather than synthetic data generated from a known, ground truth model. Various resampling approaches, such as cross-validation, which involve splitting the data into training and test sets, are commonly employed. The key question in such strategies is how well the error can be estimated under limited sample sizes. For example, the resubstitution error estimate is computed on the same data as what is used to estimate (train) the model. Thus, the estimated error is often biased low (i.e., overly optimistic) owing to overfitting of the model. This can seriously confound network inference and validation from experimental data. The performance of error estimation rules, particularly under small sample sizes, has been studied in the context of classification (see Shmulevich and Dougherty (2007)), but there appears to be an absence of the analogous study for network validation. The above discussion highlights the need of a formal framework for validation of network inference procedures. A great deal of work remains to be done to establish the properties of inference procedures under various conditions, such as the sampling procedure, model class, and validation criterion (distance function).

Chapter 4

Structural Intervention

Structural intervention in a PBN (or any gene regulatory network) refers to a perturbation of the underlying model structure to achieve a long-run alteration of network behavior as described by the stationary distribution (Shmulevich *et al.*, 2002c). We view structural intervention as permanent intervention in the sense that the change to the network is considered to be irreversible. The therapeutic aim of structural intervention is based on the observation that certain network states may represent different phenotypes or cellular functional states, such as cell invasion and quiescence, and we would like to decrease the probability that the network will end up in an undesirable set of states and increase the probability that it will end up in a desirable set of states. The idea is to develop a methodology for altering the steady-state probabilities with minimal modifications to the rule-based structure. Besides beneficially altering the steady-state distribution, an additional goal might be to change as few functions as possible. Such changes to the regulatory rules may be possible by the introduction of a factor or drug that alters the extant behavior.

We know that women can age much faster after menopause. In developed countries, estrogen is sometimes taken by women after menopause to alter this trend; however, the dose of estrogen is important because an overdose may increase the probabilities of developing breast and ovarian cancers. Although the mechanism is not clear yet, it is conceivable that this phenomenon has its basis in gene regulation. Estrogen binds its receptors, the complex getting transported into the nucleus to bind the enhancer element (a short stretch of regulatory DNA sequence) on the target genes, and functions as transcriptional factors affecting genes such as the preproenkephalin (PENK) gene (Vasudevan *et al.*, 2001). Interestingly, there are several different estrogen receptors that compete with each other for binding estrogen as well as for a coactivator, which is also required for efficient transcriptional regulation by estrogen (Zhang and Teng, 2001). It can be envisioned that estrogen binds one receptor better than another and that these complexes bind DNA and the coactivator with opposite efficiency. That is, complex C1 binds DNA better than complex C2, but complex C2 binds the coactivator better than complex C1. Thus, under low estrogen conditions, when there is not much competition for DNA binding, there would be sufficient binding of C2 to DNA so as to turn on the downstream target gene. However, when estrogen is present at high concentration, both complexes exist at very high levels and complex C2, taking up most of the coactivator away from C1, would have little chance to bind to DNA. Consequently, the better DNA binding complex (C1) would not have the necessary coactivator to activate the target gene. If the target gene plays a role in tumor suppression, for instance, this could explain why high levels of estrogen have a tumorigenic effect. Thus, by changing the concentration of estrogen, one is able to alter the rule determining the value of a gene (e.g., PENK) in terms of the levels of estrogen receptor complexes C1 and C2. For example, under a low estrogen condition, assuming Boolean values for all genes, PENK can be expressed as PENK = C1 $\vee$ C2; that is, the presence of at least one complex (C1 or C2) would be sufficient to turn on PENK; however, under a high estrogen condition, in view of the above squelching effect, PENK = C1 $\oplus$ C2, where $\oplus$ denotes exclusive OR; that is, when either C1 or C2 is individually present, with no competition from the other, PENK is turned on, but when both C1 and C2 are present together, PENK becomes turned off. The ability to alter such rules would provide a means of at least partially controlling the steady-state behavior of the network.

4.1 Impact of Function Perturbation on State Transitions

We begin by characterizing the effects of function perturbation on the state transitions and attractors of a BN. Since the attractors of a PBN are, by definition, the attractors of its constituent BNs, the theory applies to PBNs. We focus on function perturbation in the form of a one-bit change of the truth table and explore its impact on the attractors. Since state transitions completely characterize the network dynamics and define the attractors and their basins of attraction, we concentrate on two issues: (1) impact on state transitions; and (2) impact on attractors, namely, which attractors will be invariant to the perturbations and which nonattractor states will become new attractors (Xiao and Dougherty, 2007). If, in a BN, the function f_i determining the value of gene i depends on k_i input variables, $x_{i1}, x_{i2}, \ldots, x_{ik_i}$, then we will denote the evaluated input vector on row j $(1 \le j \le 2^{k_i})$ of the truth table by $\mathbf{a}^i_j \in \{0,1\}^{k_i}$, the rows being labeled $1, 2, \ldots, 2^{k_i}$.

In a BN, the state transition $\mathbf{s} \to \mathbf{w}$ depends on the vector function $\mathbf{f} = (f_1, f_2, \ldots, f_n)$. Restricting our attention to f_i means considering the ith mapping $\mathbf{s} \to w_i$. Since f_i depends on $(x_{i1}, x_{i2}, \ldots, x_{ik_i})$, the mapping depends on $(s_{i1}, s_{i2}, \ldots, s_{ik_i})$ only. If $\mathbf{u} \neq \mathbf{s}$ but $(u_{i1}, u_{i2}, \ldots, u_{ik_i}) = (s_{i1}, s_{i2}, \ldots, s_{ik_i})$, then $\mathbf{u}$ and $\mathbf{s}$ both map to w_i under f_i. If we let $\mathrm{In}_i(\mathbf{s}) = (s_{i1}, s_{i2}, \ldots, s_{ik_i})$ denote the input vector for function f_i as it operates on state $\mathbf{s}$, then $\mathrm{In}_i(\mathbf{u}) = \mathrm{In}_i(\mathbf{s})$ implies $f_i(\mathbf{u}) = f_i(\mathbf{s})$. For instance, if $n = 5$, $\mathbf{s} = 01001$, $\mathbf{u} = 00011$, and $x_{i1}x_{i2}\cdots x_{ik_i} = x_1x_3x_5$, then $\mathrm{In}_i(\mathbf{u}) = \mathrm{In}_i(\mathbf{s}) = 001$ and $f_i(\mathbf{u}) = f_i(\mathbf{s}) = f_i(001)$.

A one-bit perturbation occurs when, for a single function f_i, there is a one-bit change of its truth table resulting from flipping the value on the jth entry $(1 \le j \le 2^{k_i})$, that is, change 0 to 1 or change 1 to 0. We denote the new function by $f_i^{(j)}$, so that the one-bit perturbation on row j takes the form $f_i \to f_i^{(j)}$, where $f_i^{(j)}(\mathbf{a}^i_j) = 1 - f_i(\mathbf{a}^i_j)$. For any state $\mathbf{s} = (s_1, s_2, \ldots, s_n)$, we define $\mathbf{s}^{(i)} = (s_1, \ldots, s_{i-1}, 1 - s_i, s_{i+1}, \ldots, s_n)$.

Theorem 4.1 (Xiao and Dougherty, 2007). *The state transition* $\mathbf{s} \to \mathbf{w}$ *is affected by the one-bit perturbation* $f_i \to f_i^{(j)}$ *if and only if* $\mathrm{In}_i(\mathbf{s}) = \mathbf{a}^i_j$. *If the state transition is affected, then the new state transition will be* $\mathbf{s} \to \mathbf{w}^{(i)}$.

Proof. The first statement follows at once from the fact that the ith mapping transition $\mathbf{s} \to w_i$ depends only on $\mathrm{In}_i(\mathbf{s})$, and this is affected by the perturbation if and only if $\mathrm{In}_i(\mathbf{s}) =$

$\mathbf{a}_j^i$. Next, suppose the transition is affected by the perturbation. Then, absent perturbation, the ith mapping is given by

$$\mathbf{s} \rightarrow f_i(\mathrm{In}_i(\mathbf{s})) = f_i(\mathbf{a}_j^i) = w_i; \tag{4.1}$$

with perturbation, the ith mapping is

$$\mathbf{s} \rightarrow f_i^{(j)}(\mathrm{In}_i(\mathbf{s})) = f_i^{(j)}(\mathbf{a}_j^i) = 1 - f_i(\mathbf{a}_j^i) = 1 - w_i. \tag{4.2}$$

Since the other $n-1$ mappings remain unchanged, following the perturbation, state $\mathbf{s}$ will transit to $\mathbf{w}^{(i)}$. □

Corollary 4.2. *If $|\mathbf{a}_j^i| = k_i$, then the one-bit perturbation $f_i \rightarrow f_i^{(j)}$ will result in 2^{n-k_i} changed state transitions (altered edges) in the state transition diagram.*

Proof. According to the theorem, the transition of a state $\mathbf{s}$ is affected by the perturbation $f_i \rightarrow f_i^{(j)}$ if and only if $\mathrm{In}_i(\mathbf{s}) = \mathbf{a}_j^i$. Among the 2^n states $\mathbf{s}$, $\mathrm{In}_i(\mathbf{s}) = \mathbf{a}_j^i$ for exactly 2^{n-k_i} of them. □

Corollary 4.3. *A singleton attractor $\mathbf{s}$ will no longer be a singleton attractor following the one-bit perturbation $f_i \rightarrow f_i^{(j)}$ if and only if $\mathrm{In}_i(\mathbf{s}) = \mathbf{a}_j^i$.*

Proof. According to the theorem, subsequent to the perturbation, $\mathbf{s} \rightarrow \mathbf{s}^{(i)}$ if $\mathrm{In}_i(\mathbf{s}) = \mathbf{a}_j^i$, in which case it is no longer a singleton attractor, and $\mathbf{s} \rightarrow \mathbf{s}$ if $\mathrm{In}_i(\mathbf{s}) \neq \mathbf{a}_j^i$, in which case it remains a singleton attractor. □

Corollary 4.4. *A nonsingleton-attractor state $\mathbf{s}$ becomes a singleton attractor as a result of the one-bit perturbation $f_i \rightarrow f_i^{(j)}$ if and only if the following are true:* (1) $\mathrm{In}_i(\mathbf{s}) = \mathbf{a}_j^i$, *and* (2) *absent the perturbation, $\mathbf{s} \rightarrow \mathbf{s}^{(i)}$.*

If a singleton attractor $\mathbf{s}$ is lost on account of a one-bit perturbation, can a second perturbation restore it? If the perturbation $f_i \rightarrow f_i^{(j)}$ causes $\mathbf{s}$ to no longer be a singleton attractor, then the new transition of $\mathbf{s}$ must be $\mathbf{s} \rightarrow \mathbf{s}^{(i)}$. According to the previous corollary, $\mathbf{s}$ will again become a singleton attractor as a result of the one-bit perturbation $f_k \rightarrow f_k^{(l)}$ if and only if $\mathrm{In}_k(\mathbf{s}) = \mathbf{a}_l^k$, and, absent the perturbation, $\mathbf{s} \rightarrow \mathbf{s}^{(k)}$. From the second condition, since we know that $\mathbf{s} \rightarrow \mathbf{s}^{(i)}$, we must have $k = i$. Hence, from the first condition we must have $\mathrm{In}_i(\mathbf{s}) = \mathbf{a}_l^i$, but from the fact that $\mathbf{s}$ has been affected by the perturbation $f_i \rightarrow f_i^{(j)}$, we know that $\mathrm{In}_i(\mathbf{s}) = \mathbf{a}_j^i$. Hence, $\mathbf{s}$ is restored to being a singleton attractor by the same one-bit perturbation $f_i \rightarrow f_i^{(j)}$ that caused it to cease being a singleton attractor, and no other. This means that the original BN is restored.

According to Corollary 4.3, a singleton attractor $\mathbf{s}$ is no longer a singleton attractor following a one-bit perturbation $f_i \rightarrow f_i^{(j)}$ if $\mathrm{In}_i(\mathbf{s}) = \mathbf{a}_j^i$, but could it remain an attractor state as part of an attractor cycle following perturbation? Indeed it could. Consider the following situation for 3 nodes: 000 is a singleton attractor, 001 $\rightarrow$ 010, and 010 $\rightarrow$ 000, so that 001 and 010 are in the basin of 000. The three functions are defined by $f_1(x_2,x_3)=0$ for all x_2,x_3; $f_2(x_1,x_3)=0$ for all x_1,x_3, except for $f_2(0,1)=1$; and $f_3(x_2,x_3)=0$ for all x_2,x_3. Consider the one-bit perturbation $f_3 \rightarrow f_3^{(00)}$. Following the perturbation, the third function becomes $f_3^{(00)}(x_2,x_3)=0$ for all x_2,x_3, except for $f_3^{(00)}(0,0)=1$. This leads to the following transitions, 000 $\rightarrow$ 001, 001 $\rightarrow$ 010, and 010 $\rightarrow$ 000, and hence the attractor cycle 000 $\rightarrow$ 001 $\rightarrow$ 010 $\rightarrow$ 000.

A one-bit perturbation can result in a singleton attractor becoming a member of a multiple-state attractor cycle; on the other hand, a one-bit perturbation can affect a multiple-state attractor cycle, meaning that exact cycle is not an attractor cycle following perturbation. Suppose $\mathbf{s}_1 \rightarrow \mathbf{s}_2 \rightarrow \cdots \rightarrow \mathbf{s}_m \rightarrow \mathbf{s}_1$ is an m-state attractor cycle. It follows from Theorem 4.1 that this cycle will be affected by the one-bit perturbation $f_i \rightarrow f_i^{(j)}$ if and only if

$$\mathbf{a}_j^i \in \{\mathrm{In}_i(\mathbf{s}_1), \mathrm{In}_i(\mathbf{s}_2), \ldots, \mathrm{In}_i(\mathbf{s}_m)\}. \tag{4.3}$$

For instance, suppose $\mathbf{a}_j^i = \mathrm{In}_i(\mathbf{s}_1)$. Then, following perturbation, $\mathbf{s}_1 \rightarrow \mathbf{s}_2^{(i)} \neq \mathbf{s}_2$. Of course, $\mathbf{s}_1$ might still be an attractor state as part of some other attractor cycle.

Corollary 4.2 puts an upper bound on the number of singleton attractors that can be lost owing to a one-bit perturbation $f_i \rightarrow f_i^{(j)}$, namely, $\min\{2^{n-k_i}, N\}$, where $|\mathbf{a}_j^i| = k_i$ and N is the number of singleton attractors. Taking a network view, the total number of singleton attractors lost is bounded by

$$\min\left\{2^n \sum_{i=1}^{n} 2^{-k_i}, N\right\} \leq \min\{n2^{n-k_{\min}}, N\}, \tag{4.4}$$

where $k_{\min}$ is the minimum connectivity among the nodes. Increased connectivity provides greater robustness relative to the loss of singleton attractors via one-bit perturbations.

Increasing the number of one-bit perturbations increases the complexity of the problem; indeed, any BN on the same variables can be obtained from any other via a sufficiently long sequence of one-bit perturbations. If we consider two one-bit perturbations, then there are two cases: (1) the same function is changed and a flip occurs on two rows; and (2) two functions are changed. The two cases can be expressed as (1) $f_i \rightarrow f_i^{(j)}$ and $f_i^{(j)} \rightarrow f_i^{(j,l)}$, $j \neq l$; (2) $f_i \rightarrow f_i^{(j)}$ and $f_k \rightarrow f_k^{(l)}$, $i \neq k$. To extend Theorem 4.1 to two one-bit perturbations, let $\mathbf{w}^{(i,k)}$ denote the state obtained from $\mathbf{w}$ by flipping the ith and kth nodes. We state separate extensions for the two cases: (1) The state transition $\mathbf{s} \rightarrow \mathbf{w}$ is affected by the two-bit perturbation $f_i \rightarrow f_i^{(j,l)}$, $j \neq l$ if and only if $\mathrm{In}_i(\mathbf{s}) \in \{\mathbf{a}_j^i, \mathbf{a}_l^i\}$, and if it is affected, then $\mathbf{s} \rightarrow \mathbf{w}^{(i)}$. (2) The state transition $\mathbf{s} \rightarrow \mathbf{w}$ is affected by the two one-bit perturbations $f_i \rightarrow f_i^{(j)}$ and $f_k \rightarrow f_k^{(l)}$ ($i \neq k$) if and only if $\mathrm{In}_i(\mathbf{s}) = \mathbf{a}_j^i$ or $\mathrm{In}_k(\mathbf{s}) = \mathbf{a}_l^k$, and if it is affected, then $\mathbf{s} \rightarrow \mathbf{w}^{(i)}$ when $\mathrm{In}_i(\mathbf{s}) = \mathbf{a}_j^i$ and $\mathrm{In}_k(\mathbf{s}) \neq \mathbf{a}_l^k$, $\mathbf{s} \rightarrow \mathbf{w}^{(k)}$ when $\mathrm{In}_k(\mathbf{s}) = \mathbf{a}_l^k$ and $\mathrm{In}_i(\mathbf{s}) \neq \mathbf{a}_j^i$, and $\mathbf{s} \rightarrow \mathbf{w}^{(i,k)}$ when $\mathrm{In}_i(\mathbf{s}) = \mathbf{a}_j^i$ and $\mathrm{In}_k(\mathbf{s}) = \mathbf{a}_l^k$.

The corollaries of Theorem 4.1 can be extended to two one-bit perturbations.

Corollary 4.5. *Supposing that two one-bit perturbations take place in the BN, consider two cases:*

(a) *If* $f_i \to f_i^{(j,l)}$ *(i.e.,* f_i *is perturbed on rows* j *and* l, $j \neq l$*), then there will be* 2^{n-k_i+1} *changed state transitions;*

(b) *If* $f_i \to f_i^{(l_1)}$ *and* $f_j \to f_j^{(l_2)}$ $(i \neq j)$*, assuming* $|\mathbf{a}_{l_1}^i| = k_i$, $|\mathbf{a}_{l_2}^j| = k_j$*, and* f_i *and* f_j *have* k_{ij} *input variables in common, then* (b1) *if each of the* k_{ij} *variables takes the same value in* $\mathbf{a}_{l_1}^i$ *and in* $\mathbf{a}_{l_2}^j$*, there will be* $2^{n-k_i} + 2^{n-k_j} - 2^{n-k_i-k_j+k_{ij}}$ *changed state transitions, and* (b2) *otherwise, there will be* $2^{n-k_i} + 2^{n-k_j}$ *changed state transitions.*

Corollary 4.6. *Suppose state* $\mathbf{s}$ *is a singleton attractor. It will no longer be a singleton attractor following the two-bit perturbation* $f_i \to f_i^{(j,l)}$ *if and only if* $\mathrm{In}_i(\mathbf{s}) = \mathbf{a}_j^i$ *or* $\mathbf{a}_l^i$. $\mathbf{s}$ *will cease to be a singleton attractor following the two one-bit perturbations* $f_i \to f_i^{(l_1)}$ *and* $f_j \to f_j^{(l_2)}$ *if and only if* $\mathrm{In}_i(\mathbf{s}) = \mathbf{a}_{l_1}^i$ *or* $\mathrm{In}_j(\mathbf{s}) = \mathbf{a}_{l_2}^j$.

Corollary 4.7. *A nonsingleton-attractor state* $\mathbf{s}$ *becomes a singleton attractor as a result of the two-bit perturbation* $f_i \to f_i^{(j,l)}$ *if and only if the following is true:* $\mathrm{In}_i(\mathbf{s}) = \mathbf{a}_j^i$ *or* $\mathbf{a}_l^i$*, and absent the perturbation,* $\mathbf{s} \to \mathbf{s}^{(i)}$. $\mathbf{s}$ *will become a singleton attractor following the two one-bit perturbations* $f_i \to f_i^{(l_1)}$ *and* $f_j \to f_j^{(l_2)}$ *if and only if one of the following is true:*

(1) $\mathrm{In}_i(\mathbf{s}) = \mathbf{a}_{l_1}^i$ *and* $\mathrm{In}_j(\mathbf{s}) = \mathbf{a}_{l_2}^j$*, and absent the perturbation,* $\mathbf{s} \to \mathbf{s}^{(i,j)}$*;*

(2) $\mathrm{In}_i(\mathbf{s}) = \mathbf{a}_{l_1}^i$ *and* $\mathrm{In}_j(\mathbf{s}) \neq \mathbf{a}_{l_2}^j$*, and absent the perturbation,* $\mathbf{s} \to \mathbf{s}^{(i)}$*;*

(3) $\mathrm{In}_i(\mathbf{s}) \neq \mathbf{a}_{l_1}^i$ *and* $\mathrm{In}_j(\mathbf{s}) = \mathbf{a}_{l_2}^j$*, and absent the perturbation,* $\mathbf{s} \to \mathbf{s}^{(j)}$.

From the extension of Theorem 4.1 and its corollaries to two one-bit perturbations, it is clear how to extend them to more than two one-bit perturbations, albeit with an increased number of cases.

4.1.1 Identifying function perturbations

Suppose a function perturbation occurs unbeknownst to us except that we observe the new state transitions. By comparing the state transitions before and after perturbation, we may ask two questions: (1) Which Boolean function is perturbed? (2) On which row of the truth table is the perturbation? These identification problems may be useful in diagnosing changes in gene regulatory networks, such as changes caused by a disease, radiation therapy, drug treatment, etc. We now present a general procedure for identifying function perturbations in a BN (Xiao and Dougherty, 2007):

1. Identify perturbed function(s).

   Let $\mathbf{s}_0, \mathbf{s}_1, \ldots, \mathbf{s}_{2^n-1}$ be the successor states of $00\cdots0, 00\cdots1, \ldots, 11\cdots1$, respectively, in the original network and $\mathbf{s}'_0, \mathbf{s}'_1, \ldots, \mathbf{s}'_{2^n-1}$ be the corresponding successor states in the perturbed network. According to Theorem 4.1, if a state transition is affected by a one-bit perturbation on Boolean function f_i, then the new successor state differs from the old by the value of node x_i. Thus, by comparing the two lists, one can tell which function(s) is (are) perturbed.

2. Locate the flipped entries of the perturbed function(s).

   For the simple case of a one-bit perturbation in one function f_i, recall from Corollary 4.2 that 2^{n-k_i} states will have changed successors. Moreover, those states share the same values on nodes $x_{i1},\ldots,x_{ik_i}$. Assume the differences between the state transition rules before and after perturbation are given by the states $\mathbf{s}_{d1},\mathbf{s}_{d2},\ldots,\mathbf{s}_{dm_i}$ versus the states $\mathbf{s}'_{d1},\mathbf{s}'_{d2},\ldots,\mathbf{s}'_{dm_i}$ and k_i is unknown. We can find k_i by computing $k_i = n - \log_2 m_i$. Knowing that those states are the successor states of $(d1)_2,(d2)_2,\ldots,(dm_i)_2$, which are the length-$n$ binary representations of decimal numbers $d1,d2,\ldots,dm_i$ (e.g., for a 3-node BN, $(6)_2 = 110$), we may compare the states $(d1)_2,(d2)_2,\ldots,(dm_i)_2$ to find the common bits in order to identify the parent nodes of x_i, which are $x_{i1},\ldots,x_{ik_i}$. Moreover, if

   $$\mathrm{In}_i((d1)_2) = \mathrm{In}_i((d2)_2) = \cdots = \mathrm{In}_i((dm_i)_2) = (u_1,\ldots,u_{k_i}) = \mathbf{a}^i_j,$$

   then we can conclude that the perturbation on f_i takes place on the jth row of its truth table.

3. For the case of two-bit perturbations, we can refer to Corollaries 4.5, 4.6, and 4.7 for a similar analysis. Likewise, we can treat more complex perturbations, albeit with increased difficulty.

We will apply the perturbation identification strategy to a *Drosophila melanogaster* segmentation polarity gene network having eight nodes (genes), wg_1, wg_2, wg_3, wg_4, PTC_1, PTC_2, PTC_3, and PTC_4, and Boolean functions

$$\begin{aligned}
wg_1 &= wg_1 \cdot \overline{wg_2} \cdot \overline{wg_4},\\
wg_2 &= wg_2 \cdot \overline{wg_1} \cdot \overline{wg_3},\\
wg_3 &= wg_1 + wg_3,\\
wg_4 &= wg_2 + wg_4,\\
PTC_1 &= \overline{wg_2} \cdot \overline{wg_4} + PTC_1 \cdot \overline{wg_1} \cdot \overline{wg_3},\\
PTC_2 &= \overline{wg_1} \cdot \overline{wg_3} + PTC_2 \cdot \overline{wg_2} \cdot \overline{wg_4},\\
PTC_3 &= 1,\\
PTC_4 &= 1
\end{aligned}$$

(Albert and Othmer, 2003). The network has ten singleton attractors:

00001111(15), 00010111(23), 00011111(31), 00101011(43), 00101111(47),
00110011(51), 01010111(87), 01011111(95), 10101011(171), 10101111(175)

(the parentheses following each attractor giving the corresponding decimal number).

Among the attractors, 23 and 31 lead to a wild-type pattern and a variant of a wild-type pattern, respectively. 43 and 47 lead to patterns without parasegment. Those patterns are well known experimentally. 87 and 95 lead to patterns similar to wild types, and 171 and 175 lead to patterns similar to nonparasegment patterns, but these patterns are not observed experimentally. Assume the network is modified so that the attractors 87, 95, 171, and 175 disappear, leaving six attractors. Letting BOA denote the basin of attraction, under the modification, suppose state 87 $\in$ BOA{23}, 95 $\in$ BOA{31}, 171 $\in$ BOA{43}, and

175 ∈ BOA{47}, each reaching its attractor in one step. Suppose we have no knowledge about other changes in the network. Based on this partial knowledge, we wish to find out how the network has been modified.

States 87 (01010111) and 23 (00010111) differ by the 2nd bit. States 95 (01011111) and 31 (00011111) also differ by the 2nd bit. States 171 and 43 differ by the 1st bit. States 175 and 47 differ by the 1st bit also. According to Theorem 4.1, perturbations occur on the functions for genes wg_1 and wg_2. The former function has genes $wg1$, wg_2, and wg_3 as inputs, and in both states 87 and 95, $(wg_1, wg_2, wg_3) = 010 = \mathbf{a}_3^2$, so the 3rd row of the truth table is flipped. The latter function has genes wg_1, wg_2, and wg_4 as inputs, and in both states 171 and 175, $(wg_1, wg_2, wg_4) = 100 = \mathbf{a}_5^1$, so the 5th row of the truth table is flipped. The new definitions for the two functions are

$$wg_1 = \begin{cases} wg_1 \cdot \overline{wg_2} \cdot \overline{wg_4} & \text{if } (wg_1, wg_2, wg_4) \neq 100, \\ 1 - wg_1 \cdot \overline{wg_2} \cdot \overline{wg_4} & \text{if } (wg_1, wg_2, wg_4) = 100, \end{cases}$$

$$wg_2 = \begin{cases} wg_2 \cdot \overline{wg_1} \cdot \overline{wg_3} & \text{if } (wg_1, wg_2, wg_3) \neq 010, \\ 1 - wg_2 \cdot \overline{wg_1} \cdot \overline{wg_3} & \text{if } (wg_1, wg_2, wg_3) = 010. \end{cases}$$

Simulation results of the new BN with the above modifications agree with our conclusion.

To make the above identification, we do not require complete knowledge of the state transitions. This is because even a 1-bit difference in a single function can result in 2^{n-k_i} changes in state transitions, and when $n > k_i$, there is redundant information.

4.2 Intervention via Constructive Function Perturbation

In this section, we design intervention strategies to alter the long-run dynamics of a network by directly analyzing its predictor functions. We begin with a simple example to illustrate the concept of optimal structural intervention.

4.2.1 Concept of structural intervention

To illustrate structural intervention, following Shmulevich *et al.* (2002c) we consider two sets of states, A and B, and suppose we wish to alter the stationary probabilities of these two sets in such a way that the stationary probability, $\pi(A)$, of A is close to a given value λ_A and the stationary probability, $\pi(B)$, of B is close to a given value λ_B. To achieve this goal, we may alter function f_{ik}, the predictor of gene i in context k, by replacing it with a new function g_{ik}, all other network parameters remaining the same. The result will be a new PBN whose stationary distribution we denote by μ. We pose the following optimization problem: Given sets A and B, of states, and nonnegative values λ_A and λ_B such that $\lambda_A + \lambda_B \leq 1$, select a context k, a gene i, and a function g_{ik} to replace f_{ik} such that

$$\varepsilon(A, B) = |\mu(A) - \lambda_A| + |\mu(B) - \lambda_B| \tag{4.5}$$

is minimum among all contexts, genes, and predictor functions. Additional constraints can be added, for instance, that g_{ik} has no more essential variables than f_{ik} or that g_{ik} and f_{ik} differ for at most some maximum number of states. Moreover, some other objective functions besides the one given can be used. In this scenario, we are allowing only the alteration of one predictor function. More generally, the PBNs may be allowed to differ at most at some preselected number of predictor functions.

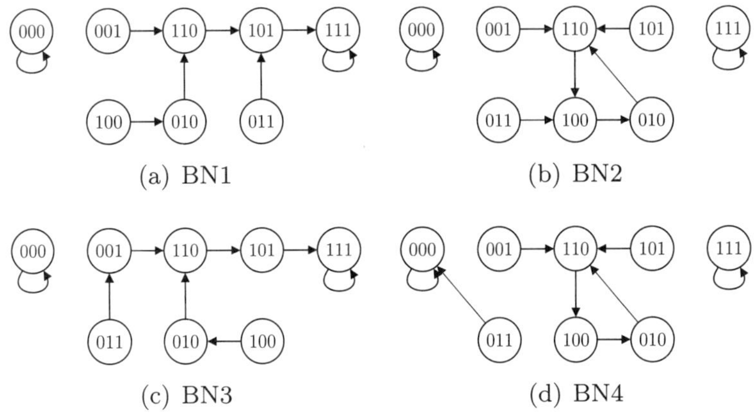

Figure 4.1. *State transition diagrams for the four constituent BNs.*

To illustrate these ideas, we consider an instantaneously random PBN with three genes (nodes) whose truth table and selection probabilities are shown in Example 2.1. Assuming no perturbations ($p = 0$), the state transition diagram is shown in Fig. 2.3. As can be seen from this figure, 000 and 111 are absorbing states. Figure 4.1 shows the four constituent BNs of the PBN. Note that 000 and 111 are attractors in all four networks, but two of the networks have an additional attractor cycle: $110 \rightarrow 100 \rightarrow 010 \rightarrow 110$. Table 4.1 gives the network functions, where $f_{ij} = f_j^{(i)}$ in Example 2.1.

Suppose that 111 corresponds to cell invasion (and rapid proliferation) and state 000 corresponds to quiescence. Assuming perturbation probability $p = 0.01$, the steady-state probabilities of 000 and 111 are 0.0752 and 0.7310, respectively. In the long run, the network will be in quiescence only 7% of the time and will be in proliferation 73% of the time. Suppose we wish to alter this imbalance and require the two steady-state probabilities to be approximately 0.4 for both 000 and 111. In the framework of the above optimization problem, $A = \{111\}$, $B = \{000\}$, $\pi(A) = 0.7310$, $\pi(B) = 0.0752$, and $\mu(A) = \mu(B) = 0.4$. Finally, suppose we are allowed to change only one predictor function. In the truth tables shown in Example 2.1, this corresponds to changing only one column while keeping the selection probabilities unchanged. Thus, there are 5 possible columns (predictors) and 256 possibilities for each.

All $5 \times 256 = 1280$ possible alterations have been generated and, for each, the steady-state probabilities $\mu(000)$ and $\mu(111)$ computed. The objective function to be minimized is

$$\varepsilon(\{000\}, \{111\}) = |\mu(000) - 0.4| + |\mu(111) - 0.4|.$$

Optimal structural intervention is achieved by altering the predictor f_{12} for gene 1 (column 2 in the truth table) and the truth table of the new predictor is $(00010101)^T$. This predictor achieves the steady-state probabilities $\mu(000) = 0.4068$ and $\mu(111) = 0.4128$, which are quite close to the desired probabilities.

We have used brute force to consider all allowable interventions, this approach being feasible owing to the small size of the network. In Shmulevich *et al.* (2002c), the use of genetic algorithms is explored.

Table 4.1. *The four components of the PBN.*

Network i	Network function f_i	Probability c_i
BN1	(f_{11}, f_{21}, f_{31})	0.3
BN2	(f_{11}, f_{21}, f_{32})	0.3
BN3	(f_{12}, f_{21}, f_{31})	0.2
BN4	(f_{12}, f_{21}, f_{32})	0.2

4.2.2 Method for constructive function-based intervention

We now provide a constructive algorithm for structural intervention (Xiao and Dougherty, 2007). Referring to Fig. 4.1, we approach intervention by considering the basins of attraction, BOA$\{x_1x_2x_3\}$ denoting the basin of attraction of $x_1x_2x_3$ and $|\text{BOA}\{x_1x_2x_3\}|$ denoting its size. In both BN1 and BN3 in Table 4.1, $|\text{BOA}\{000\}| = 1$ and $|\text{BOA}\{111\}| = 7$. According to Table 4.1, in BN2, $|\text{BOA}\{000\}| = |\text{BOA}\{111\}| = 1$ and in BN4, $|\text{BOA}\{000\}| = 2$ and $|\text{BOA}\{111\}| = 1$. Since, in the optimization problem just considered, the sum of the stationary probabilities of 000 and 111 must remain approximately unchanged, we need to increase the BOA of 000 and decrease the BOA of 111. Since function f_{21} is used in all four BNs in Table 4.1, its perturbation will result in changes in all BNs. On the other hand, if we perturb any of the other four functions, only two BNs will be affected. So we prefer not to perturb f_{21} unless necessary. Recalling Corollary 4.2, in this example, $n = k_i = 3$ and $2^{n-k_i} = 1$, so that perturbing one row of a function truth table will affect only one state transition.

First, consider BN1 and BN3 in Table 4.1. We can increase the BOA of 000 by finding any state whose successor state differs from 000 by only one bit (preferably on the 1st and 3rd bit), that state being 011 of BN3. However, a one-bit perturbation of function f_{31} to let $011 \to 000$ in BN3 increases $|\text{BOA}\{000\}|$ by only 1. This perturbation also affects BN1 ($011 \to 100$) but has no effect on BOA$\{000\}$ or BOA$\{111\}$. This approach does not provide much benefit.

Now consider BN2 and BN4 in Table 4.1. We can increase $|\text{BOA}\{000\}|$ by a one-bit function perturbation (preferably not f_{21}) by changing the state transition $110 \to 100$ to $110 \to 000$ in BN2 and BN4, i.e., perturb f_{11} or f_{12}. Another possibility is to change the state transition $100 \to 010$ to $100 \to 011$ in BN4, namely perturb f_{32} to $f_{32}^{(5)}$. Consider the following choices:

Choice 1: Perturb f_{11} to $f_{11}^{(7)}$, where $\mathbf{a}_7^{11} = 110$. As a consequence, in BN1, $110 \to 001$ and $|\text{BOA}\{111\}|$ decreases to 3; in BN2, $110 \to 000$ and $|\text{BOA}\{000\}|$ increases to 7. This is a candidate.

Choice 2: Perturb f_{12} to $f_{12}^{(7)}$. As a consequence, in BN3, $110 \to 001$ and $|\text{BOA}\{111\}|$ decreases to 2; in BN4, $110 \to 000$ and $|\text{BOA}\{000\}|$ increases to 7. This is a candidate.

Choice 3: Perturb f_{32} to $f_{32}^{(5)}$, where $\mathbf{a}_5^{32} = 100$. As a consequence, in BN4, $100 \to 011$ and $|\text{BOA}\{000\}|$ increases to 7; in BN2, $100 \to 011$, which does not affect BOA$\{111\}$ or BOA$\{000\}$. Since BN1 and BN3 adopt f_{31} rather than f_{32}, they are unaffected. Overall, this perturbation increases only $|\text{BOA}\{000\}|$ in BN4, without affecting BOA$\{111\}$. Thus, it cannot achieve the desired goal, which requires increasing $|\text{BOA}\{000\}|$ and decreasing $|\text{BOA}\{111\}|$. This choice is not a candidate.

Simulations show that Choice 1 yields steady-state probabilities $\mu(000) = 0.6$ and $\mu(111) = 0.25$. Choice 2 yields $\mu(000) = 0.43$ and $\mu(111) = 0.41$, which are close to the goal. Hence, we adopt Choice 2. Compare this solution to the exhaustive search in the preceding subsection, where the function f_{12} is perturbed from $0,1,1,0,0,1,1,1$ to $0,0,0,1,0,1,0,1$ (4-bit perturbation, on rows 2, 3, 4, and 7), and the resulting stationary probabilities are $\mu(000) = 0.4068$ and $\mu(111) = 0.4128$. The new solution is close to optimal with only a single-bit perturbation and it does not require an exhaustive search.

Using the preceding example as a guide, we have the following general procedure for structural intervention utilizing only a one-bit perturbation:

1. Formulate the goal in terms of the probability masses of the attractors. In the preceding example, the goal is to have equal steady-state probability mass for the target states 000 and 111, while the sum of the probability masses remains unchanged. Therefore, we must increase the probability mass of 000 and decrease that of 111.

2. Determine the priority of perturbation for the functions. For instance, in a PBN, if some of the functions are common in two or more constituent networks, it is preferred to perturb a function that affects as few networks as possible. In a BN, it is more favorable to perturb a function that results in fewer changes in the state transitions.

3. Plot the state transition diagrams. Analyze the BOAs of the target states by taking into consideration the BOA sizes and the selection probabilities. To increase the BOA of a target state $\mathbf{s}$ by a one bit perturbation, find a candidate state outside BOA$\{\mathbf{s}\}$ whose next state differs from a state within BOA$\{\mathbf{s}\}$ by only one bit. Find all such candidates. To decrease the BOA of a target state $\mathbf{s}$ by a one-bit perturbation, find a candidate state in BOA$\{\mathbf{s}\}$ whose next state differs from a state outside BOA$\{\mathbf{s}\}$ by only one bit. Notice that in a PBN, perturbation of one function can result in changes in two or more constituent networks.

4. List all the options of perturbation, from the highest priority to the lowest. For each option, draw a new state transition diagram and analyze the BOAs of the target states again. Throw away the options that are far from the goal. Notice that the steady-state probability mass of an attractor state is mainly affected by the size of its BOA but also has to do with the BOA structure and the perturbation probability p. Therefore, the BOA sizes can be used to estimate (but not determine) the probability masses of the target states.

5. For the remaining options, make computations either by simulation or by direct computation through Markov chain analysis, and pick the option closest to the goal. If two or more options are equally good, pick the one with the highest priority (e.g., perturbed function is present in the least number of BNs, or carries the least weight in a predefined importance rank, etc.).

4.2.3 Intervention in a WNT5A network

To pragmatically discuss optimal intervention, one must work with a network for which the goal of intervention can be quantitatively specified. Here we consider a gene regulatory network developed from data collected in a study of metastatic melanoma in which

Table 4.2. *Definitions of Boolean functions for the WNT5A BNp.*

Function	Input variables	Output
f_1	x_6	10
f_2	x_2, x_4, x_6	00010111
f_3	x_3, x_4, x_7	10101010
f_4	x_4, x_6, x_7	00001111
f_5	x_2, x_5, x_7	10101111
f_6	x_2, x_3, x_4	01110111
f_7	x_2, x_7	1101

the abundance of messenger RNA for the gene WNT5A was found to be a highly discriminating factor between cells with properties typically associated with high metastatic competence versus those with low metastatic competence (Bittner *et al.*, 2000). These findings were validated and expanded in a second study (Weeraratna *et al.*, 2002). In this study, experimentally increasing the levels of the Wnt5a protein secreted by a melanoma cell line via genetic engineering methods directly altered the metastatic competence of that cell as measured by the standard in vitro assays for metastasis. A further finding of interest in the current study was that an intervention that blocked the Wnt5a protein from activating its receptor, the use of an antibody that binds the Wnt5a protein, could substantially reduce Wnt5a's ability to induce a metastatic phenotype. This suggests a study of control based on interventions that alter the contribution of the WNT5A gene's action to biological regulation, since the available data suggest that disruption of this influence could reduce the chance of a melanoma metastasizing, a desirable outcome.

We consider the BN used in (Xiao and Dougherty, 2007) consisting of the seven genes WNT5A, pirin, S100P, RET1, MART1, HADHB, and STC2, which we label as $x_1, x_2, \ldots, x_7$, respectively. The regulatory rules are given in Table 4.2, where the ith bit of the output binary string represents the output value for the ith input predictor in binary representation. For instance, in the last row of Table 4.2, 1101 means that, for the input predictors 00, 01, 10, 11, f_7 has the output 1, 1, 0, and 1, respectively.

This BN has four attractors: 0101111, 0110110, 0111110, and 1000001. Their BOA sizes are 48, 4, 16, and 60, respectively. The last attractor, $\mathbf{s} = 1000001$, is undesirable, because the WNT5A gene is up-regulated. Moreover, $\mathbf{s}$ has a large BOA (consisting of nearly 50% of the total number of states). Our objective is to eliminate this attractor or minimize the size of its BOA if elimination is impossible. We will achieve this goal through function perturbations, with two constraints: (1) to perturb as few bits in the functions as possible; and (2) to affect as few state transitions as possible. For constraint (2), recall that the number of affected state transitions by one-bit perturbation is 2^{n-k_i}, and it is preferable to choose from the following functions for perturbation: f_2, f_3, f_4, f_5, and f_6. As a result, 16 state transitions (1/8 of the total) will be affected. To eliminate attractor $\mathbf{s}$, we will choose from among these five functions for one-bit perturbation according to the following steps.

(i) To eliminate attractor $\mathbf{s}$ by a one-bit perturbation of a function, we wish to change the state transition from $\mathbf{s} \to \mathbf{s}$ to $\mathbf{s} \to \mathbf{u}$, such that $\mathbf{u}$ differs from $\mathbf{s}$ by exactly one bit. Given the five preferred functions, the candidate states will be $\mathbf{u}_1 = 1100001$, $\mathbf{u}_2 = 1010001$, $\mathbf{u}_3 = 1001001$, $\mathbf{u}_4 = 1000101$, and $\mathbf{u}_5 = 1000011$.

(ii) Consider the five options.

Option 1 : Change the state transition to $\mathbf{s} \to \mathbf{u}_1$. Since $\text{In}_2(\mathbf{s}) = \{000\}$, we can achieve the change through the perturbation $f_2 \to f_2^{(1)}$. By applying the results of Section 4.1, we find that perturbation does not affect the other three attractors and that now $\mathbf{s} \to 1100001 \leftrightarrow 1000101$. A new attractor cycle is formed and its states have $x_1 = 1$, which is undesirable.

Option 2 : Change the state transition to $\mathbf{s} \to \mathbf{u}_2$ by $f_3 \to f_3^{(2)}$. As a result, the other three attractors remain the same, and a new attractor cycle, $1000011 \leftrightarrow 0010001$, will be formed, the first constituent state being undesirable ($x_1 = 1$).

Option 3 : Change the state transition to $\mathbf{s} \to \mathbf{u}_3$ by $f_4 \to f_4^{(2)}$. As a result, the other three attractors are still the same, while $\mathbf{s}$ disappears. This is a viable option.

Option 4 : Change the state transition to $\mathbf{s} \to \mathbf{u}_4$ by $f_5 \to f_5^{(2)}$. The result is a new undesirable attractor cycle, $\mathbf{s} \leftrightarrow 1000101$, while the other three attractors do not change.

Option 5 : Change the state transition to $\mathbf{s} \to \mathbf{u}_5$ by $f_6 \to f_6^{(2)}$. The result is a new attractor 0000011, while the other three attractors do not change. This is a viable option.

(iii) Options 3 and 5 can eliminate the undesirable attractor without creating a new undesirable attractor; however, Option 3 does not create any new attractor, while Option 5 creates a new attractor 0000011. By comparison, Option 3 has less impact on the original network, so it is selected.

By following the outlined procedure, we are able to eliminate the undesirable attractor associated with high competence of cellular metastasis with a one-bit function perturbation, with no other attractors affected or any new attractor created. Moreover, the perturbation is chosen so that a minimum number of state transitions will be affected.

4.3 Impact of Structural Perturbation on the Steady-State Distribution

In the previous section we considered the impact of function perturbation on state transitions and attractors by directly analyzing the effects of changes on the transitions and then considering these effects when the network is in an attractor. While such an approach has the benefit of directly relating logical changes with attractor states, it does not provide a direct computational framework for characterizing changes in the steady-state distribution arising from structural perturbation—where we implicitly assume the PBN has positive perturbation probability p and therefore possesses a steady-state distribution. In the next section we will turn to Markov chain perturbation theory to achieve a closed-form characterization for optimal structural intervention. In this section we will lay the necessary groundwork by providing the background perturbation theory and specializing it to PBNs.

We address the following question: What is the long-run effect if we alter the truth table governing a PBN, keeping $p > 0$ unchanged? Following function perturbation, we still have an irreducible Markov chain, but the probability transition matrix P and steady-state distribution π are changed to $\tilde{P}$ and $\tilde{\pi}$, respectively. Our task is to find the difference between $\tilde{\pi}$ and π based on P and $\tilde{P}$. One approach is to consider norm-type bounds of the form $||\tilde{\pi} - \pi|| \leq \kappa$. For these, we refer to Cho and Meyer (2001). Here, we provide an explicit expression of $\tilde{\pi}$ for the given P and $\tilde{P}$.

Following Qian and Dougherty (2008), we utilize the general perturbation theory for finite Markov chains (Cho and Meyer, 2001; Hunter, 1986, 1988, 1990, 2005, 2006; Kemeny and Snell, 1960; Schweitzer, 1968) and adapt the theory to PBNs, in particular, deriving $\tilde{\pi}$. Hunter (1986) derived results using generalized inverses. These were extended using special cases, including g-inverses, group inverses (Cho and Meyer, 2001), and the fundamental matrix (Schweitzer, 1968). In particular, a *g-inverse* of a matrix A is any matrix A^- such that $AA^-A = A$.

For the original and perturbed network, we have $\pi^T P = \pi^T$ and $\tilde{\pi}^T \tilde{P} = \tilde{\pi}^T$. Since the Markov chain for the original network is irreducible, the time-average transition matrix is defined by

$$P^{\infty} = \lim_{j \to \infty} [P + P^2 + \cdots + P^j]/j = e\pi^T, \tag{4.6}$$

where e is a column vector whose components are all unity. Letting $\tilde{P} = P + E$, the steady-state distribution change is

$$\tilde{\pi}^T - \pi^T = \tilde{\pi}^T \tilde{P} - \pi^T P = (\tilde{\pi} - \pi)^T P + \tilde{\pi}^T E, \tag{4.7}$$

implying that

$$(\tilde{\pi} - \pi)^T (I - P) = \tilde{\pi}^T E. \tag{4.8}$$

Theorem 4.8 (Hunter, 1986). *If P and π are the transition matrix and the stationary distribution, respectively, of a finite irreducible Markov chain, and t and u are any vectors such that $\pi^T t \neq 0$ and $u^T e \neq 0$, then*

(1) $I - P + tu^T$ is nonsingular;

(2) $[I - P + tu^T]^{-1}$ is a g-inverse of $I - P$;

(3) all g-inverses of $I - P$ are of the form $[I - P + tu^T]^{-1} + ef^T + g\pi^T$ for arbitrary f and g.

Several properties follow from the proof given in Hunter (1986):

$$(I - P)(I - P + tu^T)^{-1}(I - P) = I - P, \tag{4.9}$$

$$\pi^T = \frac{u^T [I - P + tu^T]^{-1}}{u^T [I - P + tu^T]^{-1} e}, \tag{4.10}$$

$$[I - P + tu^T]^{-1} t = \frac{e}{u^T e}. \tag{4.11}$$

We write the fundamental matrix of a PBN as

$$Z = [I - P + e\pi^T]^{-1}. \tag{4.12}$$

Since any PBN is an ergodic chain, the fundamental matrix always exists (Kemeny and Snell, 1960). After perturbation, the distance between $\tilde{P}$ and P can be measured by the *difference matrix*

$$U = (\tilde{P} - P)Z. \tag{4.13}$$

Theorem 4.9 (Schweitzer, 1968). *For a Markov chain containing a single irreducible set of states, $(I-U)^{-1}$ exists,*

$$(I-U)^{-1} = Z^{-1}(I - P^{\infty} + \tilde{P}^{\infty})\tilde{Z}, \tag{4.14}$$

and

$$\tilde{\pi}^T = \pi^T (I-U)^{-1}. \tag{4.15}$$

Although this theorem expresses $\tilde{\pi}$ in terms of π, since $(I-U)^{-1}$ depends on the perturbed fundamental matrix $\tilde{Z}$, whose computation is proportional to computing $\tilde{\pi}$, as it stands, it does not fully accomplish our aim.

4.3.1 Rank-one perturbations

We first consider a rank-one perturbation, in which case the perturbed Markov chain has the transition matrix

$$\tilde{P} = P + ab^T, \tag{4.16}$$

where a,b are two arbitrary vectors satisfying $b^T e = 0$, and ab^T represents a rank-one perturbation to the original Markov chain P. From Eq. (4.10), we obtain

$$\tilde{\pi}^T = \frac{\tilde{u}^T [I - \tilde{P} + \tilde{t}\tilde{u}^T]^{-1}}{\tilde{u}^T [I - \tilde{P} + \tilde{t}\tilde{u}^T]^{-1} e}. \tag{4.17}$$

The next theorem provides an expression for $\tilde{\pi}$ directly in terms of P.

Theorem 4.10 (Hunter, 2005). *If t and u are any vectors such that $\pi^T t \neq 0$ and $u^T e \neq 0$, then*

$$\tilde{\pi}^T = \frac{(\pi^T a)\beta^T + (1 - \beta^T a)\pi^T}{(\pi^T a)(\beta^T e) + 1 - \beta^T a}, \tag{4.18}$$

where

$$\beta^T = b^T [I - P + tu^T]^{-1}. \tag{4.19}$$

Proof. If $\tilde{t}\tilde{u}^T - tu^T = ab^T$, then

$$(I - \tilde{P} + \tilde{t}\tilde{u}^T)^{-1} = (I - P + tu^T)^{-1} \tag{4.20}$$

and the g-inverses of the original and perturbed Markov chains are equal, so that we need not compute the g-inverse after perturbation. This condition is satisfied if we let $\tilde{t} = t = a$ and $\tilde{u} = (b+u)$. Plugging this into Eq. (4.17) yields

$$\tilde{\pi}^T = \frac{(b+u)^T [I - P + au^T]^{-1}}{(b+u)^T [I - P + au^T]^{-1} e}. \tag{4.21}$$

To proceed further, we use the Sherman–Morrison identity:

$$(X+cd^T)^{-1} = X^{-1} - \frac{(X^{-1}c)(d^T X^{-1})}{1+d^T X^{-1}c}, \tag{4.22}$$

where c,d are two arbitrary vectors. Now, as we require an arbitrary t, where $\pi^T t \neq 0$, instead of a to compute the g-inverse of $I-P$ for the original transition matrix, based on Eq. (4.22) we relate $(I-P+au^T)^{-1}$ and $(I-P+tu^T)^{-1}$ according to

$$\begin{aligned}(I-P+au^T)^{-1} &= (I-P+tu^T+(a-t)u^T)^{-1} \\ &= (I-P+tu^T)^{-1}\left[I - \frac{(a-t)u^T(I-P+tu^T)^{-1}}{1+u^T(I-P+tu^T)^{-1}(a-t)}\right].\end{aligned} \tag{4.23}$$

Since $\pi^T(I-P+tu^T) = \pi^T t u^T$, we have

$$u^T(I-P+tu^T)^{-1} = \frac{\pi^T}{\pi^T t}. \tag{4.24}$$

Therefore,

$$\begin{aligned}(I-P+au^T)^{-1} &= (I-P+tu^T)^{-1}\left[I - \frac{\frac{(a-t)\pi^T}{\pi^T t}}{1+\frac{\pi^T(a-t)}{\pi^T t}}\right] \\ &= (I-P+tu^T)^{-1}\left[I - \frac{(a-t)\pi^T}{\pi^T a}\right] \\ &= (I-P+tu^T)^{-1}\left[I - \frac{a\pi^T}{\pi^T a}\right] + \frac{(I-P+tu^T)^{-1}t\pi^T}{\pi^T a}.\end{aligned} \tag{4.25}$$

From Eq. (4.11),

$$(I-P+au^T)^{-1} = (I-P+tu^T)^{-1}\left[I - \frac{a\pi^T}{\pi^T a}\right] + \frac{e\pi^T}{\pi^T a u^T e}. \tag{4.26}$$

Plugging this into Eq. (4.21) gives

$$\begin{aligned}\tilde{\pi}^T &= \frac{(b+u)^T[I-P+au^T]^{-1}}{(b+u)^T[I-P+au^T]^{-1}e} \\ &= \frac{(b+u)^T[(I-P+tu^T)^{-1}(I-\frac{a\pi^T}{\pi^T a}) + \frac{e\pi^T}{\pi^T a u^T e}]}{(b+u)^T[(I-P+tu^T)^{-1}(I-\frac{a\pi^T}{\pi^T a}) + \frac{e\pi^T}{\pi^T a u^T e}]e}.\end{aligned} \tag{4.27}$$

Here,

$$\begin{aligned} & u^T\left[(I-P+tu^T)^{-1}\left(I-\frac{a\pi^T}{\pi^T a}\right) + \frac{e\pi^T}{\pi^T a u^T e}\right] \\ \overset{u^T(I-P+tu^T)^{-1}=\frac{\pi^T}{\pi^T t}}{=} & \frac{\pi^T}{\pi^T t}\left(I - \frac{a\pi^T}{\pi^T a}\right) + \frac{u^T e\pi^T}{\pi^T a u^T e} \\ = & \frac{\pi^T}{\pi^T t} - \frac{\pi^T a\pi^T}{\pi^T t\pi^T a} + \frac{\pi^T}{\pi^T a} \\ = & \frac{\pi^T}{\pi^T a}.\end{aligned} \tag{4.28}$$

And

$$b^T\left[(I-P+tu^T)^{-1}\left(I-\frac{a\pi^T}{\pi^T a}\right)+\frac{e\pi^T}{\pi^T a u^T e}\right] \tag{4.29}$$

$$= b^T(I-P+tu^T)^{-1} - b^T(I-P+tu^T)^{-1}\frac{a\pi^T}{\pi^T a} + \frac{b^T e\pi^T}{\pi^T a u^T e}$$

$$\overset{b^T e=0}{=} b^T(I-P+tu^T)^{-1} - b^T(I-P+tu^T)^{-1}\frac{a\pi^T}{\pi^T a}.$$

Letting $\beta^T = b^T(I-P+tu^T)^{-1}$ yields

$$b^T\left[(I-P+tu^T)^{-1}\left(I-\frac{a\pi^T}{\pi^T a}\right)+\frac{e\pi^T}{\pi^T a u^T e}\right] = \beta^T - \frac{\beta^T a\pi^T}{\pi^T a}. \tag{4.30}$$

Integrating all of these results gives

$$\tilde{\pi}^T = \frac{(b+u)^T[I-P+au^T]^{-1}}{(b+u)^T[I-P+au^T]^{-1}e} \tag{4.31}$$

$$= \frac{\beta^T - \frac{\beta^T a\pi^T}{\pi^T a} + \frac{\pi^T}{\pi^T a}}{(\beta^T - \frac{\beta^T a\pi^T}{\pi^T a} + \frac{\pi^T}{\pi^T a})e}$$

$$= \frac{\pi^T + \pi^T a\beta^T - \beta^T a\pi^T}{(\pi^T + \pi^T a\beta^T - \beta^T a\pi^T)e}$$

$$= \frac{(\pi^T a)\beta^T + (1-\beta^T a)\pi^T}{(\pi^T a)(\beta^T e) + 1 - \beta^T a},$$

and the theorem is proved. □

If we let $t = e$ in Eq. (4.18), then $\beta^T e = 0$ according to Eq. (4.11) and we obtain

$$\tilde{\pi} = \pi + \frac{\pi^T a}{1-\beta^T a}\beta. \tag{4.32}$$

With $u = \pi$, $\beta^T = b^T Z$, and therefore the steady-state distribution of the rank-one perturbation is expressed in terms of π and Z, the steady-state distribution and fundamental matrix of the original network. The next theorem states this expression explicitly and also provides expressions for the fundamental matrix $\tilde{Z}$ and a g-inverse for $I - \tilde{P}$ in terms of π and Z.

Theorem 4.11 (Hunter, 2005; Qian and Dougherty, 2008). *The steady-state distribution, fundamental matrix, and a g-inverse for $I - \tilde{P}$ for the rank-one perturbed network are given by*

$$\tilde{\pi}^T = \pi^T + \frac{\pi^T a}{1-b^T Za}b^T Z, \tag{4.33}$$

$$\tilde{Z} = \left[I - \frac{(\pi^T a)eb^T Z}{1-b^T Za}\right]\left[Z + \frac{Zab^T Z}{1-b^T Za}\right], \tag{4.34}$$

$$[I-\tilde{P}+a(\pi+b)^T]^{-1} = Z\left[I+(e-a)\frac{\pi^T}{\pi^T a}\right]. \tag{4.35}$$

Proof. We start with the derivation of Eq. (4.33). From Eqs. (4.19) and (4.11), choosing $t = e$, we obtain

$$\beta^T e = \beta^T t = b^T (I - P + tu^T)^{-1} t = \frac{b^T e}{u^T e} = 0. \tag{4.36}$$

Since $E = ab^T$ is the perturbation matrix, it must be that $b^T e = 0$ to be sure that $P + E$ is a transition matrix. Thus, $\beta^T e = 0$. Substituting this into Eq. (4.18) yields

$$\tilde{\pi}^T = \pi^T + \frac{\pi^T a}{1 - \beta^T a} \beta^T. \tag{4.37}$$

Further, choosing $u = \pi$ gives

$$(I - P + tu^T)^{-1} = (I - P + e\pi^T)^{-1} = Z \tag{4.38}$$

and $\beta^T = b^T Z$. Substituting β^T by $b^T Z$ in Eq. (4.37) results in Eq. (4.33).

For the derivation of the updating scheme (4.34) for the fundamental matrix, we need to use $(I - U)^{-1}$ in Theorem 4.9. For the rank-one perturbation, the Sherman–Morrison identity yields

$$(I - U)^{-1} = [I - (\tilde{P} - P)Z]^{-1} = [I - ab^T Z]^{-1} = I + \frac{ab^T Z}{1 - b^T Za}. \tag{4.39}$$

Letting $H = (I - U)^{-1}$, by Theorem 4.9,

$$ZH = (I - e\pi^T + e\pi^T H)\tilde{Z}. \tag{4.40}$$

Thus,

$$\begin{aligned}
\tilde{Z} &= (I - e\pi^T + e\pi^T H)^{-1} ZH = [I - e\pi^T (I - H)]^{-1} ZH \\
&\overset{SM\ Identity}{=} \left(I - \frac{e\pi^T (H - I)}{1 + \pi^T (H - I)e} \right) ZH \\
&= \left(I - \frac{e\pi^T \frac{ab^T Z}{1 - b^T Za}}{1 + \pi^T \frac{ab^T Z}{1 - b^T Za} e} \right) ZH \\
&\overset{b^T Ze = \beta^T e = 0}{=} \left(I - \frac{(\pi^T a) e b^T Z}{1 - b^T Za} \right) \left(Z + \frac{Zab^T Z}{1 - b^T Za} \right),
\end{aligned} \tag{4.41}$$

which is Eq. (4.34).

We now consider the g-inverses of $I - P$ and $I - \tilde{P}$. Letting $t = e$ and $u = \pi$ gives Eq. (4.38). For $(I - \tilde{P} + \tilde{t}\tilde{u}^T)^{-1}$, we have the following relationship with $\tilde{t} = a$ and $\tilde{u} = u + b$ from the derivation of Theorem 4.10:

$$I - \tilde{P} + \tilde{t}\tilde{u}^T = I - P - ab^T + \tilde{t}(u + b)^T = I - P + au^T = I - P + tu^T + (\tilde{t} - t)u^T. \tag{4.42}$$

Taking $t = e$ and $u = \pi$, from the Sherman–Morrison identity we have

$$(I - \tilde{P} + \tilde{t}\tilde{u}^T)^{-1} = Z\left[I + (t - \tilde{t}) \frac{u^T Z}{1 + u^T Z(t - \tilde{t})} \right] = Z\left[I + (t - \tilde{t}) \frac{\pi^T}{\pi^T t + \pi^T (t - \tilde{t})} \right], \tag{4.43}$$

implying

$$[I-\tilde{P}+a(\pi+b)^T]^{-1}=Z\left[I+(e-a)\frac{\pi^T}{\pi^T a}\right], \tag{4.44}$$

which gives Eq. (4.35) and completes the proof. □

4.3.2 Perturbation in the same row

An important special case occurs when the transition mechanisms before and after perturbation differ only in one state, say the kth state. Then $E=e_k b^T$ has nonzero values only in its kth row, where e_k is the elementary vector with a 1 in the kth position and 0's elsewhere. Substituting this into Eq. (4.33) yields

$$\tilde{\pi}^T=\pi^T+\frac{\pi^T e_k}{1-b^T Z e_k}b^T Z=\pi^T+\frac{\pi_k}{1-\beta_k}\beta \tag{4.45}$$

(Schweitzer, 1968). For the ith state,

$$\tilde{\pi}_i=\pi_i+\frac{\pi_k\beta_i}{1-\beta_k}. \tag{4.46}$$

If we let $b=\Delta p_k$ be the difference vector for the kth rows of the transition matrices, then

$$\beta_i=\sum_j \Delta p_k^j z_{ji}, \tag{4.47}$$

where $\sum_j \Delta p_k^j=0$. Using the relationship between the mean first passage time and the fundamental matrix, $z_{ij}=z_{jj}-\pi_j m_{ij}$, yields

$$\begin{aligned}\beta_i&=\sum_{j\neq i}\Delta p_k^j(z_{ii}-\pi_i m_{ji})+\Delta p_k^i z_{ii}\\&=z_{ii}\sum_j \Delta p_k^j-\pi_i\sum_{j\neq i}\Delta p_k^j m_{ji}\\&=-\pi_i\sum_{j\neq i}\Delta p_k^j m_{ji}.\end{aligned} \tag{4.48}$$

Substituting this into Eq. (4.46) yields

$$\tilde{\pi}_i=\pi_i-\frac{\pi_k\pi_i\sum_{j\neq i}\Delta p_k^j m_{ji}}{1+\pi_k\sum_{j\neq i}\Delta p_k^j m_{jk}}. \tag{4.49}$$

This expression does not depend on any parameters of the perturbed Markov chain and therefore gives a desirable solution.

For the most special case, where Δp_k has only two nonzero components for u and v states, $\tilde{p}_{ku}=p_{ku}-\epsilon$ and $\tilde{p}_{kv}=p_{kv}+\epsilon$, where ϵ is the only perturbation. In this case,

$$\tilde{\pi}_i=\pi_i+\frac{\pi_k\epsilon(z_{vi}-z_{ui})}{1-\epsilon(z_{vk}-z_{uk})}=\pi_i+\frac{\pi_i\pi_k\epsilon(m_{ui}-m_{vi})}{1-\pi_k\epsilon(m_{uk}-m_{vk})}. \tag{4.50}$$

4.3.3 Extension to multiple rows by iterative computation

To extend the results for these special cases to arbitrary types of perturbations, we can compute the steady-state distributions of arbitrarily perturbed Markov chains in an iterative fashion. Let P^w, Z^w, and π^w be the transition matrix, fundamental matrix, and steady-state distribution, respectively, after w single-row (or more general rank-one) perturbations. Let

$$P^{w+1} = P^w + e_{w+1}b_{w+1}^T \tag{4.51}$$

be the transition matrix with the next single-row perturbation. We can apply Eqs. (4.33) and (4.34), or Eq. (4.35), to update the new steady-state distribution and fundamental matrix iteratively. Both approaches are discussed in Qian and Dougherty (2008). Here we restrict ourselves to the simpler approach, using Eq. (4.35), and explicitly write out the expression for two-row perturbations.

$\hat{P}$ is defined as the new transition matrix with differences in two rows and $\tilde{P}$ is the same as before, with a difference in the kth row. $\hat{P} - \tilde{P}$ has nonzero values in its lth row. By Eq. (4.35), with $\beta^T = \Delta P_k^T Z$ and $\beta^{'T} = \Delta p_l^T Z$

$$\begin{aligned} \hat{\beta}^T &= \Delta p_l^T [I - \tilde{P} + \tilde{\pi}\tilde{u}^T]^{-1} \\ &= \Delta p_l^T Z \left[I + (e - e_k)\frac{\pi^T}{\pi_k} \right] \\ &= \beta^{'T} - \frac{\beta_k^{'}}{\pi_k}\pi^T, \end{aligned} \tag{4.52}$$

since $\beta^{'T} e = 0$ according to Eq. (4.11). Plugging this into Eq. (4.18) yields

$$\begin{aligned} \hat{\pi} &= \frac{\tilde{\pi}_l(\beta' - \frac{\beta_k^{'}\pi}{\pi_k}) + (1 - \beta_l^{'} + \frac{\beta_k^{'}\pi_l}{\pi_k})\tilde{\pi}}{\tilde{\pi}_l(-\frac{\beta_k^{'}}{\pi_k}) + 1 - \beta_l^{'} + \frac{\beta_k^{'}\pi_l}{\pi_k}} \\ &= \frac{\pi_l\beta' + \frac{\pi_k\beta_l}{1-\beta_k}\beta' - \frac{\beta_k^{'}\beta_l}{1-\beta_k}\pi + \pi + \frac{\pi_k}{1-\beta_k}\beta - \beta_l^{'}\pi - \frac{\beta_l^{'}\pi_k}{1-\beta_k}\beta + \frac{\beta_k^{'}\pi_l}{1-\beta_k}\beta}{1 - \beta_l^{'} - \frac{\beta_k^{'}\beta_l}{1-\beta_k}} \\ &= \pi + \frac{\pi_k - \beta_l^{'}\pi_k + \beta_k^{'}\pi_l}{(1-\beta_k)(1-\beta_l^{'}) - \beta_k^{'}\beta_l}\beta + \frac{\pi_l - \beta_k\pi_l + \beta_l\pi_k}{(1-\beta_k)(1-\beta_l^{'}) - \beta_k^{'}\beta_l}\beta'. \end{aligned} \tag{4.53}$$

This iterative analytic procedure for computing the perturbed steady-state distribution confronts increasing complexity as the number of perturbations increases.

4.3.4 Application to PBNs

Having expressed the relevant perturbation theory in the framework of Markov chains, we now particularize it to PBNs. We first consider a single BN with perturbation probability $p > 0$, the simplest irreducible PBN. In this case, a single function perturbation is a flip for one target gene at the response side for the predictor state **k** in the truth table. If momentarily we assume $p = 0$, then each row in the transition matrix corresponds to a single input state **i** and the row consists of all 0's except for a 1 in the column of the output state **f**(**i**). A one-bit perturbation for input state **k** means that there is a network function transformation $\mathbf{f} \to \tilde{\mathbf{f}}$ with $\mathbf{v} = \tilde{\mathbf{f}}(\mathbf{k}) \neq \mathbf{f}(\mathbf{k}) = \mathbf{u}$ and $\tilde{\mathbf{f}}(\mathbf{i}) = \mathbf{f}(\mathbf{i})$ for all $\mathbf{i} \neq \mathbf{k}$. Hence, there is now (with $\tilde{\mathbf{f}}$) a 1 in the column of the output state **v** and a 0 in the column of state **u**, whereas originally (with **f**) there was a 0 in the column of the output state **v** and a 1 in the column of state **u**, and all other entries in the matrix are unchanged. Relative to the network, this changes the transition matrix P only at p_{ku} and p_{kv}, where we use the italic font, instead of bold, to represent the relative positions in matrices when we use them as subscripts for matrix entries. Thus, $\tilde{p}_{ku} = p_{ku} - \epsilon$ and $\tilde{p}_{kv} = p_{kv} + \epsilon$, where ϵ is decided by the change to the truth table and the perturbation probability p: $\epsilon = (1-p)^n$. With Eq. (4.46), we have

$$\begin{aligned}\tilde{\pi}_i &= \pi_i + \frac{\epsilon \pi_k (z_{vi} - z_{ui})}{1 - \epsilon(z_{vk} - z_{uk})} \\ &= \pi_i + \frac{(1-p)^n \pi_k (z_{vi} - z_{ui})}{1 - (1-p)^n (z_{vk} - z_{uk})}.\end{aligned} \tag{4.54}$$

Applying the relationship between the mean first passage time and the fundamental matrix, $z_{ij} = z_{jj} - \pi_j m_{ij}$ (Kemeny and Snell, 1960), yields

$$\tilde{\pi}_i - \pi_i = \epsilon \pi_i \pi_k \frac{(m_{ui} - m_{vi}) - (m_{uk} - m_{vk})}{1 - \epsilon \pi_k (m_{uk} - m_{vk})}. \tag{4.55}$$

Hence, the steady-state perturbation depends on whether the states **k** and **i** are absorbing or transient. If **u** and **v** are either both transient states or attractor states belonging to the same attractor cycle, then the steady-state perturbation should be small. More generally, properties of the steady-state perturbation are revealed by the mean first passage times (which are related to state status in the Markov chain).

This framework can be extended to more sophisticated function perturbations by applying the results for either rank-one perturbation in Eq. (4.33) or the iterative updating scheme for multirow perturbations. For example, if we now consider function perturbation to a single regulatory function f_i for gene i having K_i genes in its regulatory set, then there will be changes to 2^{n-K_i} rows of the transition matrix P. Although the results for f_i change only the value for one target gene, the output states could change between multiple states. In this case, the iterative scheme has to be implemented. But when the output states change only from **u** to **v**, then this is actually a rank-one perturbation with

$$E = e_{K_i} \xi^T . \xi^T = [0, \ldots, 0, \underbrace{-\epsilon}_{u}, 0, \ldots, 0, \underbrace{\epsilon}_{v}, 0, \ldots, 0], \tag{4.56}$$

where e_{K_i} is a vector with 2^{n-K_i} 1's for the corresponding entries and 0's for the others. Thus, the perturbed steady-state distribution can be written as

$$\tilde{\pi}^T = \pi^T + \frac{\pi^T e_{K_i}}{1 - \xi^T Z e_{K_i}} \xi^T Z. \tag{4.57}$$

Now let us consider an instantaneously random PBN consisting of a number of constituent networks. Again momentarily assuming $p = 0$, each row in the transition matrix corresponds to a single input state **i** and that row contains a number of positive values summing to 1, with the remaining values being 0. The value in column $\mathbf{j}$ is the sum of all context-selection probabilities c_r such that $\mathbf{f}_r(\mathbf{i}) = \mathbf{j}$, where $\mathbf{f}_r$ is the network function for the rth constituent BN. A one-bit perturbation for input state $\mathbf{k}$ corresponds to a change in one network function $\mathbf{f}_r$, so that exactly two columns are changed. This means that there is a network function transformation $\mathbf{f}_r \to \tilde{\mathbf{f}}_r$ with $\mathbf{v} = \tilde{\mathbf{f}}_r(\mathbf{k}) \neq \mathbf{f}_r(\mathbf{k}) = \mathbf{u}$ and $\tilde{\mathbf{f}}_r(\mathbf{i}) = \mathbf{f}_r(\mathbf{i})$ for all $\mathbf{i} \neq \mathbf{k}$. There is a gain of $c_r(1-p)^n$ in the column of the output state $\mathbf{v}$ and a loss of $c_r(1-p)^n$ in the column of state $\mathbf{u}$, and all other entries in the matrix are unchanged. Hence, we can proceed in the same manner as with a one-bit perturbation for a single BN with perturbation using Eq. (4.46).

If we are not limited to changing a single constituent network function for input state $\mathbf{k}$, then the situation becomes more complicated because there could be several changes in the transition matrix on the row corresponding to state $\mathbf{k}$. For instance, if two network functions change, say for the constituent networks s and r, then there are two transformations, $\mathbf{f}_r \to \tilde{\mathbf{f}}_r$ with $\tilde{\mathbf{f}}_r(\mathbf{k}) \neq \mathbf{f}_r(\mathbf{k})$ and $\mathbf{f}_s \to \tilde{\mathbf{f}}_s$ with $\tilde{\mathbf{f}}_s(\mathbf{k}) \neq \mathbf{f}_s(\mathbf{k})$. Consequently, there could be up to four changes for input state $\mathbf{k}$. In general this means that Δp_k can have more than two nonzero entries. Nevertheless, we can still directly compute the perturbed steady-state distribution with Eq. (4.45).

For a context-sensitive PBN, each state of the Markov chain is of the form $(r, \mathbf{i})$, where r denotes the rth constituent network and $\mathbf{i}$ the gene state vector. If there are m contexts and n genes, then the size of the transition matrix is $m2^n \times m2^n$. A one-bit perturbation for input state $(s, \mathbf{k})$ corresponds to a change in network function $\mathbf{f}_s$, so that exactly two columns are changed, both being columns of the matrix corresponding to the sth constituent network. Hence, we can proceed in the same manner as with a rank-one perturbation for a BN with perturbation using Eq. (4.33). To change d constituent network functions for input gene state $\mathbf{k}$ means one-bit perturbations to d states, $(s_1, \mathbf{k}), (s_2, \mathbf{k}), \ldots, (s_d, \mathbf{k})$, in the Markov chain. Relative to the matrix this means multiple rank-one perturbations caused by d perturbations. This can be solved analogously as in the multirow situation with a single BN with perturbation.

4.4 Structural Intervention via Markov Chain Perturbation Theory

In this section we treat optimal structural intervention by using the perturbation theory to achieve the intervention objective of keeping cells away from certain states by changing the long-run dynamics of the network via alteration of the rule structure (Qian and Dougherty, 2008). According to Eqs. (4.46) and (4.47), given a set $\mathcal{U}$ of undesirable states, a general optimization problem can be posed as a minimization:

$$\arg\min_{\Delta p_k} \sum_{i \in \mathcal{U}} \tilde{\pi}_i = \arg\min_{\Delta p_k} \sum_{i \in \mathcal{U}} \left(\pi_i + \frac{\pi_k \sum_j \Delta p_k^j z_{ji}}{1 - \sum_j \Delta p_k^j z_{jk}} \right). \tag{4.58}$$

This is equivalent to the minimization

$$\arg\min_{\Delta p_k} \sum_{i \in \mathcal{U}} \frac{\pi_k \sum_j \Delta p_k^j z_{ji}}{1 - \sum_j \Delta p_k^j z_{jk}}. \tag{4.59}$$

Analogously, we can also maximize the probability mass of a set $\mathcal{D}$ of desirable states:

$$\arg\max_{\Delta p_k} \sum_{i\in\mathcal{D}} \frac{\pi_k \sum_j \Delta p_k^j z_{ji}}{1-\sum_j \Delta p_k^j z_{jk}}. \tag{4.60}$$

$\mathcal{U}$ and $\mathcal{D}$ need not form a partition of the state space because there may be "don't care" states.

As stated, the intervention consists of a one-bit change in the regulatory rules. For a more general perturbation, we can apply Eqs. (4.33), (4.34), and (4.35). The extension to multiple changes by applying the iterative expression will be more computationally expensive.

Equations (4.59) and (4.60) are not the only ways to formulate optimal structural intervention. For instance, one might constrain the optimization by limiting the amount of mass in any given state, which in effect would mean that the basins of attraction must not be too large. One might also put limits on the change in mass $|\tilde{\pi}_i - \pi_i|$ of some states, so that there is no overconcentration of mass in certain states of the altered network.

4.4.1 A WNT5A BN

Here, we consider a 7-gene BN discussed in Section 4.2.3 but with perturbation probability $p = 0.001$. If we are interested only in steady-state mass, then the objective function is chosen to minimize that mass for the states with $x_1 = 1$ (WNT5A expressed); namely, we minimize $\sum_{x_1=1} \tilde{\pi}(\mathbf{x})$. This leads to an obvious intervention strategy: perturb the output from 1 to 0 for the predictor $x_6 = 0$ of f_1, which is a canalizing function. For this perturbation, $\sum_{x_1=1} \tilde{\pi}(\mathbf{x}) = 0.0010$. Figure 4.2(a) shows the original steady-state distribution and Fig. 4.2(b) shows the steady-state distribution resulting from the structural intervention.

While the preceding strategy appears obvious, it has the effect of significantly altering the attractor structure. The original network has four attractors possessing the following steady-state probabilities: $\pi(0101111) = 0.3645$, $\pi(0110110) = 0.0260$, $\pi(0111110) = 0.1301$, and $\pi(1000001) = 0.4662$. Since we desire the down-regulation of WNT5A, the attractor 1000001 is undesirable. This is accomplished by the intervention, with $\tilde{\pi}(1000001) = 0.0005$. But the intervention introduces a new attractor, 0000001, with $\tilde{\pi}(0000001) = 0.4677$. If we lack knowledge of this phenotype, although the overall result is beneficial from the standpoint of WNT5A, we may not want such a large mass associated with a little known phenotype.

Suppose we take a different approach. Since we want to down-regulate WNT5A, the attractor 1000001 is undesirable and we want to reduce its steady-state probability, but, in addition, suppose we do not want to introduce other attractors that might lead to unexpected cellular behavior. Then we can maximize the objective function

$$\varepsilon = \tilde{\pi}(0101111) + \tilde{\pi}(0110110) + \tilde{\pi}(0111110).$$

The optimal function perturbation in this case is to perturb the output from 0 to 1 for $f_4(011)$. The perturbed steady-state probabilities are $\tilde{\pi}(0101111) = 0.6921$, $\tilde{\pi}(0110110) = 0.0495$, $\tilde{\pi}(0111110) = 0.2472$, and $\tilde{\pi}(1000001) = 0.0007$. The steady-state mass for the rest of the states is 0.0105, which appears to be a safe intervening strategy. Figure 4.2(c) illustrates that the steady-state distribution shifts from undesirable to desirable states after

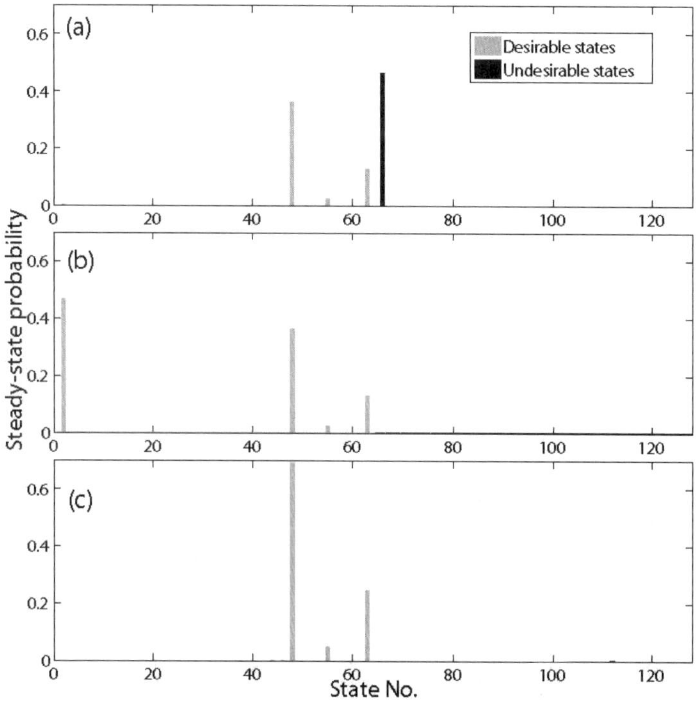

Figure 4.2. *Steady-state distribution shifts for BNps with* $p = 0.001$*:* (a) *Original steady-state distributions;* (b) *after optimal perturbation to* f_1*;* (c) *after optimal perturbation to* f_4.

this one-bit function perturbation. This shift is permanent since the perturbation changes the underlying structure of the BN.

Let us return to the original strategy, recognizing that in practice one requires a physical method to structurally intervene. To wit, not all functions can be controlled practically. Using the closed-form analytic solution provided by the Markovian perturbation theory, we can check all possible interventions and then choose the one that both minimizes the objective function and is practically implementable. For the WNT5A network, we find the optimal one-bit control policy for each f_i and list them in Table 4.3. The table shows that the perturbations to f_1, f_4, and f_6 significantly reduce the steady-state mass of the undesirable states. In practice, we can choose the biologically admissible intervening strategy among them. Intervention with f_1 greatly shifts the mass away from undesirable states, but this overall optimal solution introduces a new attractor, 0000001. The second best strategy, which is to perturb f_4, does not introduce new attractors and still moves the steady-state mass towards desirable states.

We can consider many potential intervention strategies because the analytic results lead to an efficient search for the optimal function perturbation. In addition, it is easy to find optimal practical strategies by looking into the subset of biologically admissible intervening strategies. Another point to notice is that the entire process of finding an optimal function perturbation requires computation of a single matrix inverse—the fundamental

Table 4.3. *Optimal one-bit function perturbations for all f_is in the WNT5A BN. The last column shows the stationary mass after the optimal perturbation when $p = 0.001$.*

Function	Input variables	Input values for optimal perturbation	Optimal perturbation	$\sum_{x_1=1} \tilde{\pi}(\mathbf{x})$
f_1	x_6	0	$1 \to 0$	0.0010
f_2	x_2, x_4, x_6	100	$0 \to 1$	0.3777
f_3	x_3, x_4, x_7	101	$0 \to 1$	0.2084
f_4	x_4, x_6, x_7	011	$0 \to 1$	0.0035
f_5	x_2, x_5, x_7	100	$1 \to 0$	0.4728
f_6	x_2, x_3, x_4	000	$0 \to 1$	0.0038
f_7	x_2, x_7	01	$1 \to 0$	0.2305

matrix of the original network. Thus, the efficiency of finding an optimal strategy in a one-bit function perturbation is determined by computation of a matrix inverse.

4.5 Long-Run Sensitivity

Classical sensitivity measures are based on temporary gene perturbation and measure transient effects, with research being focused on ensemble dynamical behavior for networks possessing different properties, such as connectivity, bias, and function classes. If we are concerned with network changes that alter the steady-state distribution, then we need to consider sensitivity relative to such changes. In this sense, *long-run sensitivity* is defined in terms of the difference between the steady-state distributions before and after perturbation (Qian and Dougherty, 2009a). Its particular form depends upon the change being considered—that is, change to the network structure or to other network parameters. Measuring long-run sensitivity facilitates analysis of network robustness and controllability. Therefore, it can be used to provide guidelines for network inference and control problems. To efficiently compute the long-run sensitivity with respect to structural perturbation, the change of regulatory functions, we can apply the previously studied perturbation theory to derive exact quantitative changes of steady-state distributions.

In general, both the probability transition matrix P and the steady-state distribution π for PBNs are functions of the network regulatory functions $\mathbf{F}$ and the involved probabilistic parameters p, q and selection probability vector $\mathbf{c}$. Hence, we rewrite the steady-state relationship as

$$\pi^T(\mathbf{F}, p, q, \mathbf{c}) = \pi^T(\mathbf{F}, p, q, \mathbf{c}) P(\mathbf{F}, p, q, \mathbf{c}). \tag{4.61}$$

The perturbations we can apply to network functions and probabilistic parameters have different properties. For the parameters involved, both P and π are continuous functions of $p, q, \mathbf{c}$. Hence, we can define long-run sensitivity as the partial derivative of the steady-state distribution with respect to any one of them. Since the change to regulatory rules of networks is discrete (binary in our examples), we will have a different long-run sensitivity defined by the change of steady-state distribution.

4.5.1 Long-run sensitivity with respect to probabilistic parameters

We first consider the long-run sensitivity of PBNs with respect to the parameters p, q, and $\mathbf{c}$. For these continuous parameters, the long-run sensitivity is defined via the partial derivatives of the steady-state distribution π. To obtain these derivatives, using Eq. (4.61), we take derivatives on both sides with respect to some parameter ξ:

$$\frac{\partial \pi^T(\mathbf{F},p,q,\mathbf{c})}{\partial \xi} = \frac{\partial \pi^T(\mathbf{F},p,q,\mathbf{c})}{\partial \xi} P(\mathbf{F},p,q,\mathbf{c}) + \pi^T(\mathbf{F},p,q,\mathbf{c}) \frac{\partial P(\mathbf{F},p,q,\mathbf{c})}{\partial \xi}, \quad (4.62)$$

where ξ can be any of $p,q,c_1,\dots,c_m$. After rearranging the above equation, we have

$$\frac{\partial \pi^T(\mathbf{F},p,q,\mathbf{c})}{\partial \xi}[I - P(\mathbf{F},p,q,\mathbf{c})] = \pi^T(\mathbf{F},p,q,\mathbf{c}) \frac{\partial P(\mathbf{F},p,q,\mathbf{c})}{\partial \xi}. \quad (4.63)$$

The generalized inverse of $[I - P(\mathbf{F},p,q,\mathbf{c})]$ (e.g., the fundamental matrix $Z(\mathbf{F},p,q,\mathbf{c})$) is easy to compute when we compute $\pi(\mathbf{F},p,q,\mathbf{c})$.

Definition 4.12. *The* long-run sensitivity *with respect to any probabilistic parameter* ξ *in* $\{p,q,c_1,\dots,c_m\}$ *for a given PBN defined by* $P(\mathbf{F},p,q,\mathbf{c})$ *is given by*

$$\omega_p[\xi] = \left\| \frac{\partial \pi^T(\mathbf{F},p,q,\mathbf{c})}{\partial \xi} \right\|_l = \left\| \pi^T(\mathbf{F},p,q,\mathbf{c}) \frac{\partial P(\mathbf{F},p,q,\mathbf{c})}{\partial \xi} Z(\mathbf{F},p,q,\mathbf{c}) \right\|_l, \quad (4.64)$$

where $\|\cdot\|_l$ *denotes the* l*-norm.*

To get $\omega_p[\xi]$, we need to derive $\frac{\partial P(\mathbf{F},p,q,\mathbf{c})}{\partial \xi}$ based on the transition probabilities. We now do this for BNs with perturbation (BNps), instantaneously random PBNs, and context-sensitive PBNs. For BNps,

$$P_{\mathbf{y}}(\mathbf{x}) = \mathbf{1}_{[\mathbf{f}(\mathbf{y})=\mathbf{x}]}(1-p)^n + \mathbf{1}_{[\mathbf{x}\neq\mathbf{y}]} p^{\eta(\mathbf{x},\mathbf{y})}(1-p)^{n-\eta(\mathbf{x},\mathbf{y})}, \quad (4.65)$$

where $\eta(\mathbf{x},\mathbf{y})$ is the Hamming distance between $\mathbf{x}$ and $\mathbf{y}$, and $\mathbf{1}_{[\mathbf{f}(\mathbf{y})=\mathbf{x}]}$ is the indicator function that takes value 1 if $\mathbf{f}(\mathbf{y}) = \mathbf{x}$ according to the truth table and is equal to 0 otherwise. Taking the derivative with respect to p on both sides yields

$$\begin{aligned} \frac{\partial P_{\mathbf{y}}(\mathbf{x})}{\partial p} &= -n\mathbf{1}_{[\mathbf{f}(\mathbf{y})=\mathbf{x}]}(1-p)^{n-1} \\ &\quad + \mathbf{1}_{[\mathbf{x}\neq\mathbf{y}]}\{\eta(\mathbf{x},\mathbf{y}) p^{\eta(\mathbf{x},\mathbf{y})-1}(1-p)^{n-\eta(\mathbf{x},\mathbf{y})} \\ &\quad - [n-\eta(\mathbf{x},\mathbf{y})] p^{\eta(\mathbf{x},\mathbf{y})}(1-p)^{n-\eta(\mathbf{x},\mathbf{y})-1}\}. \end{aligned} \quad (4.66)$$

For instantaneously random PBNs,

$$P_{\mathbf{y}}(\mathbf{x}) = \sum_{j=1}^{m} c_j \mathbf{1}_{[\mathbf{f}_j(\mathbf{y})=\mathbf{x}]}(1-p)^n + \mathbf{1}_{[\mathbf{x}\neq\mathbf{y}]} p^{\eta(\mathbf{x},\mathbf{y})}(1-p)^{n-\eta(\mathbf{x},\mathbf{y})}. \quad (4.67)$$

Taking the derivative with respect to p and c_j yields

$$\begin{aligned} \frac{\partial P_{\mathbf{y}}(\mathbf{x})}{\partial p} &= -n\sum_{j=1}^{m} c_j \mathbf{1}_{[\mathbf{f}_j(\mathbf{y})=\mathbf{x}]}(1-p)^{n-1} \\ &\quad + \mathbf{1}_{[\mathbf{x}\neq\mathbf{y}]}\{\eta(\mathbf{x},\mathbf{y}) p^{\eta(\mathbf{x},\mathbf{y})-1}(1-p)^{n-\eta(\mathbf{x},\mathbf{y})} \\ &\quad - [n-\eta(\mathbf{x},\mathbf{y})] p^{\eta(\mathbf{x},\mathbf{y})}(1-p)^{n-\eta(\mathbf{x},\mathbf{y})-1}\} \end{aligned} \quad (4.68)$$

and

$$\frac{\partial P_{\mathbf{y}}(\mathbf{x})}{\partial c_j} = \mathbf{1}_{[\mathbf{f}_j(\mathbf{y})=\mathbf{x}]}(1-p)^n, \tag{4.69}$$

respectively. For context-sensitive PBNs,

$$\begin{aligned} P_{s,\mathbf{y}}(r,\mathbf{x}) = \mathbf{1}_{[r=s]}(1-q+qc_s)\{\mathbf{1}_{[\mathbf{f}_s(\mathbf{y})=\mathbf{x}]}(1-p)^n + \mathbf{1}_{[\mathbf{x}\neq\mathbf{y}]}p^{\eta(\mathbf{x},\mathbf{y})}(1-p)^{n-\eta(\mathbf{x},\mathbf{y})}\} \\ + \mathbf{1}_{[r\neq s]}qc_r\{\mathbf{1}_{[\mathbf{f}_r(\mathbf{y})=\mathbf{x}]}(1-p)^n + \mathbf{1}_{[\mathbf{x}\neq\mathbf{y}]}p^{\eta(\mathbf{x},\mathbf{y})}(1-p)^{n-\eta(\mathbf{x},\mathbf{y})}\}, \end{aligned} \tag{4.70}$$

where r,s denote the rth and sth BNps, which are the BNps at time $t+1$ and t. Taking derivatives with respect to $p,q,c_1,\ldots,c_m$ yields

$$\begin{aligned} \frac{\partial P_{s,\mathbf{y}}(r,\mathbf{x})}{\partial p} =& - \mathbf{1}_{[r=s]}(1-q+qc_s)\{n\mathbf{1}_{[\mathbf{f}_s(\mathbf{y})=\mathbf{x}]}(1-p)^{n-1} \\ & - \mathbf{1}_{[\mathbf{x}\neq\mathbf{y}]}\{\eta(\mathbf{x},\mathbf{y})p^{\eta(\mathbf{x},\mathbf{y})-1}(1-p)^{n-\eta(\mathbf{x},\mathbf{y})} \\ & - [n-\eta(\mathbf{x},\mathbf{y})]p^{\eta(\mathbf{x},\mathbf{y})}(1-p)^{n-\eta(\mathbf{x},\mathbf{y})-1}\}\} \\ & - \mathbf{1}_{[r\neq s]}qc_r\{n\mathbf{1}_{[\mathbf{f}_r(\mathbf{y})=\mathbf{x}]}(1-p)^{n-1} \\ & - \mathbf{1}_{[\mathbf{x}\neq\mathbf{y}]}\{\eta(\mathbf{x},\mathbf{y})p^{\eta(\mathbf{x},\mathbf{y})-1}(1-p)^{n-\eta(\mathbf{x},\mathbf{y})} \\ & - [n-\eta(\mathbf{x},\mathbf{y})]p^{\eta(\mathbf{x},\mathbf{y})}(1-p)^{n-\eta(\mathbf{x},\mathbf{y})-1}\}\}; \end{aligned} \tag{4.71}$$

$$\begin{aligned} \frac{\partial P_{s,\mathbf{y}}(r,\mathbf{x})}{\partial q} =& -(1-c_s)\mathbf{1}_{[r=s]}\{\mathbf{1}_{[\mathbf{f}_s(\mathbf{y})=\mathbf{x}]}(1-p)^n + \mathbf{1}_{[\mathbf{x}\neq\mathbf{y}]}p^{\eta(\mathbf{x},\mathbf{y})}(1-p)^{n-\eta(\mathbf{x},\mathbf{y})}\} \\ & + \mathbf{1}_{[r\neq s]}c_r\{\mathbf{1}_{[\mathbf{f}_r(\mathbf{y})=\mathbf{x}]}(1-p)^n + \mathbf{1}_{[\mathbf{x}\neq\mathbf{y}]}p^{\eta(\mathbf{x},\mathbf{y})}(1-p)^{n-\eta(\mathbf{x},\mathbf{y})}\}; \end{aligned} \tag{4.72}$$

$$\frac{\partial P_{s,\mathbf{y}}(r,\mathbf{x})}{\partial c_j} = \begin{cases} \mathbf{1}_{[r=s]}q\{\mathbf{1}_{[\mathbf{f}_s(\mathbf{y})=\mathbf{x}]}(1-p)^n + \mathbf{1}_{[\mathbf{x}\neq\mathbf{y}]}p^{\eta(\mathbf{x},\mathbf{y})}(1-p)^{n-\eta(\mathbf{x},\mathbf{y})}\} & \text{if } j=s, \\ \mathbf{1}_{[r\neq s]}q\{\mathbf{1}_{[\mathbf{f}_r(\mathbf{y})=\mathbf{x}]}(1-p)^n + \mathbf{1}_{[\mathbf{x}\neq\mathbf{y}]}p^{\eta(\mathbf{x},\mathbf{y})}(1-p)^{n-\eta(\mathbf{x},\mathbf{y})}\} & \text{if } j=r, \\ 0 & \text{otherwise.} \end{cases}$$

By computing all the matrix entries, we can obtain all possible matrices $\frac{\partial P(\mathbf{F},p,q,\mathbf{c})}{\partial \xi}$ and then the long-run sensitivities.

4.5.2 Long-run sensitivity with respect to regulatory functions

Sensitivity analysis with respect to perturbations to regulatory functions is more complicated. Function perturbations are not continuous and therefore computation of the partial derivative of the transition matrix is not straightforward. Thus, we define the long-run sensitivity by the difference of the steady-state distributions due to perturbations to the regulatory Boolean functions. We continue to use the notation $\tilde{P} = P + E$, where E represents the perturbation. If $\mathbf{F}$ and $\tilde{\mathbf{F}}$ denote the vectors of regulatory functions for a PBN before and after perturbation, respectively, we define the sensitivity with respect to regulatory functions as follows.

Definition 4.13. *Let* $\mathbf{F}$ *denote the vector of regulatory functions for a PBN defined by* $P(\mathbf{F},p,q,\mathbf{c})$ *and* $\tilde{\mathbf{F}}$ *denote the vector of regulatory functions for the new PBN defined by* $P(\tilde{\mathbf{F}},p,q,\mathbf{c})$ *following function perturbation. The* long-run sensitivity *with respect to the perturbation mapping* $\phi : \mathbf{F} \to \tilde{\mathbf{F}}$ *is defined by*

$$\omega[\phi] = \|\tilde{\pi}^T(\tilde{\mathbf{F}},p,q,\mathbf{c}) - \pi^T(\mathbf{F},p,q,\mathbf{c})\|_l. \tag{4.73}$$

For a class of perturbation mappings $\Phi = \{\phi_\alpha : \alpha \in \mathsf{A}\}$, *where* A *is a probability space, the long-run sensitivity for the class, or* average long-run sensitivity, *is defined as the expectation* $\omega[\Phi] = \mathbb{E}_\alpha[\omega[\phi_\alpha]]$.

The preceding definition is general. Based on Eq. (4.18), for rank-one perturbations to regulatory functions we have an explicit way to compute the exact long-run sensitivity defined in Eq. (4.73). We will now consider some special cases and derive the exact analytic long-run sensitivity. Note that we have defined the average long-run sensitivity in this case as averaging over all constituent networks, target genes, and input states, but one could, if desired, average with respect to any of these.

4.5.3 One-predictor function perturbations

For a PBN with multi-output function $\mathbf{F}$, consider the multi-output Boolean function $\mathbf{f}_j$ for the jth constituent BNp. Instead of flipping gene state x_i, we alter the truth table directly by $\mathbf{f}_j(\mathbf{x} \oplus e_i)$ for a particular predictor state $\mathbf{x}$ (for BNps there is only one network function $\mathbf{f}$). This leads to the following definition.

Definition 4.14. *The* long-run sensitivity *with respect to a one-predictor function perturbation* $\mathbf{f}_j(\mathbf{x} \oplus e_i)$ *for a given state* $\mathbf{x}$ *for a PBN defined by* $P(\mathbf{F}, p, q, \mathbf{c})$ *is given by*

$$\omega_b[\mathbf{f}_j, \mathbf{x}, e_i] = \|\tilde{\pi}^T(\tilde{\mathbf{F}}^i_j[\mathbf{x}], p, q, \mathbf{c}) - \pi^T(\mathbf{F}, p, q, \mathbf{c})\|_l, \tag{4.74}$$

where $\tilde{\mathbf{F}}^i_j[\mathbf{x}]$ *differs from* $\mathbf{F}$ *only in* $\mathbf{f}_j$ *for the state* $\mathbf{x}$ *by replacing* $\mathbf{f}_j(\mathbf{x})$ *by* $\mathbf{f}_j(\mathbf{x} \oplus e_i)$. *Letting* Π *denote the class of all such perturbations, the* average long-run sensitivity *is defined by*

$$\omega_b[\Pi] = \frac{1}{n} \sum_{j=1}^{m} c_j \sum_{i=1}^{n} \mathbb{E}_\mathbf{x}[\omega_b(\mathbf{f}_j, \mathbf{x}, e_i)]. \tag{4.75}$$

A one-predictor function perturbation on $\mathbf{f}_j$ replaces the output state for the predictor state $\mathbf{x}$ with the output state for the predictor $\mathbf{x} \oplus e_i$ in the truth table. Ignoring the perturbation parts in Eqs. (4.65) and (4.67), each row in the transition matrix corresponds to a single input state $\mathbf{i}$ and the row consists of all 0's except for a 1 in the column of the output state $\mathbf{f}(\mathbf{i})$ in the BNp case, c_j's ($1 \le j \le m$) in several columns for instantaneously random PBNs, or more complicated nonzero values for context-sensitive PBNs. The one-predictor perturbation $\tilde{\mathbf{f}}_j(\mathbf{x}) = \mathbf{f}_j(\mathbf{x} \oplus e_i)$ from $\mathbf{f}_j$ for input state $\mathbf{x}$ means that $\mathbf{v} = \mathbf{f}_j(\mathbf{x} \oplus e_i) = \tilde{\mathbf{f}}_j(\mathbf{x}) \ne \mathbf{f}_j(\mathbf{x}) = \mathbf{u}$ and $\tilde{\mathbf{f}}_j(\mathbf{i}) = \mathbf{f}_j(\mathbf{i})$ for all $\mathbf{i} \ne \mathbf{x}$. This perturbation changes the transition matrix P by a special rank-one perturbation. In the next theorem, we continue to use the bold font to denote the gene states, while we write them in italic to represent the relative positions in matrices when we use them as the subscripts for matrix entries.

Theorem 4.15 (Qian and Dougherty, 2009a). *For a one-predictor function perturbation, the long-run sensitivity can be computed with a closed form for BNps, instantaneously random PBNs, and context-sensitive PBNs as follows:*

(1) *For BNps and instantaneously random PBNs,*

$$\omega_b[\mathbf{f}_j, \mathbf{x}, e_i] = \left\| \frac{\epsilon \pi_x (z_v - z_u)}{1 - \epsilon(z_{vx} - z_{ux})} \right\|_l, \tag{4.76}$$

where z_u and z_v are the uth and vth row, respectively, in the fundamental matrix Z of the original network; and $\pi_{\mathbf{x}}$ is the stationary mass for the state **x** *in the original network. For BNps, $\epsilon = (1-p)^n$; while for instantaneously random PBNs, $\epsilon = (1-p)^n c_j$.*

(2) *For context-sensitive PBNs,*

$$\omega_b[\mathbf{f}_j, \mathbf{x}, e_i] = \left\| \frac{(1-p)^n \pi^T a (z_{(j-1)2^n+v} - z_{(j-1)2^n+u})}{1-(1-p)^n (z_{(j-1)2^n+v} - z_{(j-1)2^n+u})^T a} \right\|_l, \tag{4.77}$$

where $z_{(j-1)2^n+u}$ and $z_{(j-1)2^n+v}$ again are the two rows of the fundamental matrix Z corresponding to states $(j,\mathbf{u})$ and $(j,\mathbf{v})$, respectively, in the PBN; and a is an $m2^n$-dimensional vector with

$$a[y] = \begin{cases} qc_r & \text{if } y = (r-1)2^n + x, \text{ and } r = 1,\dots,j-1,j+1,\dots,m, \\ 1-q+qc_j & \text{if } y = (r-1)2^n + x, \text{ and } r = j, \\ 0 & \text{otherwise.} \end{cases}$$

Proof. Beginning with instantaneously random PBNs, we have reasoned that a one-predictor function perturbation on $\mathbf{f}_j$ changes the output state from **u** to **v** at the predictor state **x**. In general, $\mathbf{v} = \mathbf{f}_j(\mathbf{x} \oplus e_i) = \tilde{\mathbf{f}}_j(\mathbf{x}) \neq \mathbf{f}_j(\mathbf{x}) = \mathbf{u}$; while $\tilde{\mathbf{f}}_j(\mathbf{i}) = \mathbf{f}_j(\mathbf{i})$ for all $\mathbf{i} \neq \mathbf{x}$. Hence, there is now (with $\tilde{\mathbf{f}}_j$) a 1 (for BNps) or c_j (for instantaneously random PBNs) in the column of the output state **v** and a 0 in the column of state **u**, whereas originally (with $\mathbf{f}_j$) there was a 0 in the column of the output state **v** and a 1 (for BNps) or c_j (for instantaneously random PBNs) in the column of state **u**, and all other entries in the matrix are unchanged. Consequently, there are changes to the transition matrix P only at p_{xu} and p_{xv}. Thus, $\tilde{p}_{xu} = p_{xu} - \epsilon$ and $\tilde{p}_{xv} = p_{xv} + \epsilon$, where ϵ is decided by the change to the truth table and the perturbation probability p: $\epsilon = (1-p)^n$ for BNps and $\epsilon = (1-p)^n c_j$ for instantaneously random PBNs. Hence, we write the perturbation matrix as $E = e_x \xi^T$, where

$$\xi^T = [0,\dots,0,\underbrace{-\epsilon}_{u},0,\dots,0,\underbrace{\epsilon}_{v},0,\dots,0]. \tag{4.78}$$

Applying Eq. (4.37), $a = e_x$, and $\beta^T = \xi^T Z$ gives Eq. (4.76), where z_u and z_v are the uth and vth row, respectively, in the fundamental matrix Z.

For a context-sensitive PBN, each state of the Markov chain is of the form $(r,\mathbf{i})$, where r denotes the rth constituent BNp and **i** the gene state vector. Since there are m constituent networks and n genes, the size of the transition matrix is $m2^n \times m2^n$. A one-predictor perturbation for $\mathbf{f}_j$ causes changes to multiple rows (m rows to be exact). But only two columns are changed, both being columns of the matrix corresponding to the jth constituent BN as the output changes from $(j,\mathbf{u})$ to $(j,\mathbf{v})$. Hence, we can proceed similarly using Eq. (4.37), but now the perturbation matrix is $E = a\xi^T$, where a and ξ are $m2^n$-dimensional vectors,

$$\xi^T = [0,\dots,0,\underbrace{-(1-p)^n}_{(j-1)2^n+u},0,\dots,0,\underbrace{(1-p)^n}_{(j-1)2^n+v},0,\dots,0], \tag{4.79}$$

and a is defined by (i) $a[(r-1)2^n + x] = qc_r$ for $r = 1,\dots,m, r \neq j$, (ii) $a[(r-1)2^n + x] = 1 - q + qc_j$ for $r = j$, and (iii) $a[y] = 0$ elsewhere. Applying Eq. (4.37) gives Eq. (4.77), where $z_{(j-1)2^n+u}$ and $z_{(j-1)2^n+v}$ are the two rows of the fundamental matrix Z corresponding to states $(j,\mathbf{u})$ and $(j,\mathbf{v})$, respectively, in the PBN. □

4.5.4 One-bit function perturbations

Another basic type of function perturbation is to change one bit in the truth table for the multi-output function $\mathbf{F}$. For example, we can flip the output for the Boolean function f_j^i that decides the gene state x_i in the jth constituent BNp at a particular state $\mathbf{x}$. Thus, we alter the function by $\mathbf{f}_j(\mathbf{x}) \oplus e_i$ (for BNps, there is only one network function $\mathbf{f}$). This leads to the following definition.

Definition 4.16. *The* long-run sensitivity *with respect to a one-bit function perturbation* $\mathbf{f}_j(\mathbf{x}) \oplus e_i$ *for a given state* $\mathbf{x}$ *for a PBN defined by* $P(\mathbf{F}, p, q, \mathbf{c})$ *is given by*

$$\omega_f[f_j^i, \mathbf{x}] = \|\tilde{\pi}^T(\tilde{\mathbf{F}}_j^i[\mathbf{x}], p, q, \mathbf{c}) - \pi^T(\mathbf{F}, p, q, \mathbf{c})\|_l, \tag{4.80}$$

where $\tilde{\mathbf{F}}_j^i[\mathbf{x}]$ *differs from* $\mathbf{F}$ *only in* f_j^i *for the state* $\mathbf{x}$. *Letting* Π *again denote the class of all such perturbations, the* average long-run sensitivity *is defined by*

$$\omega_f[\Pi] = \frac{1}{n}\sum_{j=1}^{m} c_j \sum_{i=1}^{n} \mathbb{E}_{\mathbf{x}}[\omega_f[f_j^i, \mathbf{x}]]. \tag{4.81}$$

A one-bit function perturbation on $\mathbf{f}_j$ flips one target gene at the response side for the predictor state $\mathbf{x}$ in the truth table. Similar as the derivation for the previous definition of the long-run sensitivity for one-predictor function perturbations, the one-bit perturbation $\tilde{\mathbf{f}}_j(\mathbf{x}) = \mathbf{f}_j(\mathbf{x}) \oplus e_i$ from $\mathbf{f}_j$ for input state $\mathbf{x}$ means that $\mathbf{v} = \tilde{\mathbf{f}}_j(\mathbf{x}) \neq \mathbf{f}_j(\mathbf{x}) = \mathbf{u}$ and $\tilde{\mathbf{f}}_j(\mathbf{i}) = \mathbf{f}_j(\mathbf{i})$ for all $\mathbf{i} \neq \mathbf{x}$. Here, $\mathbf{u}$ and $\mathbf{v}$ differ only for one bit. Hence, this type of function perturbation changes the transition matrix for either BNps, instantaneously random PBNs, or context-sensitive PBNs by either a single row perturbation or a rank-one perturbation in the exact same form as in the previous definition. Therefore, we can use the same Eqs. (4.76) and (4.77) to compute the long-run sensitivity for one-bit function perturbations.

It is straightforward to extend this definition for the long-run sensitivity for multiple-bit function perturbations and other more general perturbations.

4.5.5 Function perturbations considering one-gene regulation

Since in practice it is possible to (approximately) suppress or express certain genes in living cells, we consider network sensitivity relative to permanently blocking or expressing a gene in the network; that is, we consider perturbing $\mathbf{F}$ by setting all the output values of $\tilde{f}_j^i$ equal to either 0 or 1.

Definition 4.17. *The* long-run sensitivity *with respect to one-gene function perturbation in* $\mathbf{F}$ *for a given PBN defined by* $P(\mathbf{F}, p, q, \mathbf{c})$ *can be described by*

$$\begin{aligned}\omega_g[f_j^i] = \max\Big\{ &\|\tilde{\pi}^T(\tilde{\mathbf{F}}_j^{i,0}, p, q, \mathbf{c}) - \pi^T(\mathbf{F}, p, q, \mathbf{c})\|_l, \\ &\|\tilde{\pi}^T(\tilde{\mathbf{F}}_j^{i,1}, p, q, \mathbf{c}) - \pi^T(\mathbf{F}, p, q, \mathbf{c})\|_l \Big\},\end{aligned} \tag{4.82}$$

where $\tilde{\mathbf{F}}_j^{i,0}$ *differs from* $\mathbf{F}$ *by suppressing gene* x_i *with* $f_j^i(\mathbf{x}) = 0$ *for any* $\mathbf{x}$ *in the* j*th constituent network and* $\tilde{\mathbf{F}}_j^{i,1}$ *differs from* $\mathbf{F}$ *by expressing gene* x_i *with* $f_j^i(\mathbf{x}) = 1$ *for any* $\mathbf{x}$ *in the* j*th constituent network. Letting* Γ *denote the class of all such perturbations, the* average long-run sensitivity *is defined by*

$$\omega_g[\Gamma] = \frac{1}{n}\sum_{j=1}^{m} c_j \sum_{i=1}^{n} \omega_g[f_j^i]. \tag{4.83}$$

A one-gene function perturbation changes the transition matrix in multiple rows for BNps, instantaneously random PBNs, and context-sensitive PBNs. Equation (4.37) cannot be directly applied to obtain the analytic solution; however, the iterative procedure (4.51) discussed previously can be used to compute the steady-state distribution difference.

4.5.6 Properties of long-run sensitivity

In Qian and Dougherty (2009a) properties of the different definitions of long-run sensitivity are examined in the context of randomly generated networks possessing similar network properties, the two most important parameters being the bias (p_b) and connectivity (K). p_b is the mean of the *Bernoulli* distribution to generate the truth table of one Boolean function in a BN and K is the maximum input degree of the Boolean functions in the network. Here we provide a flavor of these results by considering average long-run sensitivity with respect to one-predictor function perturbations. We focus on BNps to illustrate the impact of different parameters, using simulations of BNs possessing 6 nodes, with 1,000 randomly generated networks. With small p, the steady-state mass is strongly concentrated in the attractor states.

Figure 4.3, parts (a) and (b), shows $\omega_b[\Pi]$ (in Definition 4.14 using the l_1-norm) as a function of p_b and K, respectively, for different p values. To gain an understanding of these graphs, we plot the average number of attractors ($|A|$) with respect to p_b and K in Fig. 4.4 and the average number of changed attractors ($|DA|$) by one-predictor function perturbation in Fig. 4.5. Since these numbers do not change with p, we plot only the results for one set of 1,000 random BNs. From these figures, it is clear that there are more attractors with increasing K or when $p_b \to 0.5$. At the same time, the average number of changed attractors also increases with increasing K or with $p_b \to 0.5$. Hence, with small p values, the steady-state distribution difference comes mainly from the mass shift for these attractors. With increasing K or $p_b \to 0.5$, the number of changed attractors increases. Thus, the steady-state distribution difference increases and $\omega_b[\Pi]$ increases.

For large p, $\omega_b[\Pi]$ behaves differently. With increasing p, the state can flip with larger probability, so that the network can jump out of attractor cycles more easily. Thus, there will be more steady-state mass in transient states and the steady-state distribution difference can be strongly affected by the difference of steady-state mass in those transient states. In addition, the change caused by changed attractors resulting from one-predictor function perturbations is relatively small (almost less than 1) as shown in Fig. 4.5. Hence, with large K or $p_b \approx 0.5$, there are fewer transient states and the resulting steady-state distribution difference is more complicated, as shown in Fig. 4.3. With increasing K or $p_b \to 0.5$, the average long-run sensitivity, in fact, decreases when p is large.

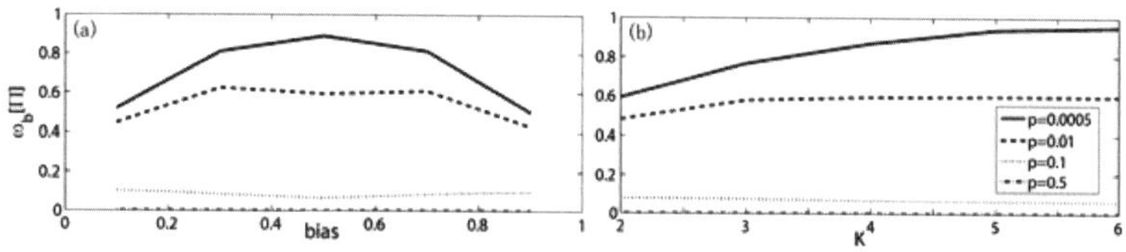

Figure 4.3. *The average long-run sensitivity ($\omega_b[\Pi]$) for* 6*-gene BNs with different network properties:* (a) *Bias (p_b);* (b) *connectivity (K).*

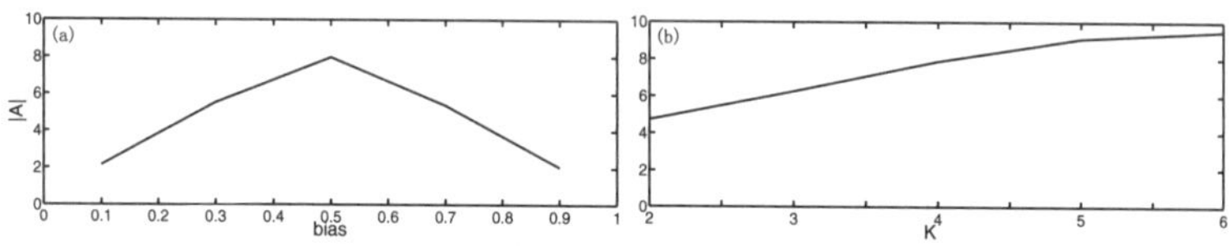

Figure 4.4. *The average number of attractors ($|A|$) for* 6*-gene BNs with different network properties:* (a) *Bias (p_b);* (b) *connectivity (K).*

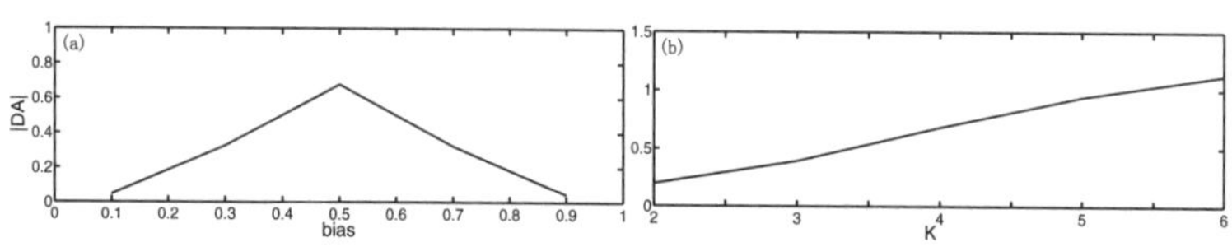

Figure 4.5. *The average number of changed attractors ($|DA|$) by one-predictor function perturbation for* 6*-gene BNs with different network properties:* (a) *Bias (p_b);* (b) *connectivity (K).*

4.5.7 Sensitivity and robustness of control

Since the l-norm distance between the steady-state distributions constitutes one measure of network closeness, and therefore a measure to quantify inference accuracy (Dougherty, 2007), the long-run sensitivity can *ipso facto* be used in the analysis of inference robustness with respect to various network parameters. We can bound this difference using a standard differential approximation:

$$\|\pi_{p'} - \pi_p\|_{l_1} \leq \left\| \frac{\partial \pi}{\partial p}(p) \right\|_{l_1} |p' - p| + \varepsilon(p' - p), \tag{4.84}$$

where $\pi_{p'}$ and π_p are the steady-state distributions with perturbation parameters p' and p, respectively, and $\frac{\varepsilon(p'-p)}{p'-p} \to 0$ as $p' - p \to 0$. Suppose now that $\pi_{p'}$ and π_p correspond to the estimated PBN and actual PBN, respectively. If we are interested in finding the one-bit optimal structural intervention, then in practice this intervention is determined on the estimated PBN.

Referring to Eq. (4.84), let $\pi_{p'}[f_j^i, \mathbf{x}]$ and $\pi_p[f_j^i, \mathbf{x}]$ denote the steady-state distributions resulting from the optimal structural intervention derived from the estimated PBN on $\pi_{p'}$ and π_p, respectively, meaning that the intervention is derived via the estimated

PBN but also applied to the original PBN. Analogously, let $\pi_{p'}[f_s^r,\mathbf{y}]$ and $\pi_p[f_s^r,\mathbf{y}]$ denote the steady-state distributions resulting from optimal structural intervention derived from the original PBN applied on $\pi_{p'}$ and π_p, respectively, We would like to know that $\|\pi_{p'}[f_j^i,\mathbf{x}]-\pi_p[f_s^r,\mathbf{y}]\|_{l_1}$ is small, because this would mean that applying the intervention on the estimated PBN yields a steady-state distribution close to the optimal one we would obtain if we actually knew the true PBN. Application of the triangle inequality yields

$$\begin{aligned}\|\pi_{p'}[f_j^i,\mathbf{x}]-\pi_p[f_s^r,\mathbf{y}\|_{l_1} &\leq \|\pi_{p'}[f_j^i,\mathbf{x}]-\pi_p[f_j^i,\mathbf{x}]\|_{l_1} \\ &\quad + \|\pi_p[f_j^i,\mathbf{x}]-\pi_p\|_{l_1}+\|\pi_p-\pi_p[f_s^r,\mathbf{y}]\|_{l_1} \\ &\leq \left\|\frac{\partial \pi_p[f_j^i,\mathbf{x}]}{\partial p}(p)\right\|_{l_1} |p'-p| \\ &\quad + \varepsilon(p'-p)+\omega_f[f_j^i,\mathbf{x}](p)+\omega_f[f_s^r,\mathbf{y}](p).\end{aligned} \tag{4.85}$$

Hence, the norm of interest is bounded in terms of long-run sensitivities for the probability parameter and one-bit function perturbations, along with the fast-converging term $\varepsilon(p'-p)$. The sensitivities can be computed as discussed previously and standard methods of calculus can be used to obtain $\varepsilon(p'-p)$. Other probabilistic parameters can be treated similarly.

4.5.8 Sensitivity in a mammalian cell-cycle network

In this section, we use a mammalian cell-cycle network to illustrate the relationship between long-run sensitivity (in particular, with respect to the perturbation probability) and the robustness of optimal intervention strategies, the expectation being that greater long-run sensitivity means fewer robust intervention strategies. We consider long-run sensitivity in a PBN for the mammalian cell cycle proposed in Faryabi *et al.* (2008b) based on the BN in Fauré *et al.* (2006). The mammalian cell cycle was first used for the study of optimal infinite-horizon control in the context of asynchronous PBNs (to be covered in the next chapter) (Faryabi *et al.*, 2008a) and then as an application of optimal structural control in Qian and Dougherty (2008, 2009a).

For a normal mammalian organism, cell division should coordinate with overall growth in a process tightly controlled via extracellular signals which indicate whether a cell should divide or remain in a resting state. The positive signals (growth factors) instigate the activation of Cyclin D (CycD), which is one of the key genes in the mammalian cell-cycle. The other two important genes are retinoblastoma (Rb) and p27. Rb is a tumor-suppressor gene expressed in the absence of the cyclins, which inhibit Rb by phosphorylation. Gene p27 is also active in the absence of the cyclins. Whenever p27 is present, it blocks the action of CycE or CycA and Rb can also be expressed, even in the presence of CycE or CycA. Hence, it stops the cell cycle. In this wild-type cell-cycle model, when p27 is active, the cell cycle can be stopped in cancerous situations. We define the logical states in which CycD, Rb, and p27 are all down-regulated as undesirable states since the cell can cycle forever even in the absence of any growth factor. The goal of intervention is to minimize the stationary mass for the undesirable states.

We construct the instantaneously random PBN of the cell-cycle network based on the Boolean functions given in Table 4.4. This PBN consists of 10 genes: CycD, Rb,

Table 4.4. *Definitions of Boolean functions for the wild-type mammalian cell-cycle PBN with* 10 *genes (reproduced from Table* 1 *in Fauré et al.* (2006)*).*

Order	Gene	Regulating function
x_1	$CycD$	extracellular signals
x_2	Rb	$(\overline{CycD} \wedge \overline{CycE} \wedge \overline{CycA} \wedge \overline{CycB}) \vee (p27 \wedge \overline{CycD} \wedge \overline{CycB})$
x_3	$p27$	$(\overline{CycD} \wedge \overline{CycE} \wedge \overline{CycA} \wedge \overline{CycB}) \vee (p27 \wedge \overline{(CycE \wedge CycA)} \wedge \overline{CycD} \wedge \overline{CycB})$
x_4	$E2F$	$(\overline{Rb} \wedge \overline{CycA} \wedge \overline{CycB}) \vee (p27 \wedge \overline{Rb} \wedge \overline{CycB})$
x_5	$CycE$	$(E2F \wedge \overline{Rb})$
x_6	$CycA$	$(E2F \wedge \overline{Rb} \wedge \overline{Cdc20} \wedge (\overline{Cdh1 \wedge UbcH10})) \vee (CycA \wedge \overline{Rb} \wedge \overline{Cdc20} \wedge (\overline{Cdh1 \wedge UbcH10}))$
x_7	$Cdc20$	$CycB$
x_8	$Cdh1$	$(\overline{CycA} \wedge \overline{CycB}) \vee Cdc20 \vee (p27 \wedge \overline{CycB})$
x_9	$UbcH10$	$\overline{Cdh1} \vee (Cdh1 \wedge UbcH10 \wedge (Cdc20 \vee CycA \vee CycB))$
x_{10}	$CycB$	$(\overline{Cdc20} \wedge \overline{Cdh1})$

p27, E2F, CycE, CycA, Cdc20, Cdh1, UbcH10, and CycB. The above order of genes is used in the binary representation of the logical states, with CycD as the most significant bit and CycB as the least significant bit. The order of genes in the logical states does not affect our analysis or intervention. We assume that the extracellular signal to the cell-cycle model is a latent variable (Faryabi *et al.*, 2008a). The growth factor is not part of the cell and its value is determined by the surrounding cells. The expression of CycD changes independently of the cell's content and reflects the state of the growth factor. Depending on the expression status of CycD, we obtain two constituent BNs. The first constituent BN is determined based on the Boolean functions in Table 4.4 when the value of CycD is equal to 0. Similarly, the second constituent BN is determined by setting the value of CycD to 1. The relationships among the genes in this PBN are shown in Figure 4.6. To completely define the PBN, we set the perturbation probability $p = 0.01$ and the probability of selecting each constituent BN $c_j = 0.5, j = 1,2$.

Table 4.5 gives various long-run sensitivities for this instantaneously random PBN with respect to the perturbation probability p, the BN selection probabilities c_1, c_2, and the one-predictor function perturbations, one-bit function perturbations, and one-gene function perturbations.

Structural intervention is applied to avoid the logical states with simultaneously down-regulated CycD, Rb, and p27, the undesirable states being those with $x_1 = 0$, $x_2 = 0$, and $x_3 = 0$. Optimality is achieved by the one-bit function perturbation that minimizes $\sum_{x_1=0,x_2=0,x_3=0} \tilde{\pi}(\mathbf{x})$, among all possible controllable functions. The objective function values with the optimal one-bit function perturbations for each f_i in the two constituent BNs are listed in Table 4.6. Note that, since CycD is determined by extracellular signals, we cannot perturb it and therefore it does not appear in the table. We see that the perturbations to the function regulating Rb significantly reduce the steady-state mass of the undesirable states. Flipping the output expression in the truth table for the second BN leads to the minimum objective function value. From Table 4.6, we see that the function regulating E2F is another possible choice.

To test the robustness of this optimal intervention strategy with respect to the perturbation probability p, the same optimization problem is solved with different perturbation probabilities: $p = 0.001$, 0.01, 0.02, 0.03, 0.04, 0.05. For $p \leq 0.04$, the same optimal intervention strategy is obtained; however, for $p = 0.05$ a different strategy is obtained.

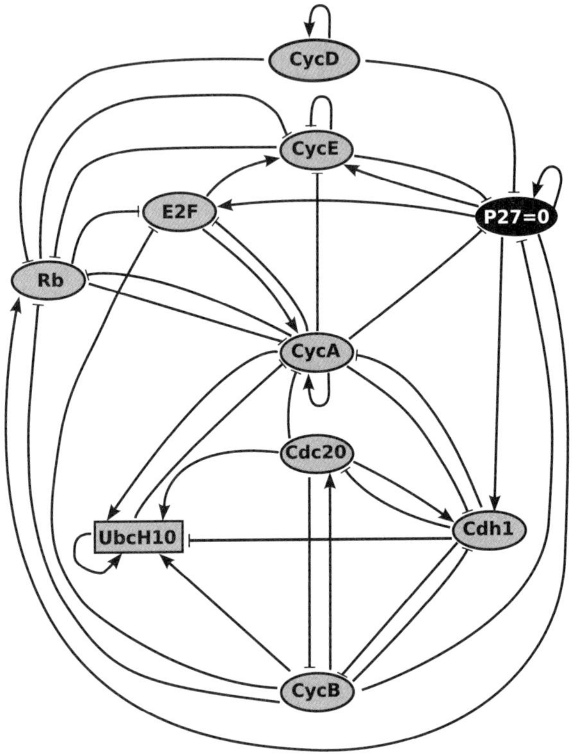

Figure 4.6. *Logical regulatory graph for the mutated mammalian cell-cycle network (Faryabi et al., 2008a). Blunt arrows stand for inhibitory effects; normal arrows stand for activations.*

Table 4.5. *Long-run sensitivities with respect to the changes to both the underlying regulatory functions and probabilistic parameters in the 10-gene cell-cycle PBN.*

$\|\partial\pi/\partial p\|_{l_1}$	$\|\partial\pi/\partial c_1\|_{l_1}$	$\|\partial\pi/\partial c_2\|_{l_1}$	ω_b	ω_f	ω_g
25.9195	1.7125	1.6379	0.1895	0.3672	1.2963

Table 4.6. $\sum_{x_1=0,x_2=0,x_3=0}\tilde{\pi}(\mathbf{x})$ *with the optimal one-bit function perturbations for all f_i's in the 10-gene cell-cycle PBN.*

Gene	Rb	p27	E2F	CycE	CycA	Cdc20	Cdh1	UbcH10	CycB
BN_1	0.1099	0.0996	0.1441	0.1436	0.1477	0.1191	0.1484	0.1467	0.1600
BN_2	0.0346	0.1329	0.0380	0.1228	0.1423	0.1723	0.1356	0.1447	0.1614

Table 4.7. *Definitions of Boolean functions for the mutated 9-gene PBN (reproduced from Table 1 in Faryabi et al. (2008a)).*

Order	Gene	Regulating function
x_1	$CycD$	extracellular signals
x_2	Rb	$(\overline{CycD} \wedge \overline{CycE} \wedge \overline{CycA} \wedge \overline{CycB})$
x_3	$E2F$	$(\overline{Rb} \wedge \overline{CycA} \wedge \overline{CycB})$
x_4	$CycE$	$(E2F \wedge \overline{Rb})$
x_5	$CycA$	$(E2F \wedge \overline{Rb} \wedge \overline{Cdc20} \wedge (\overline{Cdh1 \wedge UbcH10})) \vee (CycA \wedge \overline{Rb} \wedge \overline{Cdc20} \wedge (\overline{Cdh1 \wedge UbcH10}))$
x_6	$Cdc20$	$CycB$
x_7	$Cdh1$	$(\overline{CycA} \wedge \overline{CycB}) \vee Cdc20$
x_8	$UbcH10$	$\overline{Cdh1} \vee (Cdh1 \wedge UbcH10 \wedge (Cdc20 \vee CycA \vee CycB))$
x_9	$CycB$	$(\overline{Cdc20} \wedge \overline{Cdh1})$

Table 4.8. *Long-run sensitivities with respect to the changes to both the underlying regulatory functions and probabilistic parameters in the mutated 9-gene PBN.*

$\|\partial\pi/\partial p\|_{l_1}$	$\|\partial\pi\partial c_1\|_{l_1}$	$\|\partial\pi/\partial c_2\|_{l_1}$	ω_b	ω_f	ω_g
19.6017	1.5724	1.5541	0.0794	0.1061	1.1758

Next, we consider a PBN based on the proposed mutations in Faryabi *et al.* (2008a), in which p27 is mutated and is always off. In this cancerous scenario, p27 can never be activated. This mutation introduces a situation where both CycD and Rb might be inactive. In this mutated phenotype, the cell cycles in the absence of any growth factor and the logical states in which both Rb and CycD are down-regulated are considered undesirable. Table 4.7 summarizes the mutated Boolean functions (Faryabi *et al.*, 2008a). Using $p = 0.01$, as in the previous wild-type PBN, a new instantaneously random 9-gene PBN is constructed based on the Boolean functions in Table 4.7 with mutated p27. The relationships among the genes in this PBN are depicted in Fig. 4.6.

Table 4.8 gives the long-run sensitivities with respect to changes to both the underlying regulatory functions and probabilistic parameters in this new PBN. These values indicate that the mutated PBN is more stable (and therefore more resistant to intervention) as compared to the original PBN. In particular, the mutated PBN is less sensitive to the perturbation probability p.

Structural intervention is applied to avoid the logical states with simultaneously down-regulated CycD and Rb ($x_1 = 0$ and $x_2 = 0$) in this cancerous scenario. Relative to the objective function $\sum_{x_1=0,x_2=0} \tilde{\pi}(\mathbf{x})$, the optimal one-bit function perturbations for each f_i in the two constituent BNs are listed in Table 4.9. Again, the perturbation to the function regulating Rb in the second context BN significantly reduces the steady-state mass of the undesirable states. The function regulating E2F again is the second optimal target for applying intervention. The ranking of the values in Table 4.6 and Table 4.9 remains the same.

Regarding robustness of this optimal intervention strategy with respect to the perturbation probability p, as in the previous wild-type network we solve the same optimization

Table 4.9. $\sum_{x_1=0,x_2=0} \tilde{\pi}(\mathbf{x})$ *with the optimal one-bit function perturbations for all* f_i*'s in the mutated 9-gene PBN.*

Gene	Rb	E2F	CycE	CycA	Cdc20	Cdh1	UbcH10	CycB
BN_1	0.1350	0.2064	0.1882	0.2028	0.1703	0.2134	0.2160	0.2190
BN_2	0.0317	0.0505	0.1842	0.2042	0.2330	0.2083	0.2064	0.2221

problem with different perturbation probabilities: $p = 0.001, 0.01, 0.02, 0.03, 0.04, 0.05$. In all cases, we obtain the same optimal strategy. This differs from the wild-type case, where a different optimal strategy is obtained for $p = 0.05$. The long-run sensitivity, $\|\frac{\partial \pi}{\partial p}\|_{l_1}$, is smaller (19.6017) in the mutated network than in the wild-type network (25.9195). The greater sensitivity in the wild-type network is consistent with its being less robust with regard to optimal intervention across different perturbation probabilities.

Chapter 5

External Control

Whereas structural intervention results in a permanent change in network dynamics because it changes the underlying structure of the network, *external control* takes advantage of the fact that, since the dynamical behavior of a PBN can be modeled by a Markov chain, intervention in PBNs is amenable to the theory of Markov decision processes. Control is generally based on flipping (or not flipping) the value of a control gene. The first intervention approach of this kind involved the determination of an optimal single-gene flip to the network to drive it away from an undesirable state to a desirable state (Shmulevich *et al.*, 2002b). Following this simple, one-time intervention, attention turned to dynamic-programming-based finite-horizon external control (Datta *et al.*, 2003), in which the effects are transient and the steady-state distribution is not changed, and then to infinite-horizon stationary external control (Pal *et al.*, 2006), in which the steady-state distribution is altered. Many issues have been addressed for external control in the framework of PBNs, including computational complexity, control when there is imperfect information, suboptimal control strategies, constrained control, asynchronous control, and greedy algorithms. All of these topics will be addressed to a greater or lesser extent in the present chapter.

5.1 Intervention via One-Time Gene Perturbation

A one-time gene perturbation changes the value of one or more genes without altering the rules or probabilistic parameters of the network, which in a binary network means that the value of one or more genes is flipped at the current moment. With a one-time change of a gene's value, the network dynamics are transiently altered, but there is no change to the stationary distribution of the network. A transient perturbation having been made, the network itself is responsible for determining the values of that gene thereafter.

To illustrate one-time intervention, consider the state transition diagram in Fig. 2.3 for the Markov chain corresponding to the PBN in Example 2.1. For the predictor probabilities given in that example, the probabilities of the four possible network realizations are $c_1 = 0.3$, $c_2 = 0.3$, $c_3 = 0.2$, and $c_4 = 0.2$. For the moment, assume that the perturbation probability is zero ($p = 0$). Suppose that the network is currently in state 111 and we desire that it eventually transition to 000. With which of the three genes, x_1, x_2, or x_3, should we intervene so that the probability is greatest that the network will end up in state 000? Direct inspection of Fig. 2.3 shows that if we set $x_1 = 0$, then the network will transition into 000 with probability $c_4 = 0.2$, whereas if we set $x_2 = 0$ or $x_3 = 0$, then it will be impossible to end up in 000 and, with probability 1, the network will eventually return to 111. In other words, the network is resistant to perturbations of the second or third genes and will eventually maintain the same state. Thus, in this simple example, only by intervening with gene x_1 is there a chance of achieving the desired goal.

We will now systematically examine one-time transient perturbation based on the analysis in Shmulevich *et al.* (2002b). When $p > 0$, the Markov chain is ergodic and every state will eventually be visited. Thus, intervention can be posed in the sense of reaching a desired state as early as possible. For instance, in the preceding example, if p is very small and the network is in state 111, then it will be a long time until it reaches 000 and setting $x_1 = 0$ is much more likely to get it there faster. We are, therefore, interested in the probability, $F_k(x, y)$, that, starting in state x, the first time the PBN will reach some given state y will be at time k. This is referred to as the *first passage time* from state x to state y. The *mean first passage time* (*MFPT*) from state x to state y is defined by

$$M(x,y) = \sum_k k F_k(x,y). \tag{5.1}$$

The MFPT measures, on the average, how long it will take to get from state x to state y. $F_1(x,y) = p_{xy}$, the transition probability from x to y. For $k \geq 2$, it follows immediately from the Chapman–Kolmogorov equations that

$$F_k(x,y) = \sum_{z \in \{0,1\}^n - \{y\}} p_{xy} F_{k-1}(z,y). \tag{5.2}$$

A related measure is given by

$$H_{K_0}(x,y) = \sum_{k=1}^{K_0} F_k(x,y). \tag{5.3}$$

Because the events (the first passage time from x to y will be at time k) are disjoint for different values of $k = 1, \ldots, K_0$, $H_{K_0}(x,y)$ is equal to the probability that the network, starting in state x, will visit state y before time K_0. As a special case, when $K_0 = \infty$, this is equal to the probability that the chain *ever* visits state y, starting at state x, which in our case is equal to 1 for $p > 0$.

$H_{K_0}(x,y)$ and $M(x,y)$, the MFPT, provide two optimization criteria to decide upon which gene should be used for intervention. Generate different states $x^{(i)} = x \oplus e_i$, $i = 1, \ldots, n$, where e_i is the unit binary vector with a 1 in the ith coordinate, by perturbing each of the n genes and compute $H_{K_0}(x^{(i)}, y)$ for some desired destination state y and constant K_0. Then, the best gene for intervention is the one for which $H_{K_0}(x^{(i)}, y)$ is maximum. That is, given a fixed K_0, the optimal gene $x_{i_{\text{opt}}}$ satisfies

$$i_{\text{opt}} = \arg\max_i H_{K_0}(x^{(i)}, y). \tag{5.4}$$

Alternatively, by minimizing the MFPTs, the optimal gene satisfies

$$i_{\text{opt}} = \arg\min_i M\left(x^{(i)}, y\right). \tag{5.5}$$

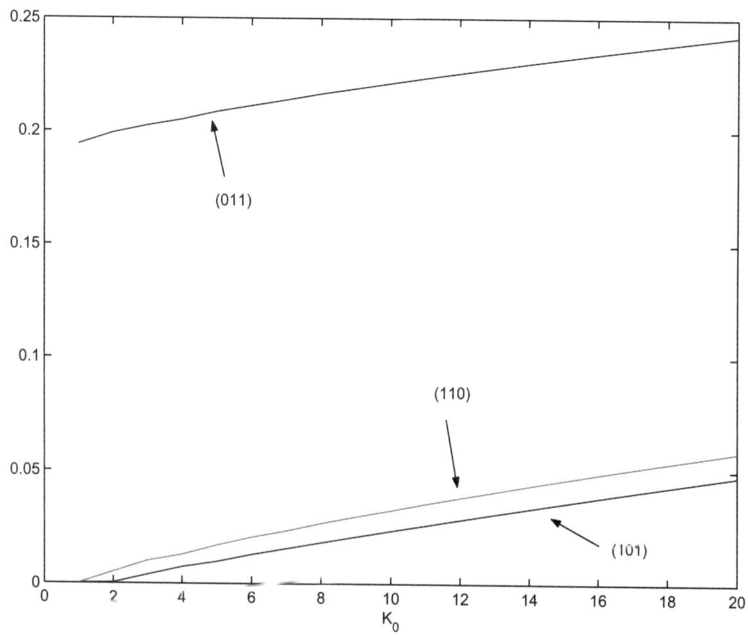

Figure 5.1. $H_{K_0^{(i)}}(x^{(i)}, y)$ *for* $K_0 = 1, \ldots, 20$ *for starting states* (011), (101), *and* (110), *corresponding to perturbations of the first, second, and third genes, respectively.*

To illustrate these notions using the preceding example, suppose $p = 0.01$. Then the steady-state distribution is given by $\{0.0752, 0.0028, 0.0371, 0.0076, 0.0367, 0.0424, 0.0672, 0.7310\}$, where the leftmost element corresponds to 000 and the rightmost to 111. As expected, the PBN spends much more time (70%) in state 111 than in any other state. Let the starting state x be 111 and the destination state y be 000, as before. With which gene should we intervene? We compute $F_k(011, 000)$, $F_k(101, 000)$, and $F_k(110, 000)$ and from these obtain $H_{K_0}(x, y)$. Figure 5.1 shows the plots of $H_{K_0}(x, y)$ for $K_0 = 1, \ldots, 20$ and for the three states of interest, namely, 011, 101, and 110. The plots indicate that, starting with state 011, there is much more likelihood of entering state 000 sooner than by starting with states 110 or 101. For example, during the first 20 steps, there is an almost 25% chance of entering 000 starting with 011, whereas starting with 110 or 101, there is only about a 5% chance. This, in turn, indicates that we should intervene with gene x_1 rather than with genes x_2 or x_3. Suppose we use the MFPT criterion. The MFPTs corresponding to the perturbations of genes x_1, x_2, and x are given by 337.51, 424.14, and 419.20, respectively. Since the first one is the smallest, this again supports the argument that gene x_1 is the best candidate for intervention.

Another approach utilizing $H_{K_0}(x, y)$ is to first fix a probability h_0 and wait until one of the $H_{K_0}(x^{(i)}, y)$ reaches it first. Owing to ergodicity, for every state $x^{(i)}$, there will always be a $K_0^{(i)}$ large enough such that $H_{K_0^{(i)}}(x^{(i)}, y) > h_0$. In that sense, the optimal gene for intervention $x_{i_{\text{opt}}}$ is one for which

$$i_{\text{opt}} = \arg\min_{i}\min_{K_0^{(i)}}\left\{K_0^{(i)} : H_{K_0^{(i)}}(x^{(i)}, y) > h_0\right\}. \tag{5.6}$$

At first glance, it might appear that both approaches, Eqs. (5.4) and (5.6), will yield the same answer, since Fig. 5.1 seems to suggest that the plots do not intersect and that if one of them is maximum for a given K_0, it will be the first to reach any fixed h_0 thereafter. While it is true that for sufficiently large K_0 the plots will not intersect, this is not in general true for smaller values of K_0.

The criteria imbedded in Eqs. (5.4) and (5.6) have different underlying interpretations. The first aims to *maximize the probability* of reaching a particular state before a certain fixed time while the second aims to *minimize the time* needed to reach a certain state with a given fixed probability. These two approaches are complementary and may be used in conjunction. Finally, the approach in Eq. (5.5) based on minimizing MFPTs is another alternative.

So far, we have discussed the notion of intervention in terms of a single starting state and a single destination state. In practice, we may be interested in the same types of questions but concerning sets of states—for example, two different sets of states may correspond to different functional cellular states, such as proliferation or quiescence. This is the kind of situation we considered with desirable and undesirable sets of states in the case of structural intervention, but now we will pose the issue in terms of starting and ending states. Given that we are in a set X of states, what gene is the best candidate for intervention if we want to end up in the set Y of states?

The question may be posed in the sense of Eq. (5.4), Eq. (5.5), or Eq. (5.6). The mathematical framework does not change substantially when considering sets of states. If $X = \{x\}$ consists of just one state, but Y is a set comprised of many states, then the first passage probabilities $F_k(x, y)$ may simply be summed over all states $y \in Y$ and we can define

$$F_k(x, Y) = \sum_{y \in Y} F_k(x, y). \tag{5.7}$$

Then, the same approaches as discussed above to find the best gene for intervention can be used.

The situation when X is comprised of a number of states is conceptually a bit more complicated, since now the starting set X, rather than having just one starting state x, represents a type of uncertainty in our knowledge of the current state of the network. We do not know which state the network is in at a particular time, but we do know that it is in a certain set of states. This may be relevant not only from an experimental perspective, as it may be difficult to determine precisely the current state at a given time, but perhaps more importantly, we may not be interested in restricting ourselves just to one state, but rather considering a whole set X of states believed to correspond to the current functional cellular state. Consequently, a gene that may be the best candidate for intervention for a particular starting state in X may not be the best for another state in X. Hence, we combine the individual results for all states $x \in X$ but weigh them by their respective probabilities of occurrence. The latter is furnished by the steady-state probabilities $\pi(x)$. We define

$$F_k(X, Y) = \frac{\sum_{x \in X}\sum_{y \in Y} F_k(x, y) \cdot \pi(x)}{\sum_{x \in X} \pi(x)} \tag{5.8}$$

to be the first passage probability from a set X to a set Y.

In addition to reaching a desired state or set of states, we may also be interested in avoiding a state or set of states. This is quite natural in terms of inducing a network not to enter into a set of states corresponding to unwanted functional cellular state (e.g., proliferation). This goal is in a sense complementary to what has been previously described in terms of reaching a desired state either as soon as possible with a given probability or with as high a probability as possible, before a given time. For example, in Eq. (5.5), the goal is to minimize the MFPT to a destination state. In order to avoid a destination state, we maximize the MFPT to that state.

As a final point, depending on the network, as well as on the starting and destination states or sets of states, it may be that not intervening with any gene is optimal in terms of the criteria given in Eqs. (5.4), (5.5), and (5.6).

5.2 Finite-Horizon Control

Rather than just consider a single flip of a gene state to alter the network trajectory, one can consider a set of auxiliary variables whose values can be chosen to make the probability distribution vector of the PBN evolve in some desirable fashion over some period of time. The use of such auxiliary variables makes sense from a biological perspective. For instance, in the case of diseases like cancer, auxiliary treatment inputs such as radiation, chemotherapy, etc., may be employed to move the state probability distribution vector away from one associated with uncontrolled cell proliferation or markedly reduced apoptosis. The auxiliary variables could include genes which serve as external master-regulators for all the genes in the network. To be consistent with the binary nature of the expression status of individual genes in a binary PBN, we will assume that the auxiliary variables (control inputs) take on the binary values 0 or 1. The values of the individual control inputs can be changed from one time step to another in an effort to make the network behave in a desirable fashion.

5.2.1 Control problem

Consider a PBN with n genes and having k control inputs $u_1, u_2, \ldots, u_k$. At any given time step t, the vector $u(t) = (u_1(t), u_2(t), \ldots, u_k(t))^T$ describes the complete status of all the control inputs and $u(t)$ can take on all binary values from $00\cdots0$ to $11\cdots1$. We can equivalently represent the control input status at time t by letting $u(t)$ be an integer ranging from 1 to 2^k and will typically take this latter approach. Letting P denote the transition probability matrix of the PBN, absent control, the probability distribution vector evolves according to $w(t+1) = w(t)P$. With control, the transition probability matrix depends on $u(t)$ and the evolution of the probability distribution vector at time t takes place according to the equation

$$w(t+1) = w(t)P(u(t)). \tag{5.9}$$

The transition probability matrix is now a function of all the control inputs, $u_1(t)$, $u_2(t), \ldots, u_k(t)$. Consequently, the evolution of the probability distribution vector of the PBN with control now depends not only on the initial distribution vector but also on the values of the control inputs at different time steps. Our objective is to make the states of the network evolve in a desirable fashion by appropriately choosing the control input at each time step.

Equation (5.9) is referred to in the control literature as a *controlled Markov chain*. Markov chains of this type occur in many applications. Given a controlled Markov chain, the objective is to come up with a sequence of control inputs, referred to as a *control strategy*, or *control policy*, such that an appropriate cost function is minimized over the entire class of allowable control strategies. To arrive at a useful solution, the cost function must capture the costs and benefits of using any control. The actual design of a "good" cost function is application dependent and is likely to require considerable expert knowledge.

In the present section, we discuss *finite-horizon control*, in which control is applied over a finite number of steps and then stopped. For instance, in the case of radiation treatment for cancer, the patient may be treated with radiation over a fixed interval of time, after which the treatment is suspended for some time as the effects are evaluated. After that, the treatment may be applied again, but in this section we are interested only in optimizing over individual treatment windows, which are assumed to be finite.

Suppose that the number of steps over which the control input is to be applied has been *a priori* determined to be M, so that we wish to control the behavior of the PBN over the interval $t = 0, 1, 2 \ldots, M-1$. Suppose, at time step t, the state of the PBN is given by $z(t)$ and the corresponding control input is $u(t)$. We define a *cost of control*, $C_t(z(t), u(t))$, as being the cost of applying the control input $u(t)$ when the state is $z(t)$. With this definition, given an initial state $z(0)$, the expected cost of control over the entire treatment horizon becomes

$$\gamma_{z(0)}(M) = E\left[\sum_{t=0}^{M-1} C_t(z(t), u(t)) | z(0)\right]. \tag{5.10}$$

Even if the network starts from a given (deterministic) initial state $z(0)$, the subsequent states will be random owing to the stochastic nature of the evolution. Consequently, $\gamma_{z(0)}(M)$ must be defined using an expectation. The cost of control $\gamma_{z(0)}(M)$ is one component of the finite-horizon cost. We now proceed to introduce the second component.

The net result of the control actions $u(0), u(1), \ldots, u(M-1)$ is that the state of the PBN will transition according to $P(u(t))$ and will end up in some state $z(M)$. Owing to the probabilistic nature of the evolution, the terminal state is a random variable that can possibly take on any of the values $1, 2, \ldots, m2^n$, where m is the number of contexts and where the states, each of which is a (context, GAP) pair, are labeled numerically and the notation z is used to denote a state. Depending on the particular PBN and the control inputs used at each step, it is possible that some states may never be reached because of noncommunicating states in the resulting Markov chain; however, since the control strategy itself has not yet been determined, it would be difficult, if not impossible, to identify and exclude such states from further consideration. Instead, we assume that all $m2^n$ terminal states are reachable and assign a penalty, or *terminal cost*, $C_M(z(M))$ to each of them.

To arrive at the terminal cost, consider the original PBN without control, divide the states into different categories depending on their desirability, and assign higher terminal costs to undesirable states. For instance, a state associated with rapid cell proliferation leading to cancer should be associated with a high terminal cost while a state associated with normal behavior should be assigned a low terminal cost. For the purposes of this section, we will assume that the assignment of terminal costs has been carried out and we have at our disposal a terminal cost $C_M(z(M))$, which is a function of the terminal state.

Recognizing that $C_M(z(M))$ is a random variable, so that its expectation is required to define an overall cost function, and taking into account the cost of control in Eq. (5.10), the *finite-horizon cost* to be minimized is defined by

$$\Gamma_{z(0)}(M) = E\left[\sum_{t=0}^{M-1} C_t(z(t),u(t)) + C_M(z(M))|z(0)\right]. \tag{5.11}$$

We assume that at time t the control input $u(t)$ is a function, μ_t, of $z(t)$, namely,

$$u(t) = \mu_t(z(t)), \tag{5.12}$$

where, for $t = 0,1,\ldots,M-1$,

$$\mu_t : \{1,2,3,\ldots,m2^n\} \rightarrow [1,2,3,\ldots,2^k] \tag{5.13}$$

is a function mapping the state space into the control space and

$$\omega = (\mu_0,\mu_1,\ldots,\mu_{M-1}) \tag{5.14}$$

defines a control policy. Under this transformation, Eq. (5.11) becomes

$$J_\omega(z(0)) = E\left[\sum_{t=0}^{M-1} C_t(z(t),\mu_t(z(t)) + C_M(z(M))|z(0)\right]. \tag{5.15}$$

The *optimal control problem* can now be stated: Given an initial state $z(0)$, find a control policy that minimizes the cost functional $J_\omega(z(0))$. Specifically, solve the optimization problem

$$\min_{\mu_0,\mu_1,\ldots,\mu_{M-1}} E\left[\sum_{t=0}^{M-1} C_t(z(t),\mu_t(z(t)) + C_M(z(M))|z(0)\right] \tag{5.16}$$

subject to the constraint

$$P(z(t+1) = j|z(t) = i) = p_{ij}(u(t)), \tag{5.17}$$

where $p_{ij}(u(t))$ is the ith row, jth column entry of the matrix $P(u(t))$.

5.2.2 Solution by dynamic programming

Optimal control problems of the type described by Eqs. (5.16) and (5.17) can be solved by using the technique of *dynamic programming*. This technique, pioneered by Bellman (1957), is based on the *principle of optimality*. To explain this principle, suppose we wish to optimize a performance index over a finite number of steps, say M. At each step, a decision is made and the objective is to find a strategy, or sequence of M decisions, that is optimal in the sense that the cumulative performance index over all the M steps is optimized. In general, such an optimal strategy may not exist. However, when such an optimal strategy does exist, the principle of optimality asserts the following: if one searches for an optimal strategy over a subset of the original number of steps, then this new optimal strategy will be given by the overall optimal strategy restricted to the steps being considered. Although intuitively obvious, the principle of optimality has significant consequences. For instance, it can be used to obtain the following proposition.

Theorem 5.1 (Bertsekas, 1976). *Let $J^*(z(0))$ be the optimal value of the cost function $J_\omega(z(0))$. Then*

$$J^*(z(0)) = J_0(z(0)), \tag{5.18}$$

where J_0 is given by the last step of the following dynamic programming algorithm, which proceeds backward in time from time step $M-1$ to time step 0:

$$J_M(z_M) = C_M(z_M), \tag{5.19}$$

$$J_t(z(t)) = \min_{u(t)\in\{1,2,\ldots,2^k\}} E[C_t(z(t),u(t)) + J_{t+1}(z(t+1))|z(t),u(t)] \tag{5.20}$$

for $t = 0,1,2,\ldots,M-1$. Furthermore, if $u_t^* = \mu_t^*(z(t))$ minimizes the right-hand side of Eq. (5.20) for each $z(t)$ and t, then the control policy $\omega^* = (\mu_0^*, \mu_1^*, \ldots, \mu_{M-1}^*)$ is optimal.

In view of Eq. (5.17),

$$E[J_{t+1}(z(t+1))|z(t),u(t)] = \sum_{j=1}^{m2^n} p_{z(t),j}(u(t)) J_{t+1}(j). \tag{5.21}$$

Thus, the dynamic programming solution to Eqs. (5.16) and (5.17) is given by

$$J_M(z_M) = C_M(z_M), \tag{5.22}$$

$$J_t(z(t)) = \min_{u(t)\in\{1,2,\ldots,2^k\}} \left[C_t(z(t),u(t)) + \sum_{j=1}^{m2^n} p_{z(t),j}(u(t)) J_{t+1}(j) \right] \tag{5.23}$$

for $t = 0,1,2,\ldots,M-1$.

In what follows, we will make two notational adjustments. To simplify the notation we will often write $z(t)$ and $u(t)$) as z_t and u_t, respectively, so that $C_t(z(t),u(t))$ becomes $C_t(z_t,u_t)$. To emphasize the state of the network or the control action, instead of $C_t(z(t),u(t))$ we will write $C_t(i,u)$, where i denotes the state at time t, u denotes the control action at time t, and the time is indicated by the subscript.

5.2.3 Illustrative example

To provide an illustrative example, from Datta *et al.* (2003), we consider the uncontrolled instantaneously random PBN whose truth table is given in Example 2.1. To introduce control, let x_1 be a control input whose value can be externally switched between 0 and 1 and let the states of the new PBN be x_2 and x_3. To be consistent with the notation in this section, the variables x_1, x_2, and x_3 will be renamed as u_1, x_1, and x_2, respectively. With this change, we have the truth table shown in Table 5.1, which also contains the values of the variables v and z corresponding to u_1 and (x_1,x_2), respectively. The values of $c_j^{(i)}$ in the table indicate that there are two constituent networks, corresponding to the function pairs $(f_1^{(1)}, f_1^{(2)})$ and $(f_1^{(1)}, f_2^{(2)})$. The selection probabilities c_1 and c_2 associated with these networks are given by $c_1 = c_2 = 0.5$.

We compute the matrices $P(1)$ and $P(2)$ corresponding to the two possible values for v. According to Table 5.1, when $v = 1$, the following transitions are associated with the network N_1 and occur with probability c_1:

$$z=1 \to z=1, \quad z=2 \to z=3, \quad z=3 \to z=3, \quad z=4 \to z=2.$$

Table 5.1. *Truth tables of the predictors for the PBN in our example. The selection probabilities of each predictor are shown in the bottom row.*

u_1	v	x_1	x_2	z	$f_1^{(1)}$	$f_1^{(2)}$	$f_2^{(2)}$
0	1	0	0	1	0	0	0
0	1	0	1	2	1	0	0
0	1	1	0	3	1	0	0
0	1	1	1	4	0	1	0
1	2	0	0	1	1	0	0
1	2	0	1	2	1	1	0
1	2	1	0	3	0	1	0
1	2	1	1	4	1	1	1
$c_j^{(i)}$					1	0.5	0.5

The corresponding transitions associated with network N_2 that occur with probability c_2 are given by

$$z=1 \rightarrow z=1, \quad z=2 \rightarrow z=3, \quad z=3 \rightarrow z=3, \quad z=4 \rightarrow z=1.$$

Consequently,

$$P(1) = \begin{pmatrix} 1 & 0 & 0 & 0 \\ 0 & 0 & 1 & 0 \\ 0 & 0 & 1 & 0 \\ c_2 & c_1 & 0 & 0 \end{pmatrix}.$$

We can analogously derive

$$P(2) = \begin{pmatrix} 0 & 0 & 1 & 0 \\ 0 & 0 & c_2 & c_1 \\ c_2 & c_1 & 0 & 0 \\ 0 & 0 & 0 & 1 \end{pmatrix}.$$

Since $n=2$, the variable z can take on any one of four values: $1,2,3,4$. Since $k=1$, the control variable v can take on two possible values: $1,2$. Suppose that the control action is to be carried out over 5 steps, so that $M=5$. Moreover, assume that the terminal penalties are given by $C_5(1)=0, C_5(2)=1, C_5(3)=2$, and $C_5(4)=3$. The choices of M and the values of the terminal penalties are completely arbitrary; in a real-world situation, this information would be obtained from biologists. The current choice of terminal penalties indicates that the most desirable terminal state is 1 while the least desirable terminal state is 4.

To set up the optimization problem, we need to define the function $C_t(z(t),u(t))$. For the sake of simplicity, we define $C_t(z(t),u(t)) = u_1(t)$, which captures the cost of applying the input $u_1(t)$ at the tth step. The optimization problem of Eqs. (5.19) and (5.20) is now determined and the dynamic programming algorithm resulting from Eqs. (5.22) and (5.23) becomes

$$J_5(z(5)) = C_5(z(5)),$$

$$J_t(z(t)) = \min_{v(t)\in\{1,2\}} \left[u_1(t) + \sum_{j=1}^{4} p_{z(t),j}(u(t)) J_{t+1}(j) \right]$$

for $t = 0,1,2,\ldots,4$. We proceed backwards step by step from $t = 4$ to obtain a solution. The resulting optimal control strategy for this finite-horizon problem is given by $\mu_0^*(z(0)) = \mu_0^*(z(1)) = \mu_0^*(z(2)) = \mu_0^*(z(3)) = 1$ for all $z(0)$, $z(1)$, $z(2)$, and $z(3)$, and $\mu_0^*(z(4)) = 2$ if $z(4) = 3$ and $\mu_0^*(z(4)) = 1$ otherwise. Thus the control input is applied only in the last time step provided the state of the system at that time step is equal to 3; otherwise, the optimal control strategy is to not apply any control at all.

We now consider a couple of initial states to illustrate the effects of control. If $z(0) = 1$, then the optimal control strategy is no control. According to the definition of the terminal penalties, for $z(0) = 1$ the evolution of the PBN is starting from the most desirable terminal state. Furthermore, from $P(1)$ we see that in the absence of any control, the state of the network remains at this position. Hence, the control strategy arrived at is, indeed, optimal and the value of the optimal cost is 0. Next suppose $z(0) = 4$, In this case, the evolution of the PBN is starting from the most undesirable terminal state. Moreover, from $P(2)$, note that if the control input were kept turned ON over the entire control horizon, then the state would continue to remain in this most undesirable position during the entire control duration. Such a control strategy cannot be optimal since not only does the network end up in the most undesirable terminal state, but also the maximum possible control cost is incurred over the entire time horizon.

To get an appreciation of the optimal control strategy, we consider the cases in which the PBN degenerates into a standard (deterministic) BN. There are two cases to consider. If $c_1 = 0$ and $c_2 = 1$, then the bottom row of $P(1)$ becomes $(1,0,0,0)$. Hence, if no control is employed, starting from $z(0) = 4$ the network will reach the state $z(1) = 1$ in one step and stay there forever. Thus, this no-control strategy is optimal and the optimal cost is 0. If $c_1 = 1$ and $c_2 = 0$, then the bottom row of $P(1)$ becomes $(0,1,0,0)$ and the second and third rows of $P(2)$ become $(0,0,0,1)$ and $(0,1,0,0)$, respectively. As derived above, the optimal control strategy is no control over the first four time steps. From $P(1)$ and $P(2)$ it follows that with $z(0) = 4$ we will have $z(1) = 2, z(2) = 3, z(3) = 3$, and $z(4) = 3$. For the optimal control, at the last time step the control input is turned ON and from $P(1)$ and $P(2)$ the resulting state is $z(5) = 2$. The optimal cost is given by 2 (the sum of the terminal cost and the cost of control).

5.2.4 Finite-horizon control in a melanoma network

Once again we turn to a WNT5A network, this time a ternary 7-gene network (WNT5A, pirin, S100P, RET1, MART1, HADHB, and STC2) considered in Datta *et al.* (2003), the expression status of each gene being quantized to one of three possible levels, -1 (down-regulated), 0 (unchanged), and 1. The GAP at each time step is ternary; nonetheless, the control strategy is developed exactly as described herein, the only difference being that now, for an n-gene network, there are 3^n states instead of 2^n states.

Because WNT5A ceasing to be down-regulated is strongly predictive of the onset of metastasis, the control objective for this 7-gene network is to externally down-regulate the WNT5A gene. The optimal control problem is completely specified by choosing (1) the treatment window, (2) the terminal penalty, and (3) the types of controls and the costs associated with them. For the treatment window, we arbitrarily choose a window of length 5; control inputs can be applied only at time steps 0, 1, 2, 3, and 4. The terminal penalty at time step 5 is chosen as follows. Since the objective is to ensure that WNT5A is down-regulated, the following penalties are assigned: 0 to all states for which WNT5A $= -1$, 3 to all states for which WNT5A $= 0$, and 6 to all states for which WNT5A $= 1$. The choice of the numbers 3 and 6 is arbitrary, but they reflect the intuitive notion that states where WNT5A $= 1$ are less desirable than those where WNT5A $= 0$. In real-world practice, these costs would have to be set by a biologist or physician.

Based on biological input, pirin is selected as the control variable, the control objective being to keep WNT5A down-regulated. The control action consists of either forcing pirin to -1 (corresponding to a control input of 1) or letting it remain wherever it is (corresponding to a control input of 0). A control cost of 1 is incurred at time step t if and only if pirin is forcibly reset to -1 at t. Using the resulting optimal controls, in Datta *et al.* (2003) the evolution of the state probability distribution vectors with and without control have been examined. For every possible initial state, simulations indicate that, at the final state, $P(\text{WNT5A} = -1)$ is higher with control than without control; however, in general, $P(\text{WNT5A} = -1)$ is not equal to 1 at the final time step. This is not surprising given that we are trying to control the expression status of WNT5A using another gene and the control horizon of length 5 simply may not be adequate for achieving the desired objective with such a high probability. Nevertheless, if the network starts from the state corresponding to STC2 $= -1$, HADHB $= 0$, MART-1 $= 0$, RET-1 $= 0$, S100P $= -1$, pirin $= 1$, WNT5A $= 1$ and evolves under optimal control, then $P(\text{WNT5A} = -1) = 0.673521$ at the final time step. This is quite good in view of the fact that the same probability would have been equal to 0 in the absence of any control action.

5.3 Infinite-Horizon Control

Finite-horizon control alters network dynamics over a transient period but does not alter the steady-state distribution. In this section we consider intervention via external control variables in PBNs over an infinite length of time, that is, *infinite-horizon* control. We desire a control policy that does not change from one time step to the next because implementation of such a stationary policy is simpler. Calculating the optimal stationary policy is quite involved. For finite-horizon control, we can use a backward dynamic programming algorithm and terminate it once the first stage is reached; for infinite-horizon control, this approach cannot be used. Optimization of the total cost rests on the fact that the total cost is finite for at least some control policy; however, if there is no termination state (state with zero cost), then the total cost may tend to infinity. This is the case with PBNs (ergodic Markov chains). Following Pal *et al.* (2006), we employ two approaches to avoid such divergence. The *discounted cost* shrinks the per-stage costs via multiplication by a discount factor less than 1 and yields a geometric series for the total cost that converges to a finite value for bounded costs. The *average cost* divides the total cost by the number of stages, a normalization that prevents divergence of the cost to infinity.

A PBN with control can be modeled as a stationary discrete-time dynamic system

$$z_{t+1} = f(z_t, u_t, w_t),\ t = 0, 1, \ldots, \tag{5.24}$$

where for all t, z_t is an element of a state space $\mathcal{S}$, the control input u_t is an element of a control space $\mathcal{C}$, the disturbance w_t is an element of a space $\mathcal{D}$, and $f : \mathcal{S} \times \mathcal{C} \times \mathcal{D} \mapsto \mathcal{S}$. For a PBN with n genes composed of m BNs with perturbation probability p and network transition probability q, $\mathcal{S} = \{1, 2, \ldots, m2^n\}$ and the control input u_t is constrained to take values in the space $\mathcal{C} = \{1, 2, \ldots, 2^k\}$. The disturbance w_t is manifested in terms of change of network based on the network transition probability q or change of state due to perturbation probability p. w_t is independent of prior disturbances $w_0, w_1, \ldots, w_{t-1}$.

With a finite horizon, there is a terminal state being separately penalized and the cost per stage, which we now write as $g_t(z_t, u_t)$, is assumed to depend only on the control applied and the current state. In the infinite-horizon problem, there is no terminal state or its associated terminal penalty. Consequently, the cost per stage should depend on the origin i, the destination j, and the applied control input u. In other words, $g_t(i,u)$ of the finite-horizon problem is replaced by $\tilde{g}(i,u,j)$, so that the per-stage cost takes into account the origin, the destination, and the control.

In the finite-horizon problem, the summation in Eq. (5.16) is finite and therefore the quantity being minimized is finite. If the control horizon goes to infinity, then the summation of the one-stage costs may diverge to infinity (for all controls), thereby leading to an ill-posed optimization problem. To make the optimization problem well-posed, the cost considered in Eq. (5.16) must be modified before letting the length M of the control horizon tend to infinity.

A standard approach is to assume that the cost per stage $\tilde{g}(i,u,j)$ is *bounded* for any $i, j \in \mathcal{S}$ and $u \in \mathcal{C}$ and to introduce a *discount factor* $\alpha \in (0,1)$ in the cost to ensure that the limit of the finite sums converges as $M \to \infty$. Specifically, our objective is to find a control policy $\omega = \{\mu_0, \mu_1, \ldots\}$, where $\mu_t : \mathcal{S} \to \mathcal{C}$, $t = 0, 1, \ldots$, that minimizes the cost function

$$J_\omega(z_0) = \lim_{M\to\infty} E\left[\sum_{t=0}^{M-1} \alpha^t \tilde{g}(z_t, \mu_t(z_t), w_t)|z(0)\right], \tag{5.25}$$

where the cost per stage $\tilde{g} : \mathcal{S} \times \mathcal{C} \times \mathcal{D} \to \Re$ is given and the expectation is with respect to the state and the disturbance. The problem is one of minimizing the total cost over an infinite number of stages with *discounted and bounded cost per stage*. In this formulation the cost is a function of the current state, the control, and the disturbance. This does not contradict our previous assumption that the cost for the control of PBNs is of the form $\tilde{g}(i,u,j)$, because a Markov chain can be modeled by $w_t = z_{t+1}$, in which case $\tilde{g}(z_t, \mu_t(z_t), w_t) = \tilde{g}(i,u,j)$ (Bertsekas, 1976). The inclusion of α in the cost captures the fact that costs incurred at a later time are less significant. In the case of cancer treatment, $\alpha < 1$ signifies that the condition of the patient in the initial stages of treatment is more important than the condition at a later stage, or in other words, the reward for improving the condition of the patient in the present is more significant than the reward obtained from similar improvement at a later stage.

A second way of avoiding the possibility of infinite total cost is to consider the *average cost per stage*, which is defined by

$$J_\omega(z_0) = \lim_{M\to\infty} \frac{1}{M} E\left[\sum_{t=0}^{M-1} \tilde{g}(z_t, \mu_t(z_t), w_t)|z(0)\right]. \tag{5.26}$$

A control policy $\omega = (\mu_0, \mu_1, \ldots)$ is chosen to minimize the above cost, and the problem is referred to as the *average cost per stage problem*. Minimization of the total cost is feasible if $J_\omega(z_0)$ is finite for at least some admissible policies ω and some admissible states z_0. If we consider no discounting, i.e., a *discount factor* of 1, and there is no zero-cost absorbing state (which is the case in PBNs with perturbation), then the total cost will frequently go to ∞. Hence, the *average cost per stage* formulation is essential when we are interested in the condition of the patient in the long run and equal importance is given to the patient's condition in all stages.

Since in Eqs. (5.25) and (5.26) the cost is obtained by taking the expectation with respect to the state and the disturbance, it is possible to replace $\tilde{g}(z_t, u_t, w_t)$ by an equivalent cost per stage that does not depend on the disturbance. This is accomplished by taking the expectation with respect to the disturbance and leaving only the expectation with respect to the state. This amounts to using the expected cost per stage in all calculations. More specifically, we use as cost per stage the expected cost $g(i,u)$ given by

$$g(i,u) = \sum_{j=1}^{m2^n} p_{ij}(u)\tilde{g}(i,u,j) \tag{5.27}$$

(Bertsekas, 1976), where $p_{ij}(u)$ is the transition probability under control u. In general, the cost $\tilde{g}(i,u,j)$ of moving from state i to state j under control u may depend on the starting state i; however, in the case of PBNs, we have no obvious basis for assigning different costs based on different initial states. Accordingly, we assume that the penalty $\tilde{g}(i,u,j)$ is independent of the starting state i and its value is based on the control effort and the terminal state j. The penalty is high if the end state is a bad state regardless of the starting state, and vice versa. Hence $\tilde{g}(i,u,j) = \tilde{g}(u,j)$ and Eq. (5.27) becomes

$$g(i,u) = \sum_{j=1}^{m2^n} p_{ij}(u)\tilde{g}(u,j) \tag{5.28}$$

For either total cost with discounted and bounded cost per stage or average cost per stage, we denote by Ω the set of all *admissible* policies ω, i.e., the set of all sequences of functions $\omega = (\mu_0, \mu_1, \ldots)$, with $\mu_t(z) : \mathcal{S} \to \mathcal{C}, t = 0, 1, \ldots$. For any $z \in \mathcal{S}$, the optimal cost function J^* is defined by

$$J^*(z) = \min_{\omega \in \Omega} J_\omega(z). \tag{5.29}$$

A *stationary policy* is an admissible policy of the form $\omega = (\mu, \mu, \ldots)$, and its corresponding cost function is denoted by J_μ. We say that the stationary policy $\omega = (\mu, \mu, \ldots)$ is optimal if $J_\mu(z) = J^*(z)$ for all states z.

5.3.1 Optimal control solution: Discounted and bounded cost per stage

In this section, we solve the problem of minimizing the cost in Eq. (5.25) under the assumption that the cost per stage $\tilde{g}(i,u,w)$ is bounded; that is, there exists $B > 0$ such that $|\tilde{g}(z,u,w)| \leq B$ for all $(z,u,w) \in \mathcal{S} \times \mathcal{C} \times \mathcal{D}$. This assumption holds for PBNs because the expected cost, $g(i,u)$, for state i is given by Eq. (5.28), $\sum_{j=0}^{2^n-1} p_{ij}(u) = 1$, and $\tilde{g}(u,j)$ is bounded since the control and disturbance spaces are finite.

If we set $g_M(z_M) = 0$ for all $z_M \in \mathcal{S}$ and $g_t(z_t, u_t) = \alpha^t g(z_t, u_t)$ in the finite-horizon problem of Eq. (5.16) and let $M \to \infty$, then we obtain the infinite-horizon cost function considered in Eq. (5.25). Thus, it seems reasonable that the finite-horizon solution described by Eqs. (5.22) and (5.23) in the previous section could provide a basis for arriving at the solution of the optimization problem of Eq. (5.29), where J_ω is given by Eq. (5.25). A formal derivation of this connection is given in Bertsekas (1976). Here we simply state the result and present an intuitive justification for it. Note that Eq. (5.23) in the dynamic programming algorithm basically describes how the optimal cost J_{t+1} propagates backwards in time to the optimal cost J_t in the finite-horizon problem of Eq. (5.16). For the cost function of Eq. (5.25), it is clear that the cost J_{t+1} must be discounted by the factor α while being propagated to the previous stage. Consequently, for the optimal control problem of this section, Eq. (5.16) is replaced by

$$J_t(i) = \min_{u \in C} \left[g(i,u) + \alpha \sum_{j=1}^{m2^n} p_{ij}(u) J_{t+1}(j) \right]. \tag{5.30}$$

The preceding equation motivates the introduction of two mappings. For any cost function $J : \mathcal{S} \to \Re$, define the mapping $TJ : \mathcal{S} \to \Re$ by

$$(TJ)(i) = \min_{u \in C} \left[g(i,u) + \alpha \sum_{j=1}^{m2^n} p_{ij}(u) J(j) \right]. \tag{5.31}$$

Note that TJ is the optimal cost function for the one-stage (finite-horizon) problem that has stage cost g and terminal cost αJ. Similarly for any cost function $J : \mathcal{S} \to \Re$ and control function $\mu : \mathcal{S} \to C$, define the mapping $T_\mu J : \mathcal{S} \to \Re$ by

$$(T_\mu J)(i) = g(i, \mu(i)) + \alpha \sum_{j=1}^{m2^n} p_{ij}(\mu(i)) J(j); \tag{5.32}$$

$T_\mu J$ can be viewed as the cost function associated with the policy μ for the one-stage problem that has stage cost function g and terminal cost αJ. Since the mappings T and T_μ map functions $J : \mathcal{S} \to \Re$ into new functions mapping $\mathcal{S} \to \Re$, one can define the composition of T with itself and T_μ with itself as follows:

$$(T^l J)(i) = (T(T^{l-1} J))(i), \tag{5.33}$$

$$(T^0 J)(i) = J(i), \tag{5.34}$$

and

$$(T_\mu^l J)(i) = (T_\mu(T_\mu^{l-1} J))(i) \tag{5.35}$$

$$(T_\mu^0 J)(i) = J(i) \tag{5.36}$$

for $i \in \mathcal{S}$ and $l = 1, 2, \ldots$. The mappings T and T_μ play key roles in the solution of the optimal control problem: (i) the optimal cost function J^* is the unique fixed point of the mapping T; (ii) the iteration $J_{t+1} = TJ_t$ converges to J^* as $t \to \infty$; and (iii) the mapping T_μ can be used to characterize the conditions under which a given stationary policy μ is optimal. These properties are formalized in the following three theorems adapted from Bertsekas (1976), as stated and proved in Pal *et al.* (2006).

Theorem 5.2 (Convergence of the discounted-cost algorithm). *For any bounded cost function $J : \mathcal{S} \to \Re$, the optimal cost function J^* satisfies*

$$J^*(i) = \lim_{M\to\infty} (T^M J)(i) \tag{5.37}$$

for all $i \in \mathcal{S}$.

Proof. To begin with,

$$\begin{aligned} J_\omega(z_0) &= \lim_{M\to\infty} E\left[\sum_{t=0}^{M-1} \alpha^t g(z_t, \mu_t(z_t))\right] \\ &= E\left[\sum_{t=0}^{K-1} \alpha^t g(z_t, \mu_t(z_t))\right] \\ &\quad + \lim_{M\to\infty} E\left[\sum_{t=K}^{M-1} \alpha^t g(z_t, \mu_t(z_t))\right]. \end{aligned}$$

Since $|g(z,u)| \le B$,

$$\left|\lim_{M\to\infty} E\left[\sum_{t=K}^{M-1} \alpha^t g(z_t, \mu_t(z_t))\right]\right| \le B \sum_{t=K}^{\infty} \alpha^t = \frac{\alpha^K B}{1-\alpha}.$$

Using these relations, we obtain

$$\begin{aligned} J_\omega(z_0) - \frac{\alpha^K B}{1-\alpha} - \alpha^K \max_{z\in\mathcal{S}} |J(z)| &\le E\left[\alpha^K J(z_K) + \sum_{t=0}^{K-1} \alpha^t g(z_t, \mu_t(z_t))\right] \\ &\le J_\pi(z_0) + \frac{\alpha^K B}{1-\alpha} + \alpha^K \max_{z\in\mathcal{S}} |J(z)|. \end{aligned} \tag{5.38}$$

Minimization over ω gives

$$\begin{aligned} J^*(z_0) - \frac{\alpha^K B}{1-\alpha} - \alpha^K \max_{z\in\mathcal{S}} |J(z)| &\le (T^K J)(z_0) \\ &\le J^*(z_0) + \frac{\alpha^K B}{1-\alpha} + \alpha^K \max_{z\in\mathcal{S}} |J(z)|. \end{aligned} \tag{5.39}$$

Letting $K \to \infty$ and utilizing the fact that $\alpha < 1$ gives the result. □

Theorem 5.3 (Bellman's equation). *The optimal cost function J^* satisfies*

$$J^*(i) = \min_{u\in C}\left[g(i,u) + \alpha \sum_{j=1}^{m2^n} p_{ij}(u) J^*(j)\right] \tag{5.40}$$

for all $i \in \mathcal{S}$ or, equivalently, $J^* = TJ^*$. Furthermore, J^* is the unique solution of this equation within the class of bounded functions.

Proof. From Eq. (5.39), for all $z \in \mathcal{S}$ and M,

$$J^*(z) - \frac{\alpha^M B}{1-\alpha} \le (T^M J_0)(z) \le J^*(z) + \frac{\alpha^M B}{1-\alpha}, \tag{5.41}$$

where J_0 is the zero function [$J_0(z) = 0$ for all $z \in \mathcal{S}$]. Applying the mapping T to Eq. (5.41) and using the monotonicity property of T yields

$$(TJ^*)(z) - \frac{\alpha^{M+1} B}{1-\alpha} \le (T^{M+1} J_0)(z) \le (TJ^*)(z) + \frac{\alpha^{M+1} B}{1-\alpha} \tag{5.42}$$

for all $z \in \mathcal{S}$ and M. Since $(T^{M+1} J_0)(z)$ converges to $J^*(z)$ (from convergence of the dynamic programming algorithm), letting $M \to \infty$ in Eq. (5.42) yields $J^* = TJ^*$. Uniqueness of the solution can be proved as follows: if J is bounded and satisfies $J = TJ$, then $J = \lim_{M\to\infty} T^M J$ and by convergence of the dynamic programming algorithm, $J = J^*$. □

Theorem 5.4 (Necessary and sufficient condition for optimality). *A stationary policy μ is optimal if and only if $\mu(z)$ attains the minimum in Bellman's equation* (5.40) *for each $z \in \mathcal{S}$; i.e.,*

$$TJ^* = T_\mu J^*. \tag{5.43}$$

Proof. If $TJ^* = T_\mu J^*$, then using Bellman's equation ($J^* = TJ^*$) gives $J^* = T_\mu J^*$, so by the uniqueness of the optimal solution, $J^* = J_\mu$; i.e., μ is optimal. On the other hand, if the stationary policy μ is optimal, then $J^* = J_\mu$, which yields $J^* = T_\mu J^*$. Combining this with Bellman's equation yields $TJ^* = T_\mu J^*$. □

The three preceding theorems can be used to computationally find the optimal policy. According to Theorem 5.3, the optimal cost function satisfies Bellman's equation and Theorem 5.2 states that the optimal cost function can be iteratively determined by running the recursion

$$J_{t+1} = TJ_t, \quad t = 0, 1, 2, \ldots, \tag{5.44}$$

for any bounded initial cost function $J_0 : \mathcal{S} \to \Re$. Since this iteration is guaranteed to converge to J^*, the iteration can be run until some stopping criterion is reached. The resulting policy is stationary and, by Theorem 5.3, must be optimal. The iteration described in Eq. (5.44) is referred to as the *Value Iteration* procedure since at every stage we are iterating on the values of the cost function and the optimal policy simply falls out as a by-product when the iteration converges to the optimal value of the cost function.

An alternative approach for solving the optimal control problem under discussion is referred to as *Policy Iteration* and involves the following matrices and vectors:

$$J = \begin{pmatrix} J(1) \\ \cdot \\ \cdot \\ J(m2^n) \end{pmatrix},$$

$$J_\mu = \begin{pmatrix} J_\mu(1) \\ \cdot \\ \cdot \\ J_\mu(m2^n) \end{pmatrix},$$

$$TJ = \begin{pmatrix} (TJ)(1) \\ \cdot \\ \cdot \\ (TJ)(m2^n) \end{pmatrix},$$

$$T_\mu J = \begin{pmatrix} (T_\mu J)(1) \\ \cdot \\ \cdot \\ (T_\mu J)(m2^n) \end{pmatrix}.$$

The transition probability matrix corresponding to the stationary policy μ is represented as

$$P_\mu = \begin{pmatrix} p_{11}(\mu(1)) & \dots & p_{1,m2^n}(\mu(1)) \\ \vdots & \ddots & \vdots \\ p_{m2^n,1}(\mu(m2^n)) & \dots & p_{m2^n,m2^n}(\mu(m2^n)) \end{pmatrix}$$

and g_μ represents the cost vector

$$g_\mu = \begin{pmatrix} g(1,\mu(1)) \\ \cdot \\ \cdot \\ g(m2^n,\mu(m2^n)) \end{pmatrix}.$$

It is straightforward to see that, for any stationary policy μ, Eq. (5.32) can be rewritten as

$$T_\mu J = g_\mu + \alpha P_\mu J.$$

By reasoning similar to proof of Theorem 5.3, it can be shown that the cost J_μ corresponding to the policy μ satisfies

$$J_\mu = g_\mu + \alpha P_\mu J_\mu$$

or

$$[I - \alpha P_\mu] J_\mu = g_\mu. \tag{5.45}$$

Eq. (5.45) is a system of linear equations that can be solved to calculate the cost J_μ corresponding to a given stationary policy μ. In the policy iteration algorithm, one starts with a given stationary policy, evaluates the corresponding cost using Eq. (5.45), and tries to find a policy that yields a smaller cost. The process is terminated when the iteration arrives at a fixed point of the mapping T.

The following steps are involved in the policy iteration algorithm:

Step 1: (*Initialization*) An initial policy μ^0 is selected.

Step 2: (*Policy evaluation*) Given a stationary policy μ^k, compute the corresponding cost function J_{μ^k} from the linear system of equations

$$(I - \alpha P_{\mu^k})J_{\mu^k} = g_{\mu^k}. \tag{5.46}$$

P_{μ^k} is the probability transition matrix obtained using control policy μ^k.

Step 3: (*Policy improvement*) An improved (relative to the cost J) stationary policy μ^{k+1} satisfying $T_{\mu^{k+1}} J_{\mu^k} = T J_{\mu^k}$ is obtained.

The iterations are stopped if $J_{\mu^k} = T J_{\mu^k}$; else return to Step 2 and repeat the process.

5.3.2 Optimal control solution: Average cost per stage

Minimization of the cost in Eq. (5.26) requires a policy that minimizes the limit of $\frac{J_M}{M}$ as $M \to \infty$, where J_M is the optimal finite-horizon cost over an interval of length M. Similar reasoning leads to the counterparts of Eqs. (5.30), (5.31), and (5.32), which now become

$$J_t(i) = \min_{u \in C} \left[g(i,u) + \sum_{j=1}^{m2^n} p_{ij}(u) J_{t+1}(j) \right], \tag{5.47}$$

$$(TJ)(i) = \min_{u \in C} \left[g(i,u) + \sum_{j=1}^{m2^n} p_{ij}(u) J(j) \right], \tag{5.48}$$

$$(T_\mu J)(i) = g(i,\mu(i)) + \sum_{j=1}^{m2^n} p_{ij}(\mu(i)) J(j) \tag{5.49}$$

for $i \in \mathcal{S}$. However, in this case the value iteration

$$J_{t+1}(i) = T J_t(i) \tag{5.50}$$

considered in the last section cannot be used directly because, in the absence of the discount factor, it may diverge to infinity. Thus, calculating the average cost by taking $\lim_{M\to\infty} \frac{J_M}{M}$ is not feasible.

Instead we consider a *differential cost* h_t obtained by subtracting a fixed component of J_t, say $J_t(n_1)$, from each element of J_t, namely,

$$h_t(i) = J_t(i) - J_t(n_1) \tag{5.51}$$

for any $i \in \mathcal{S}$. Clearly $h_t(n_1) = 0$. Defining $e = [1, 1, \ldots, 1]^T$, the above relationship can be rewritten as

$$h_t = J_t - J_t(n_1)e. \tag{5.52}$$

Similarly,

$$h_{t+1} = J_{t+1} - J_{t+1}(n_1)e. \tag{5.53}$$

Substituting for J_t and J_{t+1} into Eq. (5.50) yields

$$h_{t+1} + J_{t+1}(n_1)e = T(h_t + J_t(n_1)e), \tag{5.54}$$

which implies

$$h_{t+1} = Th_t - (J_{t+1}(n_1)e - J_t(n_1)e). \tag{5.55}$$

First, applying Eq. (5.50) and then Eq. (5.52) yields

$$\begin{aligned} J_{t+1}(n_1) &= T(J_t)(n_1) \\ &= T(h_t + J_t(n_1)e)(n_1) \\ &= Th_t(n_1) + J_t(n_1). \end{aligned} \tag{5.56}$$

It follows that

$$J_{t+1}(n_1)e - J_t(n_1)e = Th_t(n_1)e,$$

so that Eq. (5.55) yields

$$h_{t+1} = Th_t - (Th_t)(n_1)e \tag{5.57}$$

as the *value iteration algorithm* for the differential cost.

We next provide two theorems adapted from Bertsekas (1976), as stated and proved in Pal et al. (2006): the first formalizes the intuition that, since the average optimal cost is calculated over an infinite horizon, its value should be independent of the starting state; the second formalizes the fact that if the value iteration of Eq. (5.57) for the differential cost converges to some vector h, i.e., $(Th)(n_1)e + h = Th$, then $Th(n_1)$ is the optimal average cost per stage (which is the same for all initial states).

Theorem 5.5. *For ergodic Markov chains, the optimal average cost per stage is independent of the initial state.*

Proof. First, let us assume that some reference state z_r is recurrent for the optimal stationary policy π. Then the average cost from a state i in the communicating class containing z_r, defined by

$$J_\omega(z_0 = i) = \lim_{M\to\infty} \frac{1}{M} E\left[\sum_{t=0}^{M-1} g(z_t, \mu_t(z_t))\right],$$

can be written as

$$J_\omega(z_0 = i) = \lim_{M\to\infty} \frac{1}{M} E\left[\sum_{t=0}^{r_i-1} g(z_t, \mu_t(z_t)) + \sum_{t=r_i}^{M-1} g(z_t, \mu_t(z_t))\right],$$

where r_i is the smallest integer such that $z_{r_i} = z_r$. Recurrence of the state z_r guarantees the finiteness of r_i. Hence, when $M \to \infty$, the first term becomes negligible and we have

$$J_\omega(z_0 = i) = J_\omega(z_0 = z_r),$$

thereby showing that the optimal average cost per stage is independent of the starting state.

Next, if the application of the stationary policy breaks up the states into separate communicating classes $C_1,\ldots,C_s$, then these will have average costs $\gamma_1,\ldots,\gamma_s$, respectively. The optimal stationary policy will always drive the states to the communicating class having the lowest average cost. Hence, for the optimal stationary policy, there exists a state z_r approachable from all other states in a finite number of time steps. This state z_r can be used in the preceding argument to show that the optimal average cost per stage is independent of the starting state. □

Theorem 5.6. *If a scalar λ and an $m2^n$-dimensional vector h satisfy*

$$\lambda + h(i) = \min_{u\in C}\left[g(i,u) + \sum_{j=1}^{m2^n} p_{ij}(u)h(j)\right] \tag{5.58}$$

for any $i \in \mathcal{S}$ or, equivalently,

$$\lambda e + h = Th, \tag{5.59}$$

where e is the unitary vector $(1,1,\ldots,1)^T$ and $h = (h(1),h(2),\ldots,h(m2^n))^T$, then λ is the optimal average cost per stage $J^*(i)$ for all i, namely,

$$\lambda = \min_{\omega} J_\omega(i) = J^*(i) \tag{5.60}$$

for all $i \in \mathcal{S}$. Furthermore, if $\mu^*(i)$ attains the minimum in Eq. (5.58) for each i, then the stationary policy μ^* is optimal, i.e., $J_{\mu^*}(i) = \lambda$ for all $i \in \mathcal{S}$.

Proof. Let $\omega = (\mu_0,\mu_1,\ldots)$ be any admissible policy and M be a positive integer. By Eq. (5.58), $\lambda e + h = Th$. Therefore,

$$TT_{\mu_{M-1}}h \geq Th = \lambda e + h. \tag{5.61}$$

Applying $T_{\mu_{M-2}}$ to both sides of this relation and using the monotonicity of $T_{\mu_{M-2}}$ yields

$$T_{\mu_{M-2}}T_{\mu_{M-1}}h \geq T_{\mu_{M-2}}(\lambda e + h) = \lambda e + T_{\mu_{M-2}}h \geq 2\lambda e + h. \tag{5.62}$$

Continuing in the same manner, we finally obtain

$$T_{\mu_0}T_{\mu_1}\cdots T_{\mu_{M-1}}h \geq M\lambda e + h, \tag{5.63}$$

having equality if each μ_t, $t = 0,1,\ldots,M-1$, attains the minimum in Eq. (5.58). $T_{\mu_0}T_{\mu_1}\cdots T_{\mu_{M-1}}h(i)$ is equal to the M-stage cost corresponding to initial state i, policy $\mu_0,\mu_1,\ldots,\mu_{M-1}$, and terminal cost function h; that is,

$$T_{\mu_0}T_{\mu_1}\cdots T_{\mu_{M-1}}h(i) = E\left[h(z_M) + \sum_{t=0}^{M-1} g(z_t,\mu_t(z_t))|z_0 = i,\omega\right]. \tag{5.64}$$

Using this relation in Eq. (5.63) and dividing by M gives

$$\frac{1}{M}E\{h(z_M)|z_0 = i,\omega\} + \frac{1}{M}E\left[\sum_{t=0}^{M-1} g(z_t,\mu_t(z_t))|z_0 = i,\omega\right] \geq \lambda + \frac{1}{M}h(i) \tag{5.65}$$

for all $i \in \mathcal{S}$. Referring to Eq. (5.26) and letting $M \to \infty$, we have $J_\omega(i) \geq \lambda$, $i = 1,\ldots,n$, with equality if $\mu_t(i)$, $t = 0,1,\ldots$, attains the minimum in Eq. (5.58) for each i. Hence λ is the optimal average cost per stage and, by Theorem 5.4, λ is the same for every initial state i. □

For the average cost per stage problem, Eq. (5.58) plays the same role as Bellman's equation, Eq. (5.40), in the solution of the discount problem. Thus, we arrive at the following *policy iteration* algorithm for this case:

Step 1: (*Initialization*) An initial policy μ^0 is selected.

Step 2: (*Policy evaluation*) Given a stationary policy μ^k, we obtain the corresponding average and differential costs λ^k and $h^k(i)$ satisfying

$$\lambda^k + h^k(i) = g(i,\mu^k(i)) + \sum_{j=1}^{m2^n} p_{ij}(\mu^k(i))h^k(j) \tag{5.66}$$

for all $i \in \mathcal{S}$.

This linear system of equations can be solved utilizing the fact that $h^k(n_1) = 0$, where $n_1 \in \mathcal{S}$ is any particular reference state.

Step 3: (*Policy improvement*) An improved stationary policy μ^{k+1} satisfying

$$g(i,\mu^{k+1}(i)) + \sum_{j=1}^{m2^n} p_{ij}(\mu^{k+1}(i))h^k(j) = \min_{u \in C}\left[g(i,u) + \sum_{j=1}^{m2^n} p_{ij}(u)h^k(j) \right], \tag{5.67}$$

or equivalently, $T_{\mu^{k+1}}h^k = Th^k$, is obtained.

The iterations are stopped if $\mu^{k+1} = \mu^k$; else return to Step 2 and repeat the process.

5.3.3 Infinite-horizon control for a WNT5A network

We consider a 7-gene network containing the genes WNT5A, pirin, S100P, RET1, MART1, HADHB, and STC2. To obtain the PBN, we have used the algorithms described in Section 3.3 to construct four highly probable BNs to use as the constituent networks. There are $4 \times 2^7 = 512$ states in the Markov chain, 4 contexts, and 128 GAPs. The last two binary digits of the state number represent the constituent network and the first seven binary digits constitute the GAP. The genes are ordered as WNT5A, pirin, S100P, RET1, MART1, HADHB, and STC2, with WNT5A as the most significant bit (MSB) of the GAP profile and STC2 as the least significant bit (LSB) of the GAP profile. The probabilistic parameters of the PBN are $p = q = 0.01$ and pirin is the control gene ($u = 1$ signifying the state of pirin is reversed and $u = 0$ signifying no intervention). The cost of control is assumed to be 1 and the states are assigned penalties as follows:

$$\tilde{g}(u,j) = \begin{cases} 5 & \text{if } u = 0 \text{ and WNT5A is 1 for state } j, \\ 6 & \text{if } u = 1 \text{ and WNT5A is 1 for state } j, \\ 1 & \text{if } u = 1 \text{ and WNT5A is 0 for state } j, \\ 0 & \text{if } u = 0 \text{ and WNT5A is 0 for state } j. \end{cases} \tag{5.68}$$

Since our objective is to down-regulate the WNT5A gene, a higher penalty is assigned for destination states having WNT5A up-regulated. Also for a given WNT5A status for the

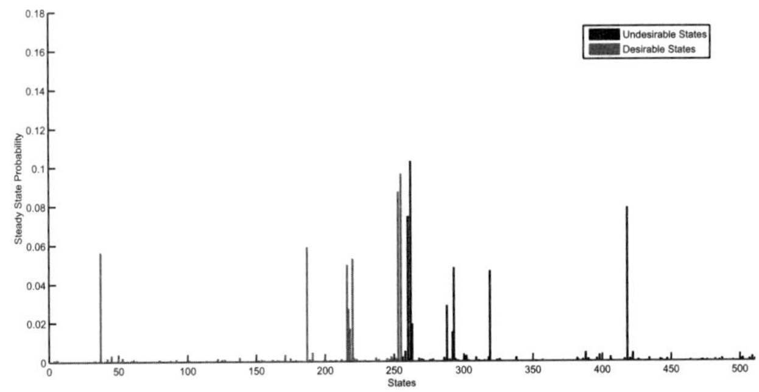

Figure 5.2. *Original steady-state distribution.*

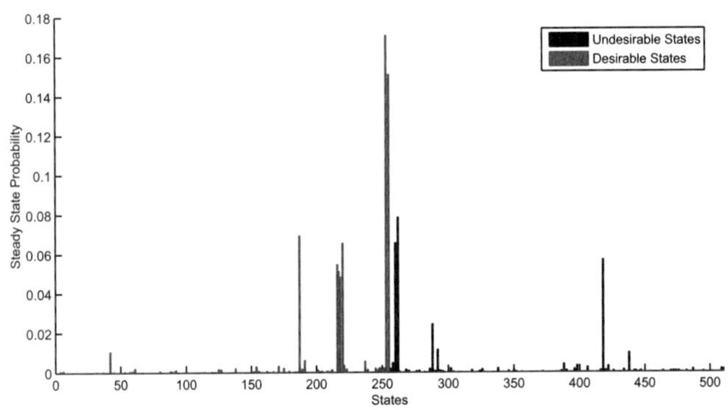

Figure 5.3. *Steady-state distribution using discounted cost stationary policy.*

destination state, a higher penalty is assigned when the control is active versus when it is not.

We first use the discounted-cost formulation with $\alpha = 0.9$ being the discount factor to design the optimal stationary policy. Figures 5.2 and 5.3 show the steady-state distributions of the original and controlled PBNs, respectively. States 1 through 256 have WNT5A = 0 (desirable) and states 257 through 512 have WNT5A = 1 (undesirable). The different steady-state distributions show that the stationary policy has shifted the probability mass from states with higher to lower metastatic competence. The probability mass of the undesirable states in the original steady-state distribution is 0.4939, while in the controlled steady-state distribution it has been reduced to 0.3061. To numerically quantify the change, we multiply the steady-state distribution by the cost vector. For the original PBN the cost vector is 0 for states 1 to 256 and 5 for states 257 through 512. For the stationary policy, the cost vector is $\tilde{g}(\mu(z), z)$, $z \in \{1, \ldots, 512\}$. The value for the stationary policy is 1.5551 as compared to 2.4695 for no control.

Using the average-cost-per-stage formulation to design the optimal stationary policy, the probability mass of the undesirable states in the steady-state distribution is 0.3052. Figure 5.4 shows the distribution. In this case, the controlled steady-state distribution multiplied by the cost vector is 1.5523.

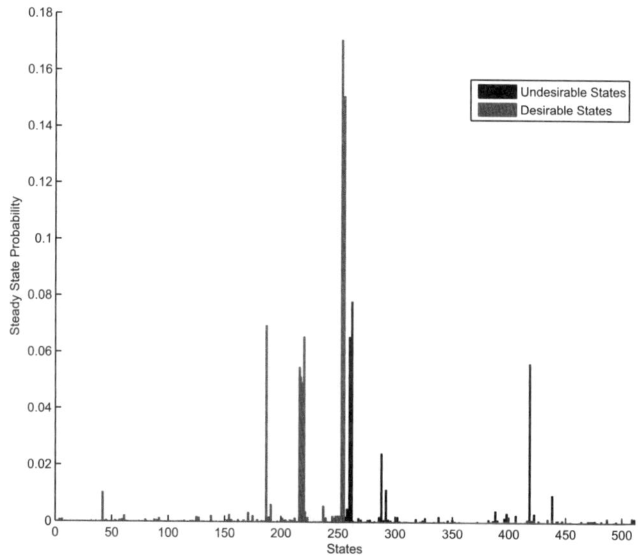

Figure 5.4. *Steady-state distribution using average cost stationary policy.*

5.4 Approximation

In this section we consider approaches to the external control of PBNs that in one way or another approximate the explicit, full-optimization methods thus far presented. We consider a linear optimization model to reduce computational complexity, control of a family of BNs in which the context probability structure of a PBN is unknown, control when the states are not observable, thereby resulting in imperfect information for state feedback, design of the control policy on an instantaneously random PBN obtained from a context-sensitive PBN via state reduction, and approximate policies designed via reinforcement learning.

5.4.1 A linear model

In general, it is well known that the direct application of optimal control methods is limited by the size of the state space—the curse of dimensionality (Bertsekas, 2005). In particular, it has been demonstrated that the finite-horizon stochastic control of a PBN is an NP-hard problem (Akutsu *et al.*, 2007). Hence, optimal methods such as those previously discussed are applicable only to models with small numbers of genes.

One way to address the complexity issue is to use a discrete linear model for controlling an instantaneously random PBN in a finite horizon:

$$w(t+1) = \alpha_t w(t)P + \beta_t u(t)B, \tag{5.69}$$

where α_t and β_t are two real-valued parameters such that $\alpha_t + \beta_t = 1$ and B is the control transition matrix (Ng *et al.*, 2006). Starting from some initial state or probability distribution, controls $u(0), u(1), \ldots, u(M-1)$ drive the probability distribution of the PBN. The problem is to find a sequence of controls to minimize the objective function of Eq. (5.16).

As discussed in Ng *et al.* (2006), we are led to the following optimization problem:

$$\min_{u(0),\ldots,u(M-1)} \sum_{t=0}^{M-1} \sum_{i=1}^{2^n} w_i(t) C_t(i, u(t)) + \sum_{i=1}^{2^n} w_i(M) C_M(i) \tag{5.70}$$

subject to Eq. (5.69), $\sum_{i=1}^{k} u_i(t) \leq 1$, and

$$\alpha_t = 1 - \beta_t \sum_{i=1}^{k} u_i(t) \tag{5.71}$$

for $t = 0, 1, \ldots, M-1$, where the last constraint assures that if there is no control, then $\alpha_t = 1$. Since this problem involves kM integer variables $u_i(t)$ and linear constraints, it is an integer programming model and therefore much more computationally tractable than the full dynamic programming solution.

5.4.2 Intervention in a family of BNs

The data may be sufficient to infer a family of BNs compatible with the data, for instance, BNs whose attractors match the data, but inference of the probability structure of a PBN is much more difficult. Assuming the family of BNs is not too small, we have the expectation that the underlying biological phenomena are reasonably modeled by at least some of the BNs in the family, so that one can design a control policy based on performance across the family (here taking a finite-horizon perspective) (Choudhary *et al.*, 2006).

Let $\mathcal{N}$ be a set of L BNs, $N_1, N_2, \ldots, N_L$, sharing the same state space $\mathcal{S}$ and the same control space $\mathcal{C}$. Associated with each network is an initial probability of it representing the underlying phenomenon. Since this information is not available, these probabilities are adaptively estimated as more transitions are observed. For each network N_l, $l = 1, 2, \ldots, L$, define $p^l_{ij}(v)$ to be the ith row, jth column entry of the matrix $P^l(v)$ of the controlled network N_l; $C^l_t(i, v)$ to be the cost of applying the control v at the tth time step in state i in network N_l; and $C^l_M(i)$ to be the terminal cost associated with state i in network N_l.

In Choudhary *et al.* (2006), a *belief vector* $\phi_t = [\phi^1_t, \phi^2_t, \ldots, \phi^L_t]$ is defined, where ϕ^l_t is the probability of network N_l being the underlying network at the tth time step. ϕ_t is the probability distribution vector for the family of networks at the tth time step. Since ϕ_t is unknown, an initial guess is made and updated as more information becomes available. The use of this vector is inspired by the *information vector* in Smallwood and Sondik (1973). Suppose i is the current state at step t, ϕ is the current estimate of the belief vector, and upon application of control v state j is observed at the next time step. Then the new belief vector is $\dot{\phi} = T(\phi, i \,|\, j, v)$, where the transformation T can be obtained by use of Bayes' theorem and the theorem of total probability,

$$\dot{\phi} = \left[\ldots, \frac{p^l_{ij}(v)\phi^l_t}{\sum_{s=1}^{L} p^s_{ij}(v)\phi^s_t}, \ldots \right]. \tag{5.72}$$

Suppose we are given an initial belief vector ϕ_0, based on prior knowledge of the system, and an initial state $z(0)$. ϕ_0 could be a function of likelihood or Bayesian scores of networks, or it could be uniform to reflect lack of prior knowledge. The objective is to find controls $v(0), v(1), \ldots, v(M-1)$ to minimize the expectation of the total cost over all networks. The cost at the tth time step ($0 \leq t < M$) is a function of the current state $z(t)$ and the updated belief vector ϕ_t. Motivated by Eq. (5.16) for the single PBN case and taking a weighted average over all networks, an average optimal cost function is defined by

$$J_t(\phi_t, i) = \min_{v \in C} \left[\sum_{l=1}^{L} \phi_t^L \left(C_t^l(i, v) + \sum_{j \in S} p_{ij}^l(v) J_{t+1}(T(\phi_t, i | j, v), j) \right) \right]. \tag{5.73}$$

The terminal cost for a state i is defined to be the average terminal cost over the entire family:

$$J_M(\phi_M, i) = \sum_{l=1}^{L} \phi_M^l C_M^l(i). \tag{5.74}$$

In addition to the optimal control being defined over a family of BNs, there are two other aspects in which Choudhary *et al.* (2006) differs from the basic finite-horizon formulation. First, rather than assign terminal penalties to states based on the expression level of certain key genes, terminal penalties are assigned based on the long-term prospective behavior of the system in the absence of control. Second, the solution to the dynamic programming problem defined by Eqs. (5.73) and (5.74) is presented as a policy tree that is optimal specific to a particular initial state and initial belief vector. An M-step policy tree has an optimal action as its root with branches for each possible observation (state) followed by $(M-1)$step policy trees (Kaelbling *et al.* 1998). Finally, for the sake of application, in Choudhary *et al.* (2006) it is assumed that all the BNs share the same attractors and that these are singletons; however, this assumption is not needed for Eqs. (5.73) and (5.74) or for the optimization.

5.4.3 Imperfect information

The finite-horizon control law that emerges from the solution of the dynamic programming problem as discussed thus far takes the form of a state feedback, as stated in Eq. (5.12). When the state vector $z(t)$ of the PBN is not available for measurement, such a control law cannot be implemented. Let us assume that when the PBN is in the state $z(t)$, it emits q measurable outputs, each of which can take on the value 0 or 1 (Datta *et al.*, 2004). Thus, the output status of the PBN at any time t is captured by a q-digit binary number or, alternatively, by an integer θ_t taking on values from 1 to 2^q as the outputs range over all possible binary values.

The design of the optimal control can use only the signals available to the controller. Specifically, at time t, the controller tries to design the control input $v_t = v(t)$ using all the available signals: $\theta_0, \theta_1, \ldots, \theta_t, v_0, v_1, \ldots, v_{t-1}$. The state $z_t = z(t)$ evolves according to the probability transition matrix of the controlled Markov chain, but it is not available for measurement; however, we assume that the output θ_t at time t is probabilistically related to the state z_t at time t and the input v_{t-1} through the known conditional probability measure $P_{\theta_t}(.|z_t, v_{t-1})$ defined by

$$P(\theta_t = j | z_t = i, v_{t-1} = v\} = r_{ij}^v. \tag{5.75}$$

Let $I_t = (\theta_0, v_0, \theta_1, v_1, \ldots, v_{t-1}, \theta_t)$ be the total information available for control at time t. I_t can be generated recursively using the equation

$$I_{t+1} = (I_t, v_t, \theta_{t+1})^T, \tag{5.76}$$

with $I_0 = \theta_0$. Since the state z_t is not available, the state feedback control of Eq. (5.12) is replaced by the information feedback control

$$v_t = \mu_t(I_t) \tag{5.77}$$

for $t = 0, 1, 2, \ldots, M-1$, and we search for the optimal μ_t over the space of all functions μ_t mapping the space of information vectors I_t into the control space $\{1, 2, 3, \ldots, 2^k\}$. If the cost function, $C_t(z_t, \mu_t(z_t))$, for the perfect-information case is replaced by the cost function $C_t(z_t, \mu_t(I_t))$ in the imperfect-information case, then the counterpart to the minimization of Eq. (5.16) becomes

$$\min_{\mu_0, \mu_1, \ldots, \mu_{M-1}} E_{z_0, \theta_0, \theta_1, \ldots, \theta_{M-1}} \left\{ \sum_{t=0}^{M-1} C_t(z_t, \mu_t(I_t)) + C_M(z_M) | z_0 \right\} \tag{5.78}$$

subject to

$$P(z_{t+1} = j | z_t = i, v_t) = p_{ij}(v_t), \tag{5.79}$$

$$I_{t+1} = [I_t^T, v_t, \theta_{t+1}]^T, \; I_0 = \theta_0 \tag{5.80}$$

(Bertsekas, 1976; Datta *et al.*, 2004). Note that $z(0)$, which is not observed, is one of the variables in Eq. (5.78) with respect to which the expectation is taken, so that the minimum is not a function of $z(0)$.

The dynamic programming algorithm for the above problem is given by (Bertsekas, 1976; Datta *et al.*, 2004)

$$J_{M-1}(I_{M-1}) = \min_{v_{M-1} \in \{1,2,\ldots,2^m\}} E_{z_{M-1}}[C_M(z_M) + C_{M-1}(z_{M-1}, v_{M-1}) | I_{M-1}, v_{M-1}], \tag{5.81}$$

$$J_t(I_t) = \min_{v_t \in \{1,2,\ldots,2^k\}} E_{\theta_{t+1}, z_t}[C_t(z_t, v_t) + J_{t+1}([I_t^T, v_t, \theta_{t+1}]^T) | I_t, v_t] \tag{5.82}$$

for $t = 0, 1, 2, \ldots, M-2$. The optimal control input is obtained from the values minimizing the right-hand sides of Eqs. (5.81) and (5.82). Using this algorithm, we will ultimately arrive at $J_0(I_0) = J_0(\theta_0)$. The optimal cost J^* can be obtained by taking the expectation of this quantity with respect to θ_0, i.e.,

$$J^* = E_{\theta_0}[J_0(\theta_0)]. \tag{5.83}$$

5.4.4 Reduction of a context-sensitive PBN

It was proposed in Pal *et al.* (2005c) that, for the purpose of control, computational complexity could be reduced for context-sensitive PBNs by using a transition probability matrix in which the context is removed from the state space and system dynamics. This reduction is accomplished by computing a weighted sum of the gene-activity profile behaviors over all the possible constituent networks. At every step, the reduced system exhibits an expected behavior by averaging over the various contexts. As such, the gene-activity profile determines the status of the approximate system and the collapsed transition probability matrix specifies its evolution. The corresponding intervention strategy is based on the approximate transition probability matrix with the collapsed state space. Such context reduction can have the added benefit that it can be used for control when the context is not observable, which is commonplace. In Pal *et al.* (2005c), the reduction was proposed for finite-horizon control and it was utilized for infinite-horizon control in Pal *et al.* (2006). Following Faryabi *et al.* (2009) we will develop the strategy for infinite-horizon control and analyze the loss of intervention performance resulting from removing the context from the state space of a context-sensitive PBN. This is done by comparing the performance of the intervention strategies derived from the full state space and the reduced state space when both are individually applied to the full state space.

The key issue is to derive the probability transition matrix for the reduced PBN, which is instantaneously random on account of context removal, beginning with the context-sensitive transition probability of Eq. (2.13). Let $(\kappa, \mathbf{x})$ denote a context-GAP pair, $p_{\mathbf{x}_1,\mathbf{x}_2}$ denote the probability of transitioning from GAP $\mathbf{x}_1$ to GAP $\mathbf{x}_2$, c_κ be the selection probability for context κ, $\eta(\mathbf{x}_1,\mathbf{x}_2)$ denote the Hamming distance between $\mathbf{x}_1$ and $\mathbf{x}_2$, $\mathbf{1}(\bullet)$ denote the indicator function, and $G_l(\mathbf{x}_1,\mathbf{x}_2) = \mathbf{1}(\mathbf{f}_l(\mathbf{x}_1) = \mathbf{x}_2)$. Under the reduction assumptions,

$$p_{\mathbf{x}_1,\mathbf{x}_2} = \sum_{\kappa_1}\sum_{\kappa_2} c_{\kappa_1} P\left(\mathbf{X}(t+1) = (\kappa_2,\mathbf{x}_2) | \mathbf{X}(t) = (\kappa_1,\mathbf{x}_1)\right). \tag{5.84}$$

Moreover, we can expand this expression as

$$\begin{aligned} p_{\mathbf{x}_1,\mathbf{x}_2} = \sum_{\kappa_1}\sum_{\kappa_2} \Big\{ & c_{\kappa_1}\mathbf{1}(\kappa_2 = \kappa_1) \\ & \times \left((1-q) + q c_{\kappa_1}\right)\left[(1-p)^n G_{\kappa_1}(\mathbf{x}_1,\mathbf{x}_2)\right. \\ & \left. + (1-p)^{(n-\eta(\mathbf{x}_1,\mathbf{x}_2))} p^{\eta(\mathbf{x}_1,\mathbf{x}_2)} \mathbf{1}(\eta(\mathbf{x}_1,\mathbf{x}_2) \neq 0)\right] \\ & + c_{\kappa_1}\mathbf{1}(\kappa_2 \neq \kappa_1) q c_{\kappa_2}\left[(1-p)^n G_{\kappa_2}(\mathbf{x}_1,\mathbf{x}_2)\right. \\ & \left. + (1-p)^{(n-\eta(\mathbf{x}_1,\mathbf{x}_2))} p^{\eta(\mathbf{x}_1,\mathbf{x}_2)} \mathbf{1}(\eta(\mathbf{x}_1,\mathbf{x}_2) \neq 0)\right]\Big\}. \end{aligned} \tag{5.85}$$

We can rewrite this as

$$p_{\mathbf{x}_1,\mathbf{x}_2} = (1-p)^n \Lambda, \tag{5.86}$$

where

$$\Lambda = \sum_{\kappa_1} c_{\kappa_1}\left((1-q)+qc_{\kappa_1}\right)\left[G_{\kappa_1}(\mathbf{x}_1,\mathbf{x}_2) + \left(\frac{p}{1-p}\right)^{\eta(\mathbf{x}_1,\mathbf{x}_2)} \mathbf{1}(\eta(\mathbf{x}_1,\mathbf{x}_2)\neq 0)\right] + \sum_{\kappa_1} c_{\kappa_1} q \left[\sum_{\substack{\kappa_2 \\ \kappa_2\neq\kappa_1}} c_{\kappa_2} G_{\kappa_2}(\mathbf{x}_1,\mathbf{x}_2) + \sum_{\substack{\kappa_2 \\ \kappa_2\neq\kappa_1}} c_{\kappa_2}\left(\frac{p}{1-p}\right)^{\eta(\mathbf{x}_1,\mathbf{x}_2)} \mathbf{1}(\eta(\mathbf{x}_1,\mathbf{x}_2)\neq 0)\right]. \tag{5.87}$$

The above expression for Λ can be further simplified as

$$\begin{aligned}\Lambda &= \sum_{\kappa_1} G_{\kappa_1}(\mathbf{x}_1,\mathbf{x}_2)\left((1-q)c_{\kappa_1}+qc_{\kappa_1}^2\right) \\ &\quad + q\sum_{\kappa_1} c_{\kappa_1} \sum_{\substack{\kappa_2 \\ \kappa_2\neq\kappa_1}} c_{\kappa_2} G_{\kappa_2}(\mathbf{x}_1,\mathbf{x}_2) \\ &\quad + \left(\frac{p}{1-p}\right)^{\eta(\mathbf{x}_1,\mathbf{x}_2)} \mathbf{1}(\eta(\mathbf{x}_1,\mathbf{x}_2)\neq 0) \\ &\quad \times \left[1-q+q\sum_{\kappa_1} c_{\kappa_1}^2 + q\sum_{\kappa_1} c_{\kappa_1} - q\sum_{\kappa_1} c_{\kappa_1}^2\right] \\ &= \left(\frac{p}{1-p}\right)^{\eta(\mathbf{x}_1,\mathbf{x}_2)} \mathbf{1}(\eta(\mathbf{x}_1,\mathbf{x}_2)\neq 0) \\ &\quad + (1-q)\sum_{\kappa_1} c_{\kappa_1} G_{\kappa_1}(\mathbf{x}_1,\mathbf{x}_2) \\ &\quad + q\sum_{\kappa_2} c_{\kappa_2} G_{\kappa_2}(\mathbf{x}_1,\mathbf{x}_2)\sum_{\kappa_1} c_{\kappa_1} \\ &= \left(\frac{p}{1-p}\right)^{\eta(\mathbf{x}_1,\mathbf{x}_2)} \mathbf{1}(\eta(\mathbf{x}_1,\mathbf{x}_2)\neq 0) \\ &\quad + \sum_{\kappa_2} c_{\kappa_2} G_{\kappa_2}(\mathbf{x}_1,\mathbf{x}_2),\end{aligned} \tag{5.88}$$

the last equality following because $\sum_{\kappa_1} r_{\kappa_1} = 1$.

Averaging over the contexts reduces the transition probabilities associated with a context-sensitive PBN to the transition probabilities arising from the corresponding instantaneously random PBN. The transition probability matrix of the corresponding instantaneously random PBN with similar parameters can be obtained from Eq. (2.13) by setting $q = 1$. Thus, the optimal and approximate control strategies perform similarly as $q \to 1$.

To analyze the performance of the approximate policy the following cost per stage has been applied in Faryabi *et al.* (2009):

$$C((\kappa,\mathbf{x}),u) = \begin{cases} 0 & \text{if } u = 0 \text{ and } (\kappa,\mathbf{x}) \in \mathcal{D}, \\ 10 & \text{if } u = 0 \text{ and } (\kappa,\mathbf{x}) \in \mathcal{U}, \\ c & \text{if } u = 1 \text{ and } (\kappa,\mathbf{x}) \in \mathcal{D}, \\ 10+c & \text{if } u = 1 \text{ and } (\kappa,\mathbf{x}) \in \mathcal{U}, \end{cases} \tag{5.89}$$

where $c = 1$ is the cost of control, and $\mathcal{U}$ and $\mathcal{D}$ are the sets of undesirable and desirable states, respectively. The target gene is chosen to be the most significant gene in the GAP and it is assumed that the up-regulation of the target gene is undesirable. Consequently, using an integer x to represent the GAP vector $\mathbf{x}$, the state space is partitioned into desirable states,

$$\mathcal{D} = \{(\kappa,x) : \kappa \in \{1,\ldots,k\}, x \in \{1,\ldots,2^{n-1}\}\},$$

and undesirable states,

$$\mathcal{U} = \{(\kappa,x) : \kappa \in \{1,\ldots,k\}, x \in \{2^{n-1}+1,\ldots,2^n\}\}.$$

The number of genes is limited to five owing to the computational complexity of dynamic programming. The cost values have been chosen in accordance with Pal *et al.* (2006). Since our objective is to down-regulate the target gene, a higher cost is assigned to destination states having an up-regulated target gene. Moreover for a given status of the target gene, a higher cost is assigned when the control is applied versus when it is not.

Synthetic context-sensitive PBNs are generated as follows. Each context-sensitive PBN consists of two constituent networks. Each is randomly generated with bias equal to 0.5, the selection probabilities are equal (0.5), and the perturbation probability is $p = 0.01$. The switching probability q is allowed to vary so its effect can be studied. For each value of q, 1000 PBNs are generated. For each PBN, the exact and approximate transition probability matrices are computed according to Eqs. (2.13) and (5.86), respectively, a control gene is randomly selected, and the optimal intervention problems for the original model and its reduced approximation are determined.

The optimal strategy, μ_g^*, for the exact transition probability matrix specifies the action that should be taken at each time step. The policy based on the reduced stochastic matrix takes only the GAP as its input. It is extended to (context, GAP) pairs by simply disregarding the context element of the state and determining the action based on its GAP element, thereby defining the approximate strategy $\hat{\mu}_g$. The average total discounted cost induced by the optimal strategy μ_g^* is estimated by generating synthetic time-course data for 1000 time steps from the transition probability matrix for the original context-sensitive PBN while intervening based on optimal strategy μ_g^*. The total cost is estimated by accumulating the discounted cost of each state given the action at that state. This procedure is repeated 10,000 times for random initial states and the average of the induced total discounted costs is computed. Following a similar procedure, the approximate strategy $\hat{\mu}_g$ is applied to the system and the average total discounted cost is computed. Finally, the average total discounted cost is estimated when no intervention is applied.

The effectiveness of an intervention strategy can be evaluated by computing the difference between its induced cost and the cost accumulated in the absence of intervention. For each set of constituent networks and a given switching probability, we compute the functions $\bar{J}_{\mu^*}$, $\bar{J}_{\hat{\mu}}$, and $\bar{J}$, these being the average total discounted cost for a given context-sensitive PBN induced by applying the optimal strategy μ_g^*, approximate strategy $\hat{\mu}_g$, and no intervention, respectively. The preceding procedure is repeated for 1000 random context-sensitive PBNs, thereby yielding 1000 values for each statistic. The effects of these strategies are compared by computing averages denoted by $E[\bar{J}_{\mu^*}]$, $E[\bar{J}_{\hat{\mu}}]$, and $E[\bar{J}]$, respectively.

The percentage of reduction in the average total discounted cost is used as a performance metric. The normalized gain obtained by each intervention strategy is taken as the immediate consequence of the intervention formulation. This metric is defined as the difference between the average discounted cost before and after intervention, normalized by the cost before intervention. The normalized gain corresponding to the optimal strategy μ_g^* is

$$\Delta J^{\mathrm{E}} = \frac{E[\bar{J}] - E[\bar{J}_{\mu^*}]}{E[\bar{J}]}, \tag{5.90}$$

and the normalized gain corresponding to the strategy derived from the approximate method $\hat{\mu}_g$ is

$$\Delta J^{\mathrm{A}} = \frac{E[\bar{J}] - E[\bar{J}_{\hat{\mu}}]}{E[\bar{J}]}. \tag{5.91}$$

Figure 5.5 shows these gains as a function of q. As $q \to 1$, the difference between normalized gains ΔJ^{A} and ΔJ^{E} decreases. The approximate method yields close to optimal performance when the switching probability is large. If one cannot obtain context knowledge or the number of contexts results in an unacceptable computational burden, then approximate method provides a practical strategy for $q = 0.01$, albeit with a 30% reduction in performance from the optimal strategy.

As a by-product of the intervention formulation, we also consider the effect of an intervention strategy on the amount of change in the steady-state probability of undesirable states before and after the intervention. For each set of constituent networks and for a given switching probability, we compute the normalized reduction in the total probability of visiting undesirable states in the long run for a given context-sensitive PBN when strategies μ_g^* and $\hat{\mu}_g$ are applied to the original system, these being defined by

$$\Delta P_{\mu^*} = \frac{\sum_{\mathbf{z}\in\mathcal{U}} \pi(\mathbf{z}) - \sum_{\mathbf{z}\in\mathcal{U}} \pi_{\mu^*}(\mathbf{z})}{\sum_{\mathbf{z}\in\mathcal{U}} \pi(\mathbf{z})} \tag{5.92}$$

and

$$\Delta P_{\hat{\mu}} = \frac{\sum_{\mathbf{z}\in\mathcal{U}} \pi(\mathbf{z}) - \sum_{\mathbf{z}\in\mathcal{U}} \pi_{\hat{\mu}}(\mathbf{z})}{\sum_{\mathbf{z}\in\mathcal{U}} \pi(\mathbf{z})}, \tag{5.93}$$

respectively, where $\pi_{\mu^*}(\mathbf{z})$, $\pi_{\hat{\mu}}(\mathbf{z})$, and $\pi(\mathbf{z})$ are the steady-state probabilities of state $\mathbf{z}$ under the optimal strategy, the approximate strategy, and when no control is applied, respectively.

The preceding procedure is repeated for 1000 random context-sensitive PBNs, thereby yielding 1000 values for each statistic. The average values for ΔP_{μ^*} and $\Delta P_{\hat{\mu}}$ are denoted by ΔP^E and ΔP^A, respectively. Figure 5.6 shows ΔP^A and ΔP^E as functions of the switching probability. The trends are similar to those observed for the normalized gains.

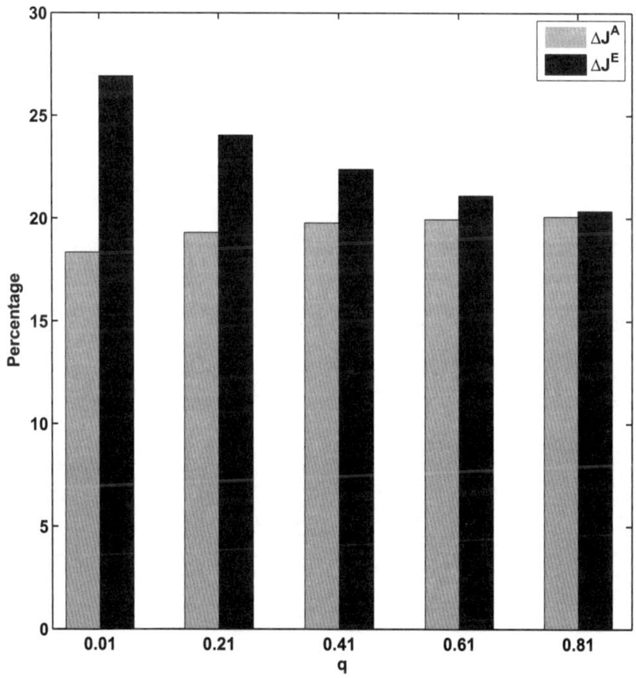

Figure 5.5. *ΔJ^A and ΔJ^E are computed for 1000 context-sensitive PBNs consisting of two contexts. The switching probability q is the parameter. The selection probability has uniform distribution $r_1 = r_2 = 0.5$.*

5.4.5 Reinforcement learning

If the system and cost structure can be simulated, repeated simulation can be used to calculate approximate transition probabilities and an expected cost, after which dynamic programming methods can be applied to find an optimal control policy. We assume that the distributions governing the PBN—the switching probability, the perturbation probability, and the selection probabilities—are known. The complexity of estimating the transition probabilities and the complexity of dynamic programming exponentially increase as the number of genes increases. Given the aforementioned distributions, a reinforcement learning algorithm progressively computes the cost function of a given policy by generating several sample trajectories of the PBN and their associated costs (Faryabi *et al.*, 2007). Hence, it eliminates the computational complexity associated with the explicit estimation of the transition probabilities.

Substituting Eq. (5.27) into Bellman's equation (5.40) yields

$$J^*(i) = \min_{u \in C} \left[\alpha \sum_{j=1}^{m2^n} p_{ij}(u)(\tilde{g}(i,u,j) + \alpha J^*(j)) \right]. \tag{5.94}$$

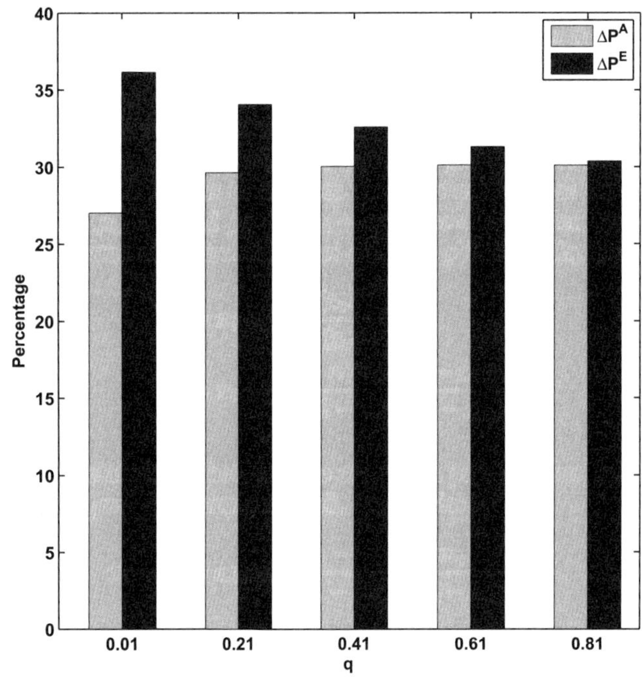

Figure 5.6. *ΔP^E and ΔP^A are computed for* 1000 *context-sensitive PBNs consisting of two contexts. The switching probability q is the parameter. The selection probability has uniform distribution* $r_1 = r_2 = 0.5$.

For each state-control pair (i,u), define the *Q-factor* by

$$Q(i,u) = \sum_{j=1}^{m2^n} p_{ij}(u)(\tilde{g}(i,u,j) + \alpha J^*(j)). \tag{5.95}$$

Then the optimal cost for state i is given by

$$J^*(i) = \min_{u \in \mathcal{C}} Q(i,u). \tag{5.96}$$

Given a state-control pair (i,u), considering state j to be random, the following expression defines a random variable as a function of j:

$$\Psi(i,u) = \tilde{g}(i,u,j) + \alpha \min_{u' \in \mathcal{C}} Q(j,u'). \tag{5.97}$$

Eq. (5.95) can be expressed as the expectation of $\Psi(i,u)$, namely,

$$Q(i,u) = E[\Psi(i,u)], \tag{5.98}$$

which can be estimated by the average

$$\overline{\Psi}_k(i,u) = \frac{\sum_{i=1}^{k} \psi_i}{k}, \tag{5.99}$$

where ψ_i is the ith sample of $\Psi(i,u)$. Upon a new observation of $\Psi(i,u)$, the value of $\overline{\Psi}_k(i,u)$ can be updated by

$$\overline{\Psi}_{k+1}(i,u) = \left(1 - \frac{1}{k+1}\right)\overline{\Psi}_k(i,u) + \frac{\psi_{k+1}}{k+1}. \tag{5.100}$$

If we generate system observations, then, given a new observation (random observation of state j), the preceding expression can be used to define an iterative updating of the Q-factor: replace $\overline{\Psi}_{k+1}$ and $\overline{\Psi}_k$ by $Q^{(k+1)}$ and $Q^{(k)}$, respectively; replace $(k+1)^{-1}$ by a small positive number λ_k; and replace ψ_{k+1} by the definition in Eq. (5.95) with Q replaced by $Q^{(k)}$. For the specific state-control pair (i,u), this leads to the iterative transformation

$$Q^{(k+1)}(i,u) \leftarrow (1-\lambda_k)Q^{(k)}(i,u) + \lambda_k\left[\tilde{g}(i,u,j) + \lambda \min_{u' \in C} Q^{(k)}(j,u')\right], \tag{5.101}$$

the right-hand side being a random variable depending on the selection of destination state j.

The revised value iteration algorithm in which the Q-factors are updated according to Eq. (5.101) is called the *Q-learning algorithm* (Bertsekas and Tsitsiklis, 1996). In Q-learning, the value of the Q-factor for a state-control pair (i,u) is updated whenever a transition from state i to state j occurs in the system's simulator, given a random selection of the control u from among all the possible controls. To specify the algorithm, an appropriate step-size, λ_k, is needed to guarantee convergence of the Q-learning algorithm to the optimal control strategy despite the asynchronous updating of the Q-factors (Faryabi *et al.*, 2007). As a general rule, the step-size should be small and diminish to zero at a suitable rate (Faryabi *et al.*, 2007). Letting $v(i,u)$ denote the number of times the state-control pair (i,u) is visited and choosing a constant $\delta \in (0,1)$, whenever the pair (i,u) is selected, the step-size is set by $\lambda_k \longleftarrow \delta/v(i,u)$. The Q-learning algorithm is initialized by setting $Q^{(0)}v(i,u) \longleftarrow 0$ and $v(i,u) \longleftarrow 0$ for all (i,u), choosing δ, and selecting the number $k_{\max}$ of iterations the algorithm will run. The value of $Q^{(k)}(i,u)$ is updated by Eq. (5.101) and the Q-learning suboptimal policy is defined by

$$\mu(i) = \arg\min_{u \in C} Q^{k_{\max}}(i,u). \tag{5.102}$$

To the extent that the transformation of Eq. (5.101) is independent of the transition probabilities of the system, the Q-learning algorithm is model-free.

In the Q-learning algorithm, the complexity of each iteration is $\mathbf{O}(2^l)$, where l is the number of binary control inputs. Q-learning runs in polynomial time complexity with respect to the number of genes in the network. Q-learning reduces the memory complexity of an optimal algorithm such as the value iteration. The values of the Q-factors are stored explicitly in a tabular form. The algorithm requires $\mathbf{O}(ld^n)$ memory units (assume a d-valued PBN). Since the number of binary control inputs is small, the memory complexity is on the order of $\mathbf{O}(d^n)$ memory units; whereas in the value iteration algorithm the required memory is $\mathbf{O}(d^{2n})$ memory units. This latter quantity stems from the fact that we must store d^n values of the value function for l control inputs along with d^{2n} entries of the transition probability matrices at each iteration of the algorithm. Given the number of binary control inputs is small, the required memory of the value iteration algorithm, $\mathbf{O}(d^{2n}+ld^n)$, has the growth of $\mathbf{O}(d^{2n})$. Consequently, the memory complexity of the value iteration algorithm is considerably reduced. The high memory complexity of the value iteration algorithm, as well as other dynamic programming algorithms, contributes to their limited direct applicability to high dimensional intervention problems.

If all the state-control pairs, (i,u), are visited infinitely often, then for each state-control pair the estimated expected value, $\overline{\Psi}_k(i,u)$, converges to its ensemble average, $E[\Psi(i,u)]$, with probability 1. Hence, the approximate stationary policy computed by the Q-learning algorithm converges to the optimal stationary policy (Faryabi *et al.*, 2007). The learning duration of the Q-learning algorithm should increase as the number of genes in the network increases in order to obtain an approximate stationary policy close to the optimal stationary policy. Therefore, the Q-learning algorithm may not be suitable for very large networks.

We illustrate Q-learning with the control of WNT5A in a 10-gene instantaneously random PBN with genes WNT5A, pirin, S100P, RET1, MMP3, PHOC, MART1, HADHB, Synuclein, and STC2. The preceding order of genes is used in the binary representation of the GAP, with WNT5A as the most significant bit and STC2 as the least significant bit. With pirin as the control gene and the down-regulation of WNT5A being the goal, the cost function of Eq. (5.68) is used. Figures 5.7, 5.8, 5.9, and 5.10 show the steady-state distributions for the original network, for the network with the optimal policy determined by value iteration, for the network with the Q-learning policy with $k_{\max} = 10^3$, and for the network with the Q-learning policy with $k_{\max} = 10^5$, respectively. As the duration of learning increases, the performance gets closer to that of the optimal control algorithm.

Let ΔP denote the percentage of reduction of the steady-state mass of the undesirable states. For the optimal policy determined by value iteration, $\Delta P = 23.1\%$. Figure 5.11 shows the value of ΔP for the optimal control policy, as well for approximate control policies derived by the Q-learning algorithm with various learning durations. The performance of approximate stochastic control converges to that of optimal stochastic control. Moreover, the time-complexity of Q-learning increases linearly with the number of iterations.

5.5 Constrained Intervention

Cancer therapy may include the use of chemotherapy, radiation therapy, targeted gene therapy, etc. All of these treatment options are directed at killing or eradicating cancerous cells. Unfortunately, cancer treatments may also damage healthy cells and it is desirable to limit the resulting side effects. This goal can be achieved by enforcing an upper bound on the number (expected number) of treatments a patient may receive during therapy, that is, by constraining optimization. The problem has been addressed in the context of finite-horizon control by placing an upper bound for the number of controls that can be applied, using a formulation in which the objective function depends on the distance between the desirable states and the terminal states (Ching *et al.*, 2009). Here we consider constraint in the infinite-horizon setting by bounding the expected number of interventions, using a discounted formulation to define both the objective cost function and a constraint cost function (Faryabi *et al.*, 2008b).

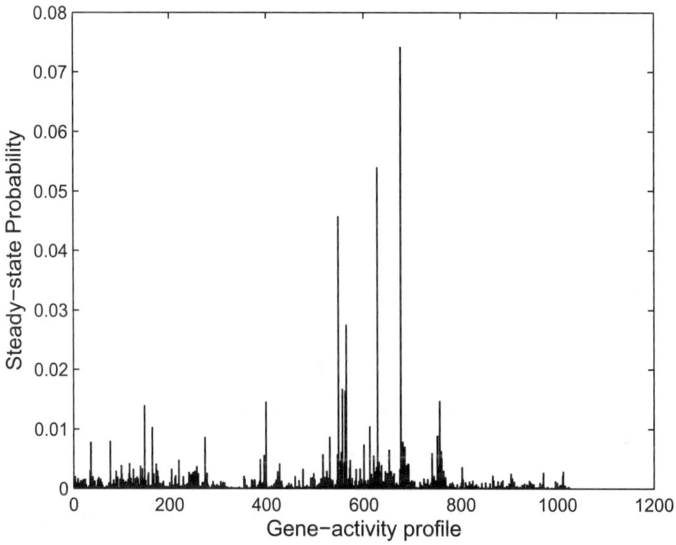

Figure 5.7. *Steady-state distribution of gene-activity profile of the* 10*-gene PBN prior to intervention.*

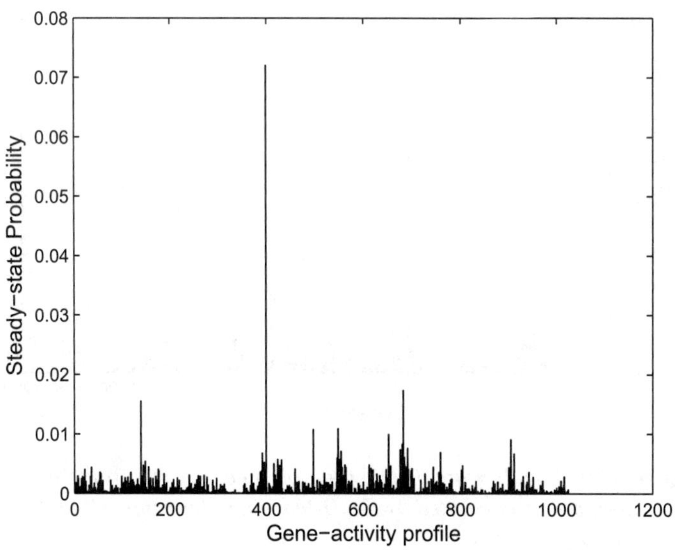

Figure 5.8. *Steady-state distribution of gene-activity profile after intervention with optimal control policy.*

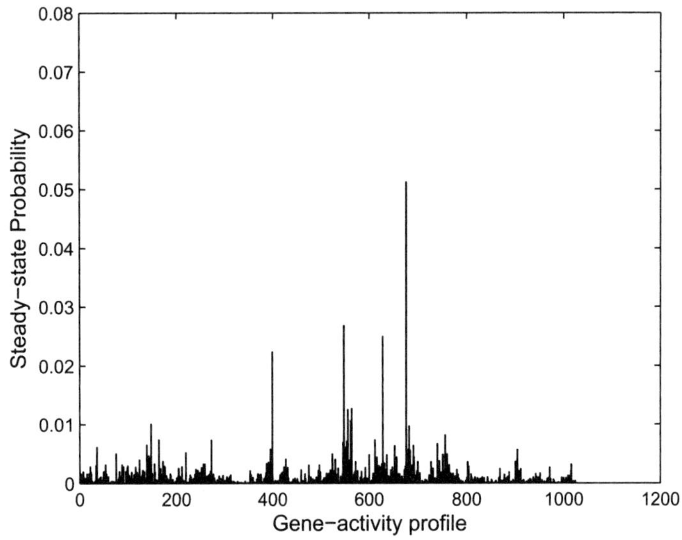

Figure 5.9. *Steady-state distribution of gene-activity profile after applying the approximate control policy computed by the Q-learning algorithm with* $k_{max} = 10^3$.

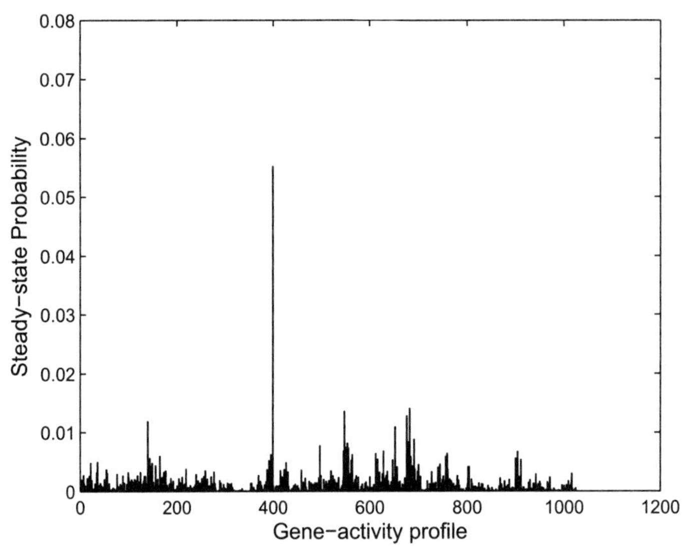

Figure 5.10. *Steady-state distribution of gene-activity profile after applying the approximate control policy computed by the Q-learning algorithm with* $k_{max} = 10^5$.

To restrict intervention frequency, for each state-action pair (z,u) in the set $\mathcal{K} = \{(z,u) : z \in \mathcal{S}, u \in \mathcal{C}\}$ of all possible state-action pairs we postulate an *associated cost per stage*, $\xi(z,u)$. The goal is to bound the discounted expected number of interventions in

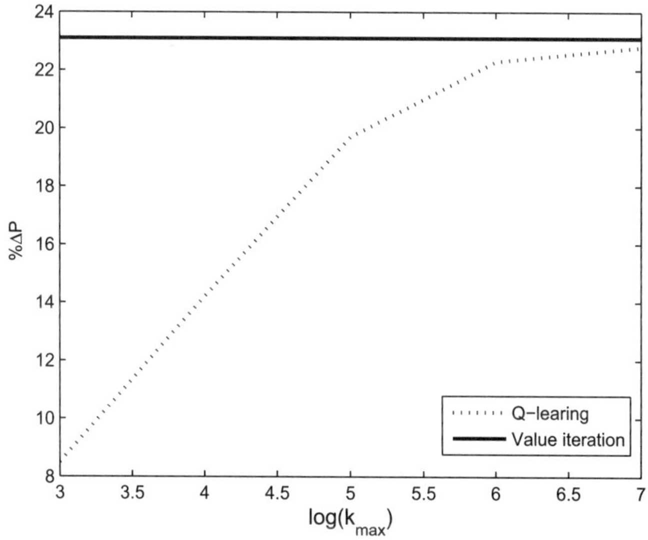

Figure 5.11. ΔP *of approximate control policy versus the optimal control policy as a function of logarithm of learning duration.*

the long run. Accordingly, the *normalized expected total discounted cost* of the constraint, given policy $\omega = (\mu_0, \mu_1, \ldots)$, initial state i, and control gene g, is denoted by

$$C_\omega(i) = (1-\alpha) \lim_{N\to\infty} E\left[\sum_{t=0}^{N-1} \alpha^t \xi(z(t), \mu_t(z(t)) | z(0) = i\right]. \tag{5.103}$$

The constrained intervention problem is now defined by the optimization of Eq. (5.25) subject to the constraint $C_\omega(i) \leq C_{\text{total}}$, where C_{total} is the upper bound on the discounted expected number of interventions in the long run and i is the initial state. Given a control gene g, the goal is to find an optimal intervention policy ω^* within the set Ω_g of admissible policies (not just Markovian policies) for control gene g satisfying the constraint $C_\omega(i) \leq C_{\text{total}}$. Thus, the constrained intervention problem takes the form

$$\begin{aligned} &\max_{\omega\in\Omega_g} J_\omega(i), \\ &\text{such that} \\ &C_\omega(i) \leq C_{\text{total}}, \end{aligned} \tag{5.104}$$

where, in this framework, we normalize J_ω in Eq. (5.25) by the multiplicative factor $1-\alpha$, as in Eq. (5.103).

Given an arbitrary policy $\omega = (\mu_0, \mu_1, \ldots)$ and starting from initial state i, the state trajectories and selected actions over time are probabilistic. Our objective is to find the expected number of times that state-action pairs (z, u) with active intervention decision, $u = 1$, occur over time. For initial state i and policy ω, we define the *occupation probability* by

$$p_t(i, \omega; z, u) = P(z(t) = z, \mu_t(z(t)) = u \,|\, z(0) = i) \tag{5.105}$$

and the *normalized discounted total expected time* spent in the state-action pair (z,u) in the long run by

$$f_\alpha(i,\omega;z,u) = (1-\alpha)\lim_{N\to\infty}\sum_{t=0}^{N-1}\alpha^t p_t(i,\omega;z,u). \tag{5.106}$$

The numbers of states and actions are finite and the discounting factor α guarantees uniform convergence of the sum. For any initial state i and policy ω, the set

$$f_\alpha(i,\omega) = \{f_\alpha(i,\omega;z,u)\,|\,(z,u)\in\mathcal{K}\} \tag{5.107}$$

determines a probability measure over $\mathcal{K}$, called an *occupation measure* (the factor $1-\alpha$ making the total probability equal to 1) (Derman, 1970). The occupation measure can be interpreted as the probability of occupying (z,u) in the long run, given that the PBN is initially in state i and policy ω is used throughout.

The normalized discounted cost function of Eq. (5.25) can be expressed as the expectation of the average cost per stage, $g(z,u)$, over the occupation probability:

$$J_\omega(i) = (1-\alpha)\lim_{N\to\infty}\sum_{t=0}^{N-1}\alpha^t\sum_{(z,u)\in\mathcal{K}} g(z(t)=z,\mu_t(z(t))=u)p_t(i,\omega;z,u), \tag{5.108}$$

which reduces to the expectation of the average cost per stage with respect to the occupation measure:

$$J_\omega(i) = \sum_{(z,u)\in\mathcal{K}} f_\alpha(i,\omega;z,u)\,g(z,u). \tag{5.109}$$

Similarly, we can express the normalized discounted objective cost corresponding to policy ω as the expectation of the associated cost per stage with respect to the occupation measure:

$$C_\omega(i) = \sum_{(z,u)\in\mathcal{K}} f_\alpha(i,\omega;z,u)\,\xi(z,u). \tag{5.110}$$

Using Eqs. (5.109) and (5.110), we can write the constrained optimization problem as

$$\begin{aligned}&\max_{\omega\in\Omega_g}\sum_{(z,u)\in\mathcal{K}} f_\alpha(i,\omega;z,u)\,g(z,u),\\ &\text{such that}\\ &\sum_{(z,u)\in\mathcal{K}} f_\alpha(i,\omega;z,u)\,\xi(z,u)\le C_{\text{total}}.\end{aligned} \tag{5.111}$$

If we define

$$\xi(z,u) = \begin{cases}0 & \text{if } u=0 \text{ and } z\in\mathcal{S},\\ 1 & \text{if } u=1 \text{ and } z\in\mathcal{S},\end{cases} \tag{5.112}$$

then the constraint in Eq. (5.111) prevents the discounted expected number of interventions in the long run from exceeding the upper-bound C_{total}. Equivalently, we can interpret $C_\omega(i)$ as the discounted frequency of applying treatments given a therapeutic strategy.

Several solutions for the constrained optimization problem of Eq. (5.111) are presented in Altman (1999). In Faryabi *et al.* (2008b), it is explained how an optimal policy satisfying the equivalent formulation of Eq. (5.111) is found by solving the following linear program:

$$\min_{\mathbf{x}=(x(1,1),x(1,2),\ldots,x(m2^n,2^k))\in\mathbb{R}^{|\mathcal{K}|}} \sum_{(z,u)\in\mathcal{K}} x(z,u)\,g(z,u),$$

such that

$$\sum_{(z,u)\in\mathcal{K}} x(z,u)(\mathbf{1}_{z=j} - \alpha\, p_{zj}(u)) = (1-\alpha)\mathbf{1}_{i=j} \tag{5.113}$$

for all $j \in \mathcal{S}$,

$$\sum_{(z,u)\in\mathcal{K}} x(z,u)\xi(z,u) \leq C_{\text{total}},$$

$$(z,u) \geq 0 \quad \text{for all } (z,u) \in \mathcal{K}.$$

This linear program is called the *primal problem.*

In Altman (1999), it is shown that an optimal stationary policy ω^* of the constrained optimization problem of Eq. (5.111) exists if and only if the primal problem has a solution $\mathbf{x}^* = \{x^*(z,u)|(z,u) \in \mathcal{K}\}$. Moreover, an optimal solution of Eq. (5.113) uniquely determines an optimal stationary policy ω^*, where an optimal stationary policy ω^* selects action u at state z with probability

$$p^*_\omega(z,u) = \frac{x^*(z,u)}{\sum_{v\in\mathcal{C}} x^*(z,v)}. \tag{5.114}$$

The optimal policy devised by Eq. (5.114) is not necessarily a deterministic policy, in contrast to the corresponding unconstrained problem.

Depending on the utilized numerical method, the computational complexity of finding a solution for the linear program in Eq. (5.113) varies. It is known that the complexity of the interior point method increases polynomially with the number of states in $\mathcal{K}$, where the exponent of the complexity polynomial is not large (Boyd and Vandenberghe, 2004). Moreover, it is known that the number of iterations required for the numerical method to converge is on the order of $O(\log(1/\epsilon))$, where ϵ is the accuracy of the outcome of the numerical method. Here, the size of $\mathcal{K}$ increases exponentially with the number n of genes and the number k of controls.

5.5.1 Constrained intervention in a mammalian cell-cycle network

We reconsider the mammalian cell-cycle network with mutated p27 (always OFF) considered in Section 4.5.8. Recall that, in this cancerous scenario, the permanent inactivation of p27 creates a situation where both CycD and Rb might be inactive, so that the cell can cycle in the absence of any growth factor. Thus, the states in which both Rb and CycD are down-regulated are undesirable states when p27 is mutated. Table 4.7 summarizes the mutated Boolean functions. Recalling Section 4.5.8, the first constituent BN is determined from Table 4.7 when CycD = 0 and the second when CycD = 1. The mutated cell cycle's PBN consists of nine genes: CycD, Rb, E2F, CycE, CycA, Cdc20, Cdh1, UbcH10, and CycB,

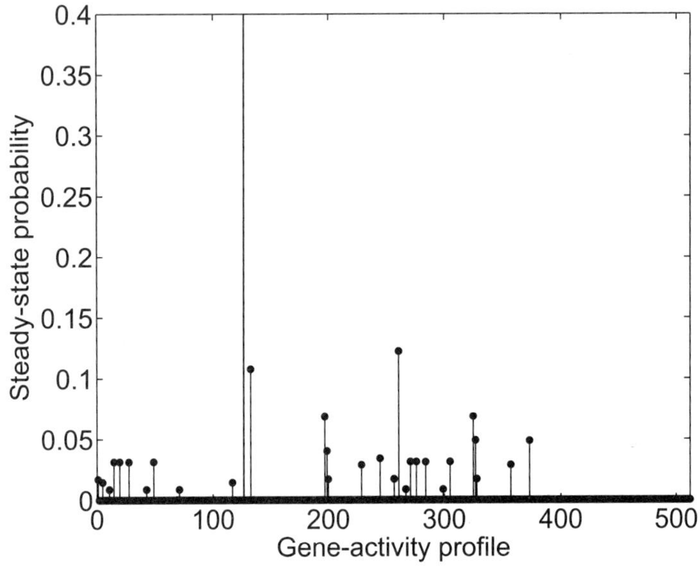

Figure 5.12. *The steady-state probability of gene-activity profile of the PBN associated with the mammalian cell-cycle network before intervention. The vertical line separates the undesirable gene-activity profiles from the desirable ones.*

with $q = 0.01$ and $p = 0.001$. This order of genes is used in the binary representation of the states, with CycD as the most significant bit and CycB as the least significant bit. This order facilitates the presentation of the results and does not affect the computed control policies. With simultaneously down-regulated CycD and Rb as our objective, we apply the constrained intervention method with various bounds on the frequency of applying control in a policy. We consider a single control, u: if $u = 1$, then the state of the control gene is reversed; if $u = 0$, then the state of the control gene remains unchanged. The control gene can be any of the genes in the model except CycD. We assume that the cost of the states with down-regulated Rb and CycD is higher than those for the states in which these two genes are not simultaneously down-regulated. We also consider the cost of applying a control action, which increases the cost of each state. We postulate the following rate-of-reward function:

$$C(z,u) = \begin{cases} 0 & \text{if } u = 0 \text{ and (CycD,Rb)} \neq (0,0) \text{ in } z, \\ 9 & \text{if } u = 0 \text{ and (CycD,Rb)} = (0,0) \text{ in } z, \\ 1 & \text{if } u = 1 \text{ and (CycD,Rb)} \neq (0,0) \text{ in } z, \\ 10 & \text{if } u = 1 \text{ and (CycD,Rb)} \neq (0,0) \text{ in } z. \end{cases} \tag{5.115}$$

Figure 5.12 shows the steady-state distribution of the gene-activity profile with no intervention, the undesirable states being to the left of the vertical line. The aggregated probability of the GAPs with simultaneously down-regulated CycD and Rb is close to 0.2; that is, the mutated cell-cycle will be in the cancerous GAPs, 0 to 127, nearly 20% of its time in the long run.

Table 5.2. *The ΔP for the intervention strategy based on various control genes and various constraint bounds.*

Control Gene	C_{total}									
	0.1	0.2	0.3	0.4	0.5	0.6	0.7	0.8	0.9	1.0
Rb	61.96	98.32	98.33	98.33	98.33	98.33	98.34	98.34	98.34	98.34
E2F	57.43	97.36	98.00	98.00	98.00	98.01	98.01	98.02	98.02	98.02
CycE	28.37	28.41	28.41	28.44	28.44	28.46	28.46	28.47	28.49	28.51
CycA	16.56	16.59	16.60	16.61	16.62	16.64	16.65	16.65	16.69	16.69
Cdc20	39.15	41.44	41.47	41.48	41.48	41.50	41.51	41.52	41.53	41.61
Cdh1	27.55	40.58	41.51	41.56	41.56	41.57	41.62	41.62	61.63	41.65
UbcH10	6.49	6.50	6.52	6.56	6.57	6.59	6.61	6.64	6.66	6.69
CycB	39.33	41.85	41.86	41.89	41.91	41.92	41.92	41.96	41.99	41.99

To determine the best intervention gene, a constrained intervention policy is determined for each gene in the network while limiting the expected number of times a control can be applied. First, we assume that the PBN's initial state is the undesirable GAP with the highest probability in the steady-state distribution of GAPs prior to intervention. Table 5.2 lists the percentage change, ΔP, in the aggregated probability of undesirable GAPs with simultaneously down-regulated CycD and Rb with and without intervention, corresponding to each gene in the network. The upper bound on the frequency of applying intervention is varied when finding the corresponding constrained policies.

Among all the genes, Rb offers the best performance when control can be applied without any constraint, based strictly on maximization of the reward function, $C_{total} = 1$. After applying the unconstrained control policy designed for Rb, the aggregated probability of undesirable GAPs is significantly altered (Fig. 5.13). Figure 5.14 indicates that by using a constrained stationary intervention policy for the control gene Rb we can reduce the aggregated probability of the undesirable states to less than 12%, while restricting the number of interventions to at most 10%. We could translate this to restricting the dose of prescribed drugs. If we wish only to limit the expected number of applied interventions to less than 20%, then we can reduce the chance of the cancerous gene-activity profiles by 98%.

According to Table 5.2, intervention policies based on gene E2F perform almost as well as Rb when the constraint is not too tight, $C_{total} \geq 0.2$. This suggests that, given the side effects of treatments, we may need to consider alternative control genes. The steady-state probability distributions of GAPs after intervention based on E2F are presented in Figs. 5.15 and 5.16. Comparing Figs. 5.13 and 5.15, one can observe that, although the final performances of intervention based on these two genes are close, the probability mass of the most probable GAPs after intervention with Rb differs from E2F-based intervention.

The results of Table 5.2 indicate that some genes are more sensitive to the bound on the frequency of control. Relaxing the constraint will not improve the result of intervention when gene UbcH10 is selected as the control gene. Genes CycB and Cdc20 perform relatively well for tightly constrained intervention policies, but relaxing the limitation on the expected number of treatments does not significantly improve the performance of the policies based on these genes.

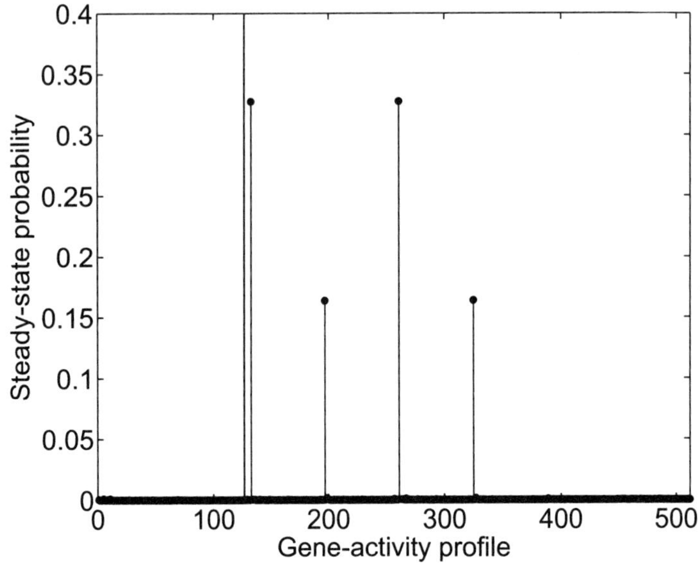

Figure 5.13. *The steady-state probability of gene-activity profile of the PBN associated with the mammalian cell-cycle network after intervention using Rb as the control gene when the frequency of applying control is unconstrained,* $C_{\text{total}} = 1.0$*. The vertical line separates the undesirable gene-activity profiles from the desirable ones.*

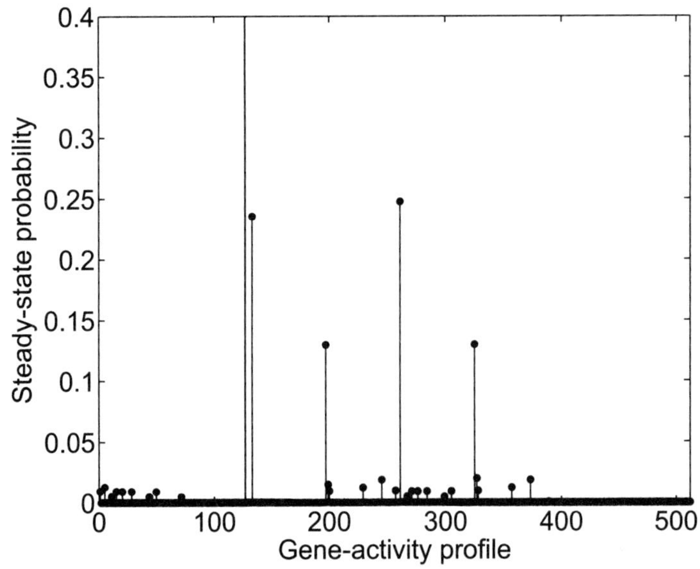

Figure 5.14. *The steady-state probability of gene-activity profile of the PBN associated with the mammalian cell-cycle network after intervention using Rb as the control gene when the frequency of applying control is upper bounded by* $C_{\text{total}} = 0.1$*. The vertical line separates the undesirable gene-activity profiles from the desirable ones.*

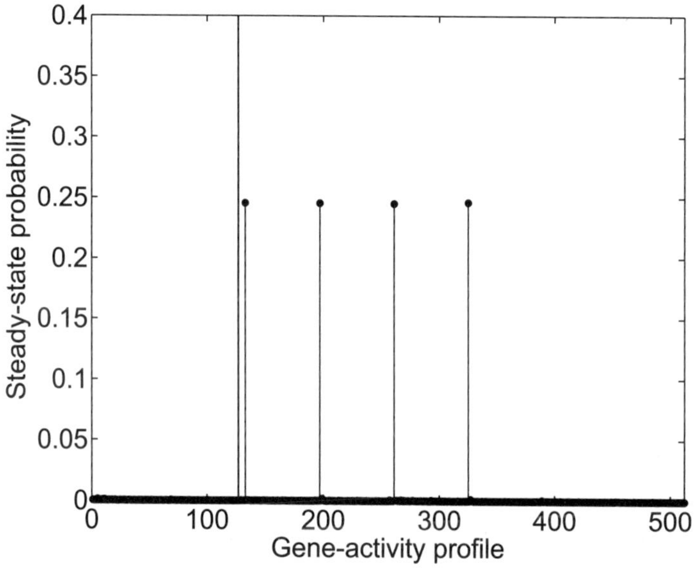

Figure 5.15. *The steady-state probability of gene-activity profile of the PBN associated with the mammalian cell-cycle network after intervention using E2F as the control gene when the frequency of applying control is unconstrained,* $C_{\text{total}} = 1.0$. *The vertical line separates the undesirable gene-activity profiles from the desirable ones.*

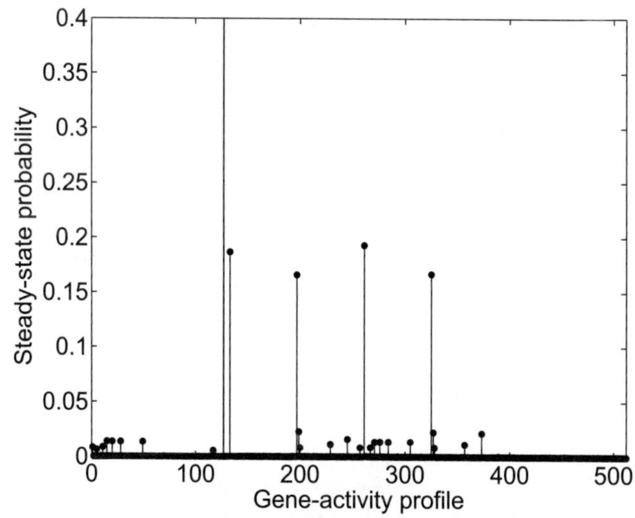

Figure 5.16. *The steady-state probability of gene-activity profile of the PBN associated with the mammalian cell-cycle network after intervention using E2F as the control gene when the frequency of applying control is upper bounded by* $C_{\text{total}} = 0.1$. *The vertical line separates the undesirable gene-activity profiles from the desirable ones.*

If we do not assume that the PBN's initial state is the undesirable GAP with the highest probability in the steady-state distribution of GAPs prior to intervention, but instead initialize the PBN from an arbitrary undesirable GAP, we observe that the policies are robust to the initial state unless the constraint is too tight. For $C_{\text{total}} \geq 0.2$, the values of ΔP do not alter significantly; the performance of the intervention policy varies more for different initial GAPs when the constraint is tight, $C_{\text{total}} = 0.1$.

5.5.2 Cyclic intervention

Cancer treatment options, such as chemotherapy, are directed at killing cells. These cells include cancer cells, which continuously divide to form more cells, as well as healthy cells that also divide quickly. Healthy cells usually recover shortly after chemotherapy is complete. To this end, chemotherapy is generally given in "cycles": each treatment is followed by a "recovery phase." In a recovery phase, the side effects tend to gradually degrade. Assuming the feasibility of extracting samples from patients during the recovery phase, we are interested in designing an effective cyclic therapeutic strategy when the treatment is permitted only after each recovery phase.

In the fixed cyclic intervention scheme proposed in Vahedi *et al.* (2009b), the objective is to design an effective intervention strategy when the treatment is permitted every W transitions, where $W \in \mathbb{N}$ denotes the length of the recovery phase. Intervention is permitted at the beginning of a treatment window. Thereafter, the system transitions $W-1$ steps without any intervention, which corresponds to a recovery phase.

In Vahedi *et al.* (2009b), it has been shown how the optimal cyclic policy can be found via dynamic programming algorithms. The objective is to find an optimal policy when it is allowed to apply control only after W transitions. To this end, in each treatment window, the average costs of the $W-1$ steps where no treatment is permitted are accumulated and added to the cost at the beginning of that treatment window, where treatment is permitted. Equation (5.116) shows how the cost at the beginning of the $(s+1)$th window affects the cost at the beginning of the (s)th window:

$$
\begin{aligned}
J_s(i) = \min_{u \in C} \Bigg\{ g(i,u) + \alpha \sum_{j=0}^{N-1} p_{ij}(u) \Bigg[\sum_{k=0}^{N-1} \left(\sum_{r=0}^{W-2} \alpha^r p_{jk}^{(r)} \right) g(k,0) \\
+ \alpha^{W-1} \sum_{k=0}^{N-1} p_{jk}^{(W-1)} J_{(s+1)}(k) \Bigg] \Bigg\},
\end{aligned}
\tag{5.116}
$$

where $p_{i,j}^{r}(u)$ is the probability of transitioning from state i to state j in r steps under control u. This equation, similar to the dynamic programming technique, ranks decisions based on the sum of the present cost and the expected future cost considering the cost of all the states where no control is allowed between two treatment points. The proof of the convergence in Eq. (5.116) can be found in Vahedi *et al.* (2009b). The optimal cost function at the beginning of each treatment window corresponding to state i satisfies

$$J^*(i) = \min_{u \in C} \left\{ g(i,u) + \alpha \sum_{j=0}^{N-1} p_{ij}(u) [\sum_{k=0}^{N-1} \left(\sum_{r=0}^{W-2} \alpha^r p_{jk}^{(r)} \right) g(k,0) \right. \\ \left. + \alpha^{W-1} \sum_{k=0}^{N-1} p_{jk}^{(W-1)} J^*(k)] \right\}. \tag{5.117}$$

The optimal cyclic control policy attains the minimum in Eq. (5.117). Proofs can be found in Vahedi *et al.* (2009b).

5.6 Robust Control

A fundamental aspect of any designed operator is the degree to which its performance degrades when it is applied to processes different from the one for which it has been designed. Qualitatively, an operator is said to be *robust* when its performance degradation is acceptable for processes close to the one for which it has been designed. Robustness is important for application because, once in practice, an operator will surely be applied in nondesign settings.

In the context of signal processing, robust filter design has been posed in the following way: given that the distribution determining the optimal filter is uncertain, in the sense that it can be in different states, then what is the best state at which to optimize the filter so as to optimize filter performance relative to the set of all possible states? It is implicit that a criterion must be posited for optimization across the state (parameter) space. Design of optimal robust filters was first treated from a minimax perspective, which does not employ distributional information regarding the states: find a linear filter having the best worst performance over all states (Kuznetsov, 1976; Poor, 1980; Verdu and Poor, 1984). Optimal robust filtering is treated in a Bayesian framework by assuming a prior distribution for the states. In this context, robustness analysis is relative to the prior distribution (Grigorian and Dougherty, 2001a, 2001b). Subsequently, robust classification has been considered from both the minimax and Bayesian perspectives (Dougherty *et al.*, 2005).

Regarding PBN intervention, to this point, control policies have been designed to minimize a cost function, with a consequence that the steady-state distribution is beneficially transformed, under the assumption that the network is exactly known. In practice the actual network is not likely to be known exactly: for instance, we may possess only an estimate of its transition probability matrix P. In this case, the control policy will be designed using a perturbed variant, $\tilde{P}$, of P. In Section 4.5 we studied the effect of perturbation on the steady-state distribution; here we are interested in the effect of perturbation on the steady state of the controlled PBN. Consequently, we would like to have perturbation bounds on the shift of the steady-state distribution resulting from the difference between $\tilde{P}$ and P. Going further, our uncertainty with respect to the network might not simply be a single alternative to the actual probability transition matrix but rather that the actual probability transition matrix belongs to an uncertainty class of matrices and we want to design a control policy that works optimally across the entire uncertainty class. There are three issues here: characterization of the uncertainty, characterization of optimality, and finding an optimal policy.

5.6.1 Perturbation bounds

Before considering perturbation bounds, we need to consider the effect of control on the probability transition matrix, that is, determination of the relationship between the probability transition matrices of the original and the controlled network, where we assume control results from a stationary infinite-horizon policy.

Suppose a PBN has binary control inputs, $u_1, u_2, \ldots, u_k$, where u_i refers to the status of the ith control gene with $u_i = 1$ signifying that the ith control gene is to be flipped. If we apply a stationary policy to the Markov chain determined by the PBN's probability transition matrix P, then the rows of the controlled transition probability matrix, P_c, will be a collection of selected rows from P. This is because flipping genes forces the Markov chain to start from another initial state.

For a concrete example, consider a network with 7 genes, three of which, namely, genes 1, 2, and 3, are control genes. Suppose that the stationary policy for state 0000001 (corresponding to the decimal number 1) is 101; i.e. flip gene 1, leave gene 2 as is, and flip gene 3. If the network is currently in state 0000001, application of the stationary policy reinitializes the state to 1010001 (corresponding to the decimal number 81). Therefore, in the controlled transition probability matrix P_c, the transition probabilities of going from state 1 to each of the other states will be the same as the transition probabilities of going from state 81 to each of those states in the original uncontrolled network with transition probability matrix P.

In general, when applying a stationary policy, P and P_c are related by $P_c = TP$, where T is a matrix having only one nonzero entry of 1 in a row. If the stationary policy is no control, then $T = I$, the identity matrix. For the example just given, the 2nd row of T consists of all 0's except for the 82nd entry, which is 1.

Let π and π_c denote the stationary distribution vectors corresponding to the transition matrices P and P_c respectively, and let $\tilde{\pi}$ and $\tilde{\pi}_c$ denote the stationary distributions of $\tilde{P}$ and $\tilde{P}_c$, respectively. The problem is to bound the difference $\tilde{\pi}_c - \pi_c$ based on the knowledge of P and some characterization of the error, $E = P - \tilde{P}$.

If, for two Markov chains with transition probability matrices P and $\tilde{P}$, and sharing a common state space, the difference between the two stationary distributions can be bounded by

$$\|\pi - \tilde{\pi}\|_q \leq k\|E\|_\infty, \tag{5.118}$$

where $q = 1$ or ∞, $k > 0$, $\|\pi - \tilde{\pi}\|_q$ denotes the q-norm of the vector $\pi - \tilde{\pi}$, and $\|E\|_\infty$ is defined as the maximum absolute row sum of E, then the constant k is referred to as a *condition number*. Several have been studied in the literature. Some condition numbers yield tighter bounds than others and a comparison of the available bounds is given in Cho and Meyer (2001). Here, we will provide a theorem for a particular condition number called the *ergodicity coefficient* and defined by

$$\tau_1(P) = \sup_{\substack{\|x_1^T\|=1 \\ x^T \mathbf{1}_n = 0}} |x^T P|_1, \tag{5.119}$$

where $\mathbf{1}_n$ denotes the n-dimensional column vector having all entries equal to 1 (Seneta,

1988). The ergodicity coefficient possesses the following two equivalent formulations:

$$\tau_1(P) = \frac{1}{2}\max_{i,j}\sum_{s=1}^{n}|p_{is} - p_{js}|, \tag{5.120}$$

$$\tau_1(P) = 1 - \min_{i,j}\sum_{s=1}^{n}\min(p_{is}, p_{js}), \tag{5.121}$$

where p_{ij} refers to the ith row and jth column entry of matrix P (Seneta, 1988). The ergodicity coefficient satisfies the relation

$$\tau_1(P_1 P_2) \leq \tau_1(P_1)\tau_1(P_2) \tag{5.122}$$

(Seneta, 1993). In Seneta (1988), the ergodicity coefficient is used to obtain a bound on the perturbation in the steady-state distribution owing to perturbations in the transition probability matrix: if $\tau_1(P) \neq 1$, then

$$\|\pi - \tilde{\pi}\|_1 \leq \frac{1}{1 - \tau_1(P)}\|E\|_\infty. \tag{5.123}$$

We extend the above result to obtain a perturbation bound on the *controlled* steady-state distributions that could result from perturbations in the uncontrolled probability transition matrix.

Theorem 5.7 (Pal *et al.*, 2008). *For stationary control via flipping genes in PBNs, if* $\tau_1(P) \neq 1$, *then*

$$\|\pi_c - \tilde{\pi}_c\|_1 \leq \frac{1}{1 - \tau_1(P)}\|E\|_\infty. \tag{5.124}$$

Proof. As discussed previously, for the class of interventions being considered, $P_c = TP$, where T is a stochastic matrix with each row containing only a single nonzero entry of 1. According to Eq. (5.122),

$$\tau_1(P_c) \leq \tau_1(T)\tau_1(P). \tag{5.125}$$

From Eq. (5.121), it is clear that the ergodicity coefficient of a stochastic matrix is less than or equal to 1 and hence

$$\tau_1(P_c) \leq \tau_1(P). \tag{5.126}$$

Thus, from Eq. (5.123), it follows that

$$\|\pi_c - \tilde{\pi}_c\|_1 \leq \frac{1}{1 - \tau_1(P)}\|E_c\|_\infty. \tag{5.127}$$

Since

$$E_c = P_c - \tilde{P}_c = TP - T\tilde{P} = T(P - \tilde{P}) = TE, \tag{5.128}$$

the rows of E_c are selected from the rows of E and therefore $\|E_c\|_\infty \leq \|E\|_\infty$. Thus, Eq. (5.124) follows from Eq. (5.127). □

The inequality in Eq. (5.126) implies that if the Markov chain corresponding to an uncontrolled genetic network has a small ergodicity coefficient bound, then the ergodicity coefficient for the corresponding controlled Markov chain is bounded by the same bound. Consequently, if a stationary policy is designed from an estimated Markov chain that is "close" to the actual one for the network, then this policy when applied to the actual network will produce results that are close to the desired outcome, as far as the steady-state behavior is concerned.

There are several condition numbers other than the ergodicity coefficient found in the literature. These perturbation bounds are mostly stated in terms of the fundamental matrix, Z, or the g-inverse, A^-, of $A = I - P$. In Cho and Meyer (2001), the following condition numbers are given for bounding the 1 and ∞ norms of the perturbations in the steady-state distributions:

$$k_1 = ||Z||_\infty, \quad q = 1, \tag{5.129}$$

$$k_2 = ||A^-||_\infty, \quad q = 1, \tag{5.130}$$

$$k_3 = \frac{\max_j (a_{jj}^- - \min_i a_{ij}^-)}{2}, \quad q = \infty, \tag{5.131}$$

$$k_4 = \max_{i,j} |a_{ij}^-|, \quad q = \infty, \tag{5.132}$$

$$k_5 = \frac{1}{1 - \tau_1(P)}, \quad q = 1, \tag{5.133}$$

$$k_6 = \tau_1(A^-) = \tau_1(Z), \quad q = 1, \tag{5.134}$$

$$k_7 = \frac{\min_j ||A_{(j)}^{-1}||_\infty}{2}, \quad q = \infty. \tag{5.135}$$

The bound k_5 involves the ergodicity coefficient. In Pal *et al.* (2008), a simulation study is carried out to analyze the effectiveness of these bounds for the steady-state distributions of controlled PBNs.

5.6.2 Minimax robust control

Minimax robustness involves finding the control policy that minimizes the maximum cost over all possible states of nature; that is, minimize the cost of the worst-case scenario. We consider the problem where robustness is relative to uncertainty in the transition probability matrix with the aim of finding the optimal minimax robust control policy for the discounted cost infinite-horizon problem under the assumption that the network is time invariant. In this case, there is an *uncertainty class* consisting of possible transition probability matrices, each characterizing the dynamics of a possible network, and the discounted cost of Eq. (5.25) now takes the form $J_{\mu,\theta}(z_0)$, where $\mu \in \Omega$, the set of all possible stationary control policies, and $\theta \in \Theta$ parameterizes the uncertainty class. (Note that in this section we will denote the control policy $(\mu, \mu, \ldots, \mu)$ by μ.) Whereas the optimal cost for a given network model is given by Eq. (5.29), the optimal cost relative to the uncertainty class is given by

$$\Phi(\Theta, \Omega, z_0) = \min_{\mu \in \Omega} \max_{\theta \in \Theta} J_{\mu,\theta}(z_0). \tag{5.136}$$

Recalling Section 5.3, the optimal cost is the unique solution of Bellman's equation, Eq. (5.40), and can be iteratively determined using value iteration. The optimal control is obtained as a by-product of the steps involved in the value iteration process. To explain the use of these results in the presence of uncertainty, following Pal *et al.* (2008), we introduce some definitions. Recall that the policy μ is a mapping, $\mu : \mathcal{S} \to \mathcal{C}$, of the form $\mu(z) = u$. For a given stationary controlled transition probability matrix P^u and any state $z \in \mathcal{S}$, define $(P^u_z)^T$ to be the row of P^u that defines the transition probabilities from state z. Furthermore, for any initial state z_0, define the value function v by $v(z_0) = \min_{\mu\in\Omega} J_{\mu,\theta}(z_0)$ and the vector $\mathbf{v}$ by $\mathbf{v} = [v(1), v(2), \ldots, v(m2^n)]^T$. According to the results of Section 5.3, the value function v satisfies the Bellman equation, which with the new definitions takes the form

$$v(z) = \min_{u\in\mathcal{C}}[g(z,u) + \alpha(P^u_z)^T \mathbf{v}] \tag{5.137}$$

for any $z \in \mathcal{S}$. Furthermore, v is the unique limit value of the convergent vector sequence defined by

$$v_{k+1}(z) = \min_{u\in\mathcal{C}}[g(z,u) + \alpha(P^u_z)^T \mathbf{v_k}] \tag{5.138}$$

for $z \in \mathcal{S}$ and $k = 1, 2, \ldots$.

Intuitively, the robust versions of the Bellman equation and the value iteration can be derived by noting that, in this case, we are focusing on the worst-case scenario and, therefore, P^u_z on the right-hand sides of Eqs. (5.137) and (5.138) must be chosen to maximize $(P^u_z)^T \mathbf{v}$ and $(P^u_z)^T \mathbf{v_k}$, respectively. These ideas have been formalized (Nilim and El Ghaoui, 2004; Iyengar, 2005) and are summarized in the following theorem.

Theorem 5.8. *For the infinite-horizon control problem* (5.136) *with stationary uncertainties on the probability transition matrices, stationary control policies, and a discounted cost function with discount factor* $\alpha \in [0,1)$, *there exists perfect duality between minimization and maximization, namely,*

$$\min_{\mu\in\Omega}\max_{\theta\in\Theta} J_{\mu,\theta}(z_0) = \max_{\theta\in\Theta}\min_{\mu\in\Omega} J_{\mu,\theta}(z_0). \tag{5.139}$$

The optimal value is given by $\Phi(\Theta, \Omega, z_0) = v(z_0)$, *where* z_0 *is the initial state and the value function satisfies the optimality condition*

$$v(z) = \min_{u\in\mathcal{C}}[g(z,u) + \alpha\sigma_{P^u_z}(\mathbf{v})] \tag{5.140}$$

for any $z \in \mathcal{S}$, *where*

$$\sigma_{P^u_z}(\mathbf{v}) = \sup_{P^u_z}[(P^u_z)^T \mathbf{v}]. \tag{5.141}$$

Furthermore, the value function is the unique limit value of the convergent vector sequence defined by

$$\mathbf{v}_{k+1}(z) = \min_{u\in\mathcal{C}}[g(z,u) + \alpha\sigma_{P^u_z}(\mathbf{v}_k)] \tag{5.142}$$

for $z \in \mathcal{S}$ *and* $k = 1, 2, \ldots$. *A stationary optimal control policy* $(\mu^*, \mu^*, \ldots)$ *is obtained as the* u *that minimizes the right-hand side of Eq.* (5.142) *once the iteration has converged.*

This theorem supplies us with a value iteration procedure to construct control policies that are optimal for the worst-case scenario. It is quite general and does not focus on a specific characterization of the uncertainty. To apply the theorem, one must repeatedly carry out the maximization in Eq. (5.141). This maximization problem can present different levels of difficulty depending on the characterization of the uncertainty. We address the solution to the maximization problem in Eq. (5.141) for three specific cases.

In the first case, referred to as *infinity-norm uncertainty*, the uncertainty is characterized by an upper bound on $||E||_\infty$, where $\tilde{P}$ is observed, P is the actual transition probability matrix, and $E = P - \tilde{P}$. This case corresponds to the situation encountered with perturbations bounds. It occurs naturally when one estimates the transition probabilities and has sufficient data to assert that $P(|p_{ij} - \tilde{p}_{ij}| \leq \varepsilon)$ is sufficiently large that, for the purposes of modeling, one can assume that the absolute values of the terms E are all bounded by ε. Note that $||E||_\infty = ||\mathbf{e}_\infty^T||_1$, where $\mathbf{e}_\infty^T$ is the row of E possessing the largest absolute sum. In the second case, with *interval uncertainty*, the transition probabilities are assumed to lie between two reference matrices; that is, there exist matrices $P^0 = (p_{ij}^0)$ and $P^1 = (p_{ij}^1)$ such that $p_{ij}^0 \leq p_{ij} \leq p_{ij}^1$ for all i and j, a situation that would occur when the selection probabilities of a PBN are imprecisely estimated. Last, with *transition-matrix uncertainty*, the probability transition matrix belongs a finite set, $\{P^1, P^2, \ldots, P^m\}$, of matrices, a situation that would occur when there is uncertainty regarding the estimation of some of the predictor functions in a PBN. The solutions to the maximization problem in Eq. (5.141) for the three cases are provided in the next three theorems, where B^* denotes the supremum.

Theorem 5.9. *The maximization problem for infinity-norm uncertainty takes the form*

$$B^* = \max_{\mathbf{p}} \mathbf{v}^T \mathbf{p}, \tag{5.143}$$

subject to the constraints $\mathbf{p}^T \mathbf{1} = 1$ *and* $\|\mathbf{p} - \mathbf{q}\|_1 \leq \beta$, *where the vector* $\mathbf{p}^T$ *represents a row of a transition probability matrix and the vector* $\mathbf{q}^T$ *represents the corresponding reference row about which the perturbation must satisfy a* 1*-norm constraint, and* β *is the limit on the infinity norm of the difference between the actual and utilized transition probability matrices. If the components of* $\mathbf{v}$ *are arranged in ascending order and denoted by* $\bar{\mathbf{v}}(i)$, $1 \leq i \leq m2^n$, *and the same order is used to sort the values of* $\mathbf{q}$ *represented as* $\bar{\mathbf{q}}$, *then*

$$B^* = \min_{1 \leq j \leq m2^n} \left[\sum_{i>j} \bar{\mathbf{q}}(i)\bar{\mathbf{v}}(i) + \bar{\mathbf{v}}(j) \left(\sum_{i \leq j} \bar{\mathbf{q}}(i) - \frac{\beta}{2} \right) + \bar{\mathbf{v}}(m2^n)\frac{\beta}{2} \right]. \tag{5.144}$$

Theorem 5.10. *The maximization problem for interval-matrix uncertainty takes the form*

$$B^* = \max_{\mathbf{p} \geq \mathbf{0}} \mathbf{v}^T \mathbf{p}, \tag{5.145}$$

subject to the constraints $\mathbf{p}^T \mathbf{1} = 1$ *and* $\mathbf{p}_1 \leq \mathbf{p} \leq \mathbf{p}_2$, *where the vector* $\mathbf{p}^T$ *represents a row of a transition probability matrix and the nonnegative vectors* $\mathbf{p}_1$ *and* $\mathbf{p}_2$ *characterize the interval uncertainty associated with that row. The corresponding solution is given by*

$$B^* = \min_{\varsigma \in \{0, v_1, \ldots, v_{m2^n}\}} \mathbf{v}^T \mathbf{p}_2 + \varsigma(1 - \mathbf{p}_2^T \mathbf{1}) + (\mathbf{p}_2 - \mathbf{p}_1)^T (\varsigma \mathbf{1} - \mathbf{v})^+, \tag{5.146}$$

where x^+ *stands for the positive part of* x.

Theorem 5.11. *In the case of transition-matrix uncertainty, for any stationary control policy* μ, *let* $\{P^{1,u}, P^{2,u}, \ldots, P^{r,u}\}$ *denote the set of transition matrices corresponding to* *the uncertainty class for P for control action u and let $\mathbf{p}_z^{l,u}$ denote the transpose of the zth row of $P^{l,u}$. Then, the maximization problem is given by*

$$B^* = \max_{\mathbf{p} \in \{\mathbf{p}_z^{1,u}, \mathbf{p}_z^{2,u}, \ldots, \mathbf{p}_z^{r,u}\}} \mathbf{v}^T p = \max_{1 \le l \le r} \mathbf{v}^T \mathbf{p}_z^{l,u}. \tag{5.147}$$

The proofs of these theorems are given in the supplementary material for Pal *et al.* (2008) and these are based on prior arguments (Nilim and El Ghaoui, 2004; Iyengar, 2005).

The complexity of each step of the Bellman recursion for the regular problem (without considering any uncertainties) is $O(2^k 2^n)$; for the robust versions, the one-step complexities are $O(2^k 2^n 2^n)$ for infinity-norm uncertainty and matrix-interval uncertainty, and $O(r 2^k 2^n)$ for transition-matrix uncertainty.

5.6.3 Bayesian robust control

Whereas the general principle of minimax robustness is to minimize the cost of the worst-case scenario, a Bayesian robust policy minimizes the average cost over the uncertainty class Θ, relative to a prior parameter distribution governing Θ. No such distribution is assumed for minimax robust control (although such a distribution might exist). Let μ_b denote the policy designed to be optimal at point b of the parameter distribution; that is, μ_b minimizes $J_{\mu,\theta_b}(z_0)$, where $\theta_b \in \Theta$.

Let $\gamma_a(\mu_b)$ denote the expected cost at point a of the parameter distribution for the intervention policy μ_b:

$$\gamma_a(\mu_b) = E_{z_0}[J_{\mu_b,\theta_a}(z_0)], \tag{5.148}$$

where the expectation over the states z_0 of the network is taken to arrive at a single value for representing the cost of a policy. The expected cost of $\gamma_a(\mu_b)$ over the distribution F_a is given by

$$E_a[\gamma_a(\mu_b)] = E_a[E_{z_0}[J_{\mu_b,\theta_a}(z_0)]], \tag{5.149}$$

where E_a denotes expectation relative to the parameter distribution. A *Bayesian robust control policy*, denoted by μ_{b^*}, is one that minimizes $E_a[\gamma_a(\mu_b)]$. The parameter value b^* that achieves the minimum is called a *maximally robust parameter value*. If the actual parameter for the distribution is b and one optimizes at b, then there is no gain, and perhaps a loss, in using the Bayesian robust policy, because $\gamma_b(\mu_b) \le \gamma_b(\mu_{b^*})$; however, in the face of uncertainty regarding the actual parameter, the gain in applying the Bayesian robust policy is $E_a[\gamma_a(\mu_b)] - E_a[\gamma_a(\mu_{b^*})]$, which is always nonnegative and can be substantial for parameter values far from the maximally robust one.

As noted in Pal *et al.* (2009), minimization is over the set of control policies that are optimal for various points in the distribution. A policy χ can exist that is not optimal for any point in the distribution but for which $E_a[\gamma_a(\chi)] \le E_a[\gamma_a(\mu_{b*})]$; however, no analytical methods exist for generation of such a policy and an exhaustive search is prohibitive due to the large number of possible policies. The problem with a dynamic programming based value iteration approach for directly calculating either χ or μ_{b*} is that the principle of optimality is not valid for the current setting.

Rather than trying to design robust optimal policies at all parameter values and then select the maximally robust policy, one can instead design a single global optimal policy by using a centrality parameter, such as the mean or median, to represent the mass of the uncertainty distribution. Here we will use the mean of the distribution as the global parameter value for designing the intervention strategy. The *global policy*, denoted by $\mu_{\bullet}$, is the optimal policy for the global parameter value.

Different approaches to robustness can be applied in different circumstances. Minimax policies are typically conservative and give too much weight to parameter values that may rarely occur in practice; however, they are useful for avoiding extremely undesirable results. Bayesian robust policy design gives better expected performance. Global policy design is advisable when one wishes to take into account the parameter distribution but has only limited knowledge of it, say, an estimate of the mean or median. Based on the application, an appropriate design strategy should be selected.

A number of cost and expected cost curves are involved in comparing the performances of different robust policies:

- $\gamma_a(\mu_b) = E_{z_0}[J_{\mu_b,\theta_a}(z_0)]$ – Cost at parameter value a using the optimal policy designed at b.
- $\gamma_b(\mu_b) = E_{z_0}[J_{\mu_b,\theta_b}(z_0)]$ – Cost at parameter value b using the optimal policy designed at b.
- $\gamma_b(\mu_{b^*}) = E_{z_0}[J_{\mu_{b^*},\theta_b}(z_0)]$ – Cost of the Bayesian robust policy at parameter value b.
- $\gamma_b(\mu_{\bullet}) = E_{z_0}[J_{\mu_{\bullet},\theta_b}(z_0)]$ – Cost of the global robust policy at parameter value b.
- $\gamma_b(\mu_{mm}) = E_{z_0}[J_{\mu_{mm},\theta_b}(z_0)]$ – Cost of the minimax robust policy at parameter value b.
- $E_a[\gamma_a(\mu_b)]$ – Expected cost of the optimal policy designed at b across the parameter space.
- $E_a[\gamma_a(\mu_{b^*})]$ – Expected cost of the Bayesian robust policy across the parameter space.
- $E_a[\gamma_a(\mu_{\bullet})]$ – Expected cost of the global robust policy across the parameter space.
- $E_a[\gamma_a(\mu_{mm})]$ – Expected cost of the minimax robust policy across the parameter space.

Following Pal *et al.* (2009), simulations involve generating r transition probability matrices $P_1, P_2, \ldots, P_r$ corresponding to r points, $b_1, b_2, \ldots, b_r$, of the parameter distribution. The Bayesian policy is selected from one of the optimal policies for points $b_1, b_2, \ldots, b_r$. To have a consistent comparison with the Bayesian approach, the set of admissible policies for the minimax policy is taken to be $\{\mu_{b_1}, \mu_{b_2}, \ldots, \mu_{b_r}\}$.

5.6.4 Uncertainty in the switching probabilities

To illustrate Bayesian robust control and to compare it to minimax robust control, we consider uncertainty in the selection probabilities of a PBN. To avoid the complication of multiple switching probabilities and thereby better graphically illustrate robustness relative to switching probabilities, we consider an artificial example of an instantaneously random PBN consisting of two constituent 4-gene BNs. The selection probability c_1 follows a beta

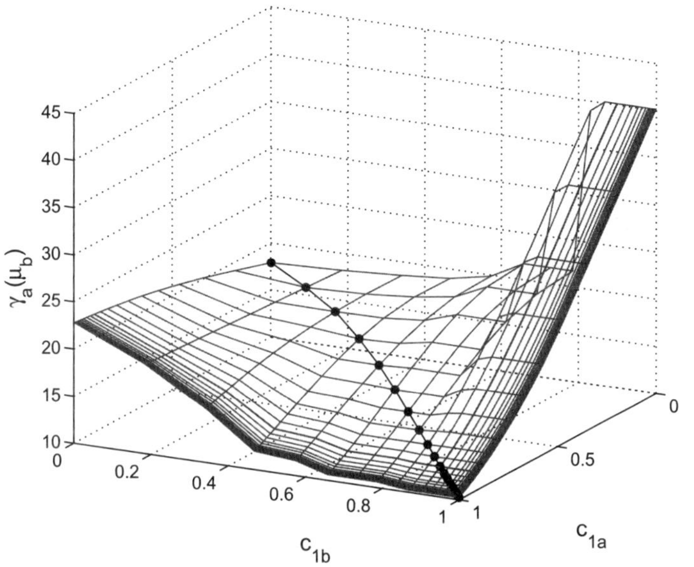

Figure 5.17. *Cost of policies designed at c_{1b} over the set of all possible c_{1a}'s.*

distribution having parameters $\alpha = 1$ and $\beta = 0.1$, and the selection probability c_2 is calculated from the equation $c_1 + c_2 = 1$. The parameter p is fixed at $p = 0.1$. The control problem for the Markov chain with dimension 16×16 is formulated with cost of control being 0.2 and the states assigned penalties as follows:

$$\tilde{g}(i,u,j) = \begin{cases} 0 & \text{if } u = 0 \text{ and } 1 \leq j \leq 8, \\ 0.2 & \text{if } u = 1 \text{ and } 1 \leq j \leq 8, \\ 5 & \text{if } u = 0 \text{ and } 9 \leq j \leq 16, \\ 5.2 & \text{if } u = 1 \text{ and } 9 \leq j \leq 16. \end{cases} \tag{5.150}$$

The penalty structure indicates that the highly desirable states have the first gene down-regulated (0) while the undesirable states have the first gene up-regulated (1). The discount factor is $\alpha = 0.9$.

Figure 5.17, the *robustness surface*, shows the cost for the different points in the distribution (c_{1a}) when the policy is designed at particular points in the distribution (c_{1b}) and applied at c_{1a}. It shows that for a network with parameter value c_{1a}, the cost is minimum when the policy designed at parameter value $c_{1b} = c_{1a}$ is applied to it. The minimum for each c_{1a} is represented by black dots in the figure. Figures 5.18 and 5.19 show the costs and expected costs, respectively, as functions of the parameter c_{1b} ranging from 0 to 1. Figure 5.18 shows that the cost for low values of c_1 is lower with the minimax policy compared to the Bayesian and global policies but that the performance of the minimax policy deteriorates for higher values of c_1. Based on the parameters of the beta distribution, the probability of occurrence of low values of c_1 is much less than that of high values of c_1. Minimax design gives equal importance to all cases irrespective of their probability of occurrence. With regard to expected cost, the minimax policy performs much worse than the

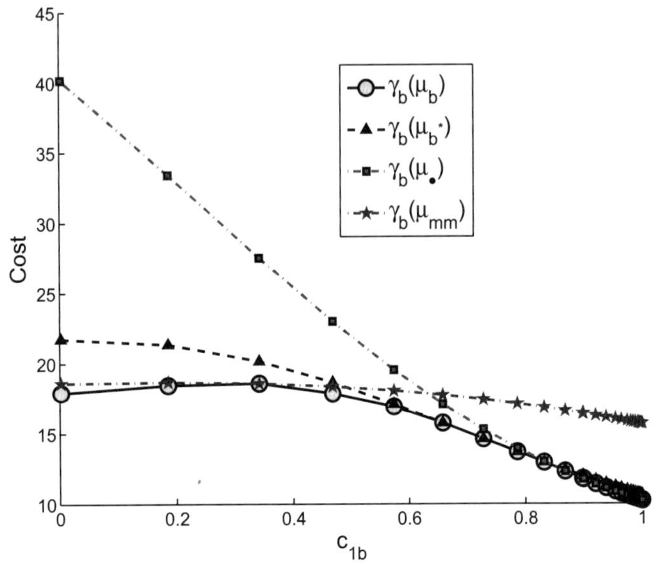

Figure 5.18. *Cost for various policies.*

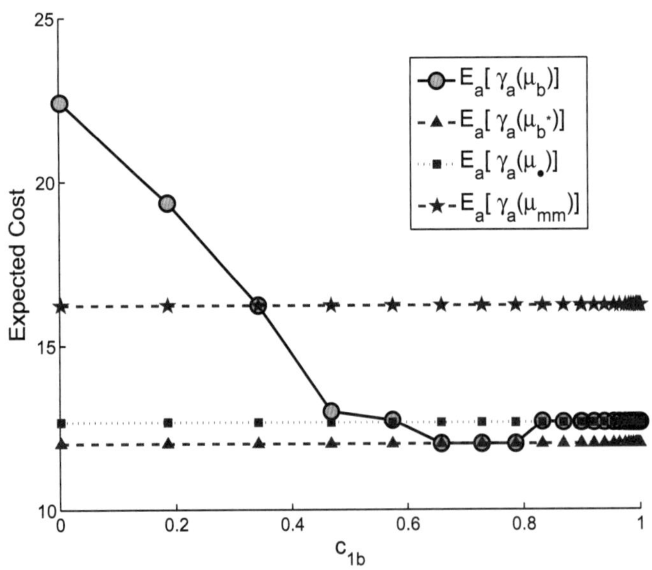

Figure 5.19. *Expected cost for various policies.*

Bayesian or global robust policies, as shown in Fig. 5.19. The maximally robust parameter value is $c_1 = 0.658$ and the globally robust parameter value is the mean of the distribution, $\alpha/(\alpha+\beta) = 1/1.1 = 0.909$.

We close this section by noting that in Pal *et al.* (2009), errors originating from data extraction are probabilistically characterized and this characterization is used as the basis for comparing robustness policies.

5.7 Adaptive Infinite-Horizon Control

One way to approach imperfect knowledge of the underlying PBN when designing a control strategy is to design an optimal robust strategy; another approach is to "tune" the intervention strategy to the actual network via on-line adaptation. This section discusses adaptive control, an approach that is feasible only if the uncertainty belongs to a specific class and prior knowledge about this class can be incorporated into the design. We assume that the underlying network is modeled by a member of a known finite family of PBNs and we have no prior knowledge about which member of that family models the actual network. The aim is to estimate the model number on-line and then use policy iteration to determine the corresponding controller. Considerable theoretical research has been aimed at showing that such *certainty equivalence* schemes can provide satisfactory performance (Ioannou and Sun, 1996; Kumar and Varaiya, 1986). We assume a family of instantaneously random PBNs parameterized by $\alpha \in A$, where $\sum_{j\in\mathcal{S}} p(i,j,u,\alpha) = 1$ for any $\alpha \in A$ and $(i,u) \in \mathcal{S} \times \mathcal{C}$, $p(i,j,u,\alpha)$ denoting the transition probability under control u and PBN α.

Following Layek *et al.* (2009), a variation of an adaptive control algorithm developed in Kumar and Lin (1982) for unknown Markov chains is utilized. While the scheme in Kumar and Lin (1982) attempts to estimate all entries of the transition probability matrix, the adaptive algorithm presented here will estimate only the model number, the underlying assumption being that the transition probabilities of the PBN are completely determined, once the model number is known. There are various ways to possess a list of PBNs and thereby be presented with the problem of adaptively determining a model number. For instance, the inference procedure discussed in Section 3.3 constructs BNs whose attractors coincide with data points and which satisfy certain constraints. PBNs are constructed from these by comparing the steady-state distributions of potentially inferred PBNs with the full set of experimental data. Alternatively, BNs can be inferred by first using a Bayesian approach to generate regulatory graphs most compatible with the data and then inferring the predictors via a nonlinear perceptron model. PBNs are then constructed from the BNs by using Bayesian scores.

The scheme in Layek *et al.* (2009) utilizes the adaptive control algorithm derived in Kumar and Lin (1982) by maximizing a modified likelihood criterion. For each $\alpha \in A$, let $J^*(\alpha)$ be the optimal long-term average cost obtained for model α and let $\phi(.,\alpha) : \mathcal{S} \to \mathcal{C}$ be the corresponding control law. Let $f : \mathbb{R} \to \mathbb{R}$, $o : \mathbb{Z} \to \mathbb{R}$, and let constant m be defined as follows: f is a strictly monotonically increasing continuous function such that $f(\inf_{\alpha\in A} J^*(a)) > 0$; o is any function such that $\lim_{t\to\infty} o(t)t^{-\theta}$ is a positive finite number for some $\theta \in (0,1)$; and m is any integer such that $m > |\mathcal{S}| + 1$. The implementation of Layek *et al.* (2009) uses f as the logarithmic function and $o(t) = 2\sqrt{(t)}$, for which $\theta = 0.5$. The adaptive controller consists of two separate operations, estimation and control:

- *Estimator:* At each time step $0, m, 2m, 3m, \ldots, km, (k+1)m, \ldots$, estimate α by

$$\hat{\alpha}_t := \operatorname*{argmax}_{\alpha\in A} \bar{D}_t(\alpha), \tag{5.151}$$

where

$$\bar{D}_t(\alpha) := K \prod_{(i,j,u)\in F^c} p(i,j,u,\alpha)^{n_t(i,j,u)}, \tag{5.152}$$

$$K = \left[\frac{1}{f\{J^*(\alpha)\}}\right]^{o(t)}, \tag{5.153}$$

and F^c is the complement of the set F of *forbidden* transitions, which is assumed to be known *a priori*. These transitions correspond to zero values for $p(i,j,u,\alpha)$. In Eq. (5.152),

$$n_t(i,j,u) = 1 + \sum_{s=0}^{t-1} \mathbf{1}(z_s = i, z_{s+1} = j, u_s = u), \tag{5.154}$$

where $\mathbf{1}(.)$ denotes the indicator function. $n_t(i,j,u)$ can be interpreted as measuring the number of times a transition occurs from i to j under control u. At time km, knowing the parameter estimate $\alpha_{\hat{km}}$, one can find the optimal cost function $J^*(\alpha_{\hat{km}})$ and the optimal control law $\phi(z_t,\alpha_{\hat{km}})$ by applying policy iteration to the estimated model. These will be used for the next m time steps. The parameter estimate is kept constant at $\alpha_{\hat{km}}$ between time steps km and $(k+1)m-1$.

- *Controller:* At each time t, the control applied is

$$u_t := \phi(z_t, \hat{\alpha}_t). \tag{5.155}$$

Although the adaptive procedure is designed for instantaneously random PBNs, it can be applied to context-sensitive PBNs by using the reduction method of Section 2.7. If so, then, as demonstrated via simulations in Layek *et al.* (2009), it can be expected that performance degradation will occur as the switching probability is reduced from 1 to 0. If, as is typically the case, the switching probability is assumed to be very small, one can proceed by considering each constituent BN to be a possible model to be identified by the estimation algorithm. Although this increases the number of possible models, as simulations in Layek *et al.* (2009) demonstrate, it results in improved performance because a small switching probability means that the constituent networks change infrequently and therefore the estimation procedure has time to identify the current BN.

Two algorithms are presented in Layek *et al.* (2009) to address these issues and a simulation study demonstrates the performance of the adaptive intervention strategy. A simulation study is important because the theoretical results in Kumar and Lin (1982) guarantee only almost sure convergence, and that in the Cesaro sense.

5.8 Mean-First-Passage-Time Stationary Control

As previously discussed, various approximations have been employed to mitigate the computational burden of full optimization when using dynamic-programming-based algorithms. In the present section we consider a greedy algorithm that bypasses altogether the specification of a cost function and dynamic programming; rather, it is based on mean first passage times (Vahedi *et al.*, 2008). Recall that the first intervention strategy proposed for PBNs utilized a single gene flip, with one possible criterion being based on mean first passage time (Shmulevich *et al.*, 2002b). Whereas that approach was transient, the mean-first-passage-time (MFPT) algorithm discussed in this section provides a stationary, infinite-horizon control strategy. the MFPT algorithm requires the designation of desirable and undesirable states, and intervention is performed by flipping the expression status of a particular gene. The MFPT algorithm is based on the following heuristic: given a control gene, when a desirable state reaches the set of undesirable states on average faster than its flipped state, it is reasonable to intervene and transition into the flipped state; if an undesirable state reaches the set of desirable states on average faster than its flipped state, it is reasonable not to intervene.

Without loss of generality we can assume that the transition probability matrix of the Markov chain is partitioned according to

$$P = \begin{pmatrix} P_{\mathcal{D},\mathcal{D}} & P_{\mathcal{D},\mathcal{U}} \\ P_{\mathcal{U},\mathcal{D}} & P_{\mathcal{U},\mathcal{U}} \end{pmatrix}, \tag{5.156}$$

where $\mathcal{D}$ and $\mathcal{U}$ are the subsets of desirable and undesirable states, respectively. The MFPT times are computed by solving the following systems of linear equations:

$$K_{\mathcal{U}} = e + P_{\mathcal{D},\mathcal{D}} K_{\mathcal{U}}, \tag{5.157}$$

$$K_{\mathcal{D}} = e + P_{\mathcal{U},\mathcal{U}} K_{\mathcal{D}}, \tag{5.158}$$

where e is a column vector of 1's with appropriate length, $K_{\mathcal{U}}$ is a vector containing the MFPTs from each state in $\mathcal{D}$ to $\mathcal{U}$, and $K_{\mathcal{D}}$ is a vector containing the MFPTs from each state in $\mathcal{U}$ to $\mathcal{D}$ (Norris, 1998).

The dynamic-programming-based methods discussed thus far involve a heuristic cost-of-control function that imposes a penalty for intervention. In the MFPT algorithm, this function is replaced by a tuning parameter γ, with higher settings of γ making the intervention condition more difficult to satisfy. Although we will not go into the matter here, in Vahedi *et al.* (2008) there is substantial discussion of the relation between γ and the cost-of-control function, and the way that relationship affects the degree to which the greedy MFPT stationary policy can be used as an approximation of an optimal dynamic-programming-based control policy. When discussing such an "approximation," one must keep in mind that both the cost-of-control function and γ are heuristics, and the goals of the algorithms are different, one to lower an overall cost and the other to move the network as quickly as possible into desirable states. Perhaps the key point is that both the cost-of-control function and γ serve similar roles.

Given γ, the MFPT algorithm selects the control policy $\hat{\mu}_g^\gamma$ for control gene g in the following manner. For any network state $\mathbf{x}$, let $\tilde{\mathbf{x}}$ be the state that is identical to $\mathbf{x}$ except that the value of g is flipped (from 0 to 1 or from 1 to 0). Suppose $\mathbf{x}$ is an undesirable state. The algorithm compares the average MFPTs from state $\mathbf{x}$ to $\mathcal{D}$ and from the flipped state $\tilde{\mathbf{x}}$ to $\mathcal{D}$, the issue being on average which one of the two states, $\mathbf{x}$ and $\tilde{\mathbf{x}}$, "hits" the set of desirable states for the first time faster than the other one. The algorithm sets $\hat{\mu}_g^\gamma(\mathbf{x}) = 1$ if

$$K_{\mathcal{D}}(\mathbf{x}) - K_{\mathcal{D}}(\tilde{\mathbf{x}}) > \gamma; \tag{5.159}$$

otherwise, $\hat{\mu}_g^\gamma(\mathbf{x}) = 0$. Analogously, if state $\mathbf{x}$ is desirable, then $\hat{\mu}_g^\gamma(\mathbf{x}) = 1$ if

$$K_{\mathcal{U}}(\tilde{\mathbf{x}}) - K_{\mathcal{U}}(\mathbf{x}) > \gamma; \tag{5.160}$$

otherwise, $\hat{\mu}_g^\gamma(\mathbf{x}) = 0$. These comparisons are repeated for all states.

As we have discussed, although an optimal control policy might result from a cost function defined for a particular setting, a general measure of its performance is the normalized amount to which it reduces the steady-state mass of undesirable states. The relevant performance measure is given by Eq. (5.92). The analogous performance measure for the MFPT control policy is given by Eq. (5.93) with $\hat{\mu}_g^\gamma$ in place of $\hat{\mu}$. To be consistent with the notation in Vahedi *et al.* (2008), in this section we denote these performance measures for the optimal and MFPT policies by ΔP_g^{opt} and $\Delta P_g^{\text{MFPT}(\gamma)}$, respectively.

The MFPT algorithm has other uses besides its direct utilization for designing a stationary control policy. One example is finding the best control gene. The best control gene g^* can be found by directly solving a dynamic programming algorithm and computing ΔP_g^{opt} for all the genes g in the network. In short, g^* is given by

$$g^* = \arg \max_{g=1,\ldots,n} \Delta P_g^{\text{opt}}; \tag{5.161}$$

however, this optimal method to find the best control gene is computationally prohibitive. On the other hand, the MFPT algorithm enables more efficient computation of the MFPT control policies $\hat{\mu}_g^\gamma$ for all the genes in the network. Taking this approach, the MFPT algorithm predicts the best control gene to be

$$\hat{g} = \arg \max_{g=1,\ldots,n} \Delta P_g^{\text{MFPT}(\gamma)}. \tag{5.162}$$

In Vahedi *et al.* (2008), it is shown via simulation that $\hat{g} = g^*$ with high probability, and that $\Delta P_{g^*}^{\text{opt}} - \Delta P_{\hat{g}}^{\text{opt}}$ is generally small when $\hat{g} \neq g^*$. Once the best gene candidate is identified, an optimal control policy can be obtained using dynamic programming algorithms.

One can use the MFPT algorithm to generate a measure of controllability by setting $\gamma = 0$, the interest here being the ability to beneficially shift the steady-state distribution absent consideration of a cost of control. For example if $\Delta P_{g^*}^{\text{MFPT}(0)}$ is very small, we conclude that the network is not controllable. If $\Delta P_{g^*}^{\text{MFPT}(0)} = 0.5$, then we conclude that it is possible to shift 50% of the probability mass of the undesirable states to desirable states in the long run, given no cost of control. While one takes a similar approach using dynamic programming by setting the cost of control to zero, such a measure would still be dependent on the cost function and not provide a universal measure. Moreover, simulations in Vahedi *et al.* (2008) indicate that $\Delta P_{g^*}^{\text{MFPT}(0)}$ is usually close to $\Delta P_{g^*}^{\text{opt}}$ when the cost of control is 0.

5.8.1 Model-free intervention

A key advantage of the MFPT algorithm is that it does not require full knowledge of the model and in this sense is considered to be *model-free*. Even if we require only estimation of the associated Markov chain of a PBN, this is still a formidable task because the complexity of estimating the transition probabilities of a Markov chain increases exponentially with the number of genes in the model. For lower numbers of observations, which correspond better to feasible experimental conditions, estimating the Markov chain yields poor results. Estimation errors may have a huge impact on finding an effective intervention strategy, which is often quite sensitive to changes in the transition probabilities (Feinberg and Shwartz, 2002). Furthermore, the complexity of estimating the transition probabilities of a Markov chain increases exponentially with the number of genes in the model, $\mathbf{O}(2^{2n})$.

This is in addition to the complexity of designing an effective intervention strategy. Hence, a procedure that can find an effective intervention strategy without having to know the transition probabilities is very attractive.

When time-course data are available, the MFPT algorithm can be implemented by directly estimating the MFPTs. This is done in a straightforward manner: when state $\mathbf{x}$ is observed, record the number of time points required to reach the set, $\mathcal{D}$ or $\mathcal{U}$, not containing $\mathbf{x}$, and from these counts estimate the required MFPTs. The estimated MFPTs are used to construct the MFPT vectors $K_{\mathcal{U}}$ and $K_{\mathcal{D}}$. The MFPT algorithm can then be applied to the estimates of $K_{\mathcal{U}}$ and $K_{\mathcal{D}}$ to devise a control policy. An advantage of the model-free approach is that the estimates of $K_{\mathcal{U}}$ and $K_{\mathcal{D}}$ can be updated whenever new time-course data become available. The possibility of updating the estimated MFPTs enables the MFPT algorithm to adapt its control policy to the status of gene interactions.

The effectiveness of the model-free MFPT control policy obtained by estimating the MFPTs can be demonstrated by estimating $K_{\mathcal{U}}$ and $K_{\mathcal{D}}$ based on synthetic time-course data and using the MFPT algorithm to find the control policy, and then using the same time-course data to build a Markov chain representing the dynamics of the model and finding the control policy based on the estimated transition probability matrix using dynamic programming. In Vahedi *et al.* (2008), this was done by generating synthetic time-course data for 100,000 time steps from an existing model, estimating the MFPTs after each 10^k time steps, for $k = 2, \ldots, 5$, and fixing the cost of control to 1. As the duration of estimating the MFPTs increases, $\Delta P_{g*}^{\mathrm{MFPT}(\gamma)}$ approaches $\Delta P_{g*}^{\mathrm{opt}}$. Figure 5.20 shows the average of $|\Delta P_{g*}^{\mathrm{opt}} - \Delta P_{g*}^{\mathrm{MFPT}(\gamma)}|$, where $\Delta P_{g*}^{\mathrm{opt}}$ is obtained from the original transition probabilities, with various estimation durations over 1000 trials. For an optimal control policy based on the Markov chain estimated from the data, the shift in the steady-state distribution is denoted by $\widehat{\Delta P^{\mathrm{opt}}}$. Figure 5.20 also shows the average of $|\Delta P_{g*}^{\mathrm{opt}} - \widehat{\Delta P_{g*}^{\mathrm{opt}}}|$ with various

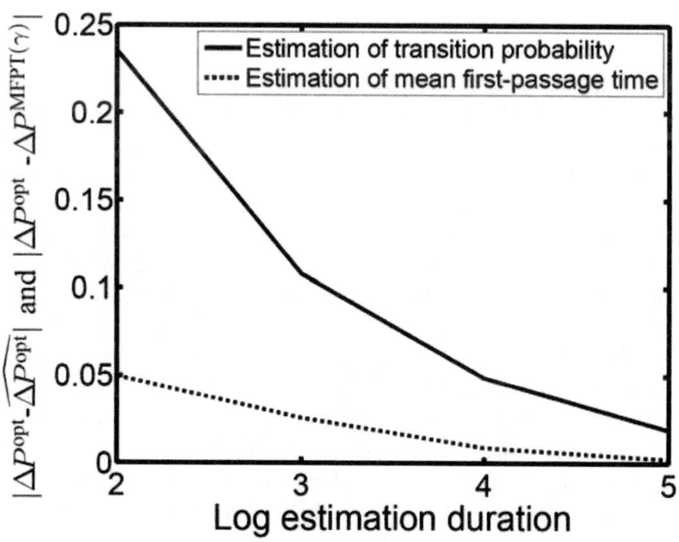

Figure 5.20. *Average of* $|\Delta P^{opt}-\widehat{\Delta P^{opt}}|$ *(solid) and* $|\Delta P^{opt} - \Delta P^{MFPT(\gamma)}|$ *(dash) over* 1000 *trials as a function of the logarithm of estimation duration.*

estimation durations over 1000 trials. The graphs demonstrate the superior performance of the model-free approach using the MFPT algorithm. In particular, for lower numbers of observations, which correspond better to feasible experimental conditions, estimating the Markov chain yields poor results, whereas the MFPT approximation performs quite well. The same experiment has been conducted with various costs of the control for networks with different numbers of genes (see the supplementary materials in Vahedi *et al.* (2008)).

Let us consider algorithm complexity. To find an optimal control policy using value or policy iteration given the full model, one should iteratively find the value (cost) function until the algorithm reaches the fixed point of the Bellman optimality equation. Once the optimal cost functions are computed, one must check which control value attains the minimum on the right-hand side of the Bellman optimality equation and this procedure should be iterated for all the states. To the best of our knowledge, there does not exist a tight upper bound on the number of iterations required to find an optimal policy using either value or policy iteration. Given the control gene, the policy iteration algorithm has complexity $\mathbf{O}(2^{3n})$ per iteration, whereas the complete complexity of the MFPT algorithm, which consists of two matrix inversions, is $\mathbf{O}(2^{3n})$. In general, it is known that the policy iteration algorithm converges, but it is not known whether "the number of iterations in policy iteration can be bounded by a polynomial in the instance size" (Blondel and Tsitsiklis, 2000). Even assuming that the number of iterations can be bounded by a polynomial in the number of states, the complexity of the MFPT algorithm is lower than the policy iteration algorithm because it is computed once and does not require iteration. Regarding the value iteration algorithm, the asymptotic upper bound on the number of iterations required to find an optimal policy using the value iteration algorithm is polynomial in the number of states (Blondel and Tsitsiklis, 2000). The degree of the polynomial is determined to exceed 2 in special cases (Madani, 2000; Tseng, 1990). Given that the complexity of each iteration in the value iteration algorithm is $\mathbf{O}(2^{2n})$, the complexity of the value iteration algorithm to find an optimal control policy is $\mathbf{O}(2^{(2+\alpha)n})$, where $\alpha > 1$. Hence, the complexity of the MFPT algorithm is also lower than the complexity of the value iteration algorithm. To find the optimal cost functions for n control genes, the complexity of a dynamic programming algorithm is k times the complexity of this algorithm for one control gene. On the other hand, once the MFPT vectors are computed, they can be used to devise MFPT control policies for all control genes.

It is important to point out that for any control gene, in addition to the above complexities, the dynamic programming and the MFPT algorithms must loop over all the states to find their corresponding control policies. In dynamic programming algorithms, to obtain the optimal control policy, one must check which control value attains the minimum on the right-hand side of the Bellman optimality equation and this procedure must be iterated for all the states. In the MFPT algorithm, one must investigate which control value leads to a more favorable MFPT and this procedure must be repeated for all the states.

It is evident from the above analysis that, like dynamic programming, the MFPT algorithm is restricted to small numbers of genes since its complexity increases exponentially with the number of genes. Nonetheless, for small numbers of genes, the MFPT algorithm has computational advantages. Figure 5.21 shows the average execution time of the value and policy iteration algorithms over 1000 randomly generated intervention problems as a function of the number n of genes, along with the execution times of the MFPT algorithm. Per this figure, the execution time of the MFPT algorithm is much smaller than the execution time of the two optimal algorithms. The direct comparison has been limited to 10-gene networks on account of the high complexity of the optimal intervention algorithms.

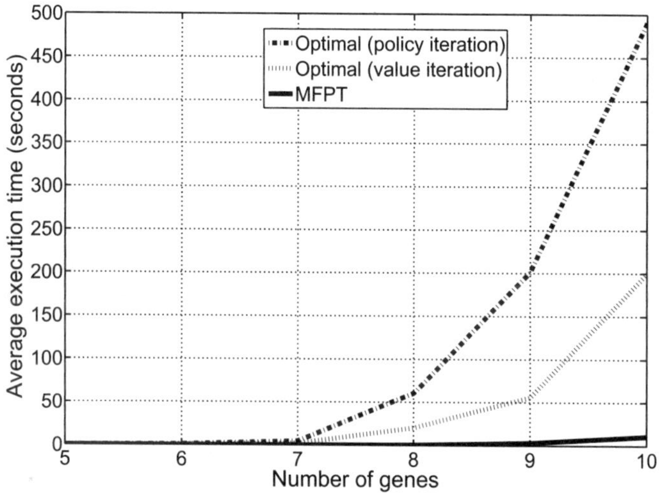

Figure 5.21. *The average execution time of the value and policy iteration algorithms over* 1000 *randomly generated intervention problems as functions of the number of genes, along with the execution times of the MFPT algorithm.*

5.9 Steady-State-Based Control Policies

The MFPT policy is a stationary control policy that circumvents the need for a cost function and works directly with the transition probabilities of the Markov chain associated with the network. Since the intervention goal is to shift the steady-state distribution, rather that using MFPTs, which do this indirectly, it is prudent to develop greedy policies based directly on long-run behavior. Three such policies have been proposed in Qian *et al.* (2009b). One is based on the distance to (un)desirable attractor states, a second is based on the shift of the steady-state distribution, and a third also uses the steady-state distribution as the criterion but gives up some computational efficiency in order to increase the certainty that applying the derived control policy will lead to the reduction of the total stationary mass for undesirable states. In this section we will address the two policies based on the steady-state distribution. The policies are developed for instantaneously random PBNs, where the states of the associated Markov chain are the same as the network GAPs. We subsequently comment on their extension to context-sensitive PBNs.

5.9.1 Steady-state-distribution greedy control policy

Control by flipping one control gene g at any given state $\mathbf{x}$ changes the original transition matrix P to the controlled transition matrix $P^{u_g(\mathbf{x})=1}$ by replacing the row in P corresponding to the state $\mathbf{x}$ by the row that corresponds to the state $\mathbf{x}_c$ with g flipped from $\mathbf{x}$. Hence, the perturbation matrix can be written as a rank-one matrix (see Theorem 4.10) and the perturbed steady-state distribution can be computed efficiently by

$$\tilde{\pi}(\mathbf{x}) = \pi - \frac{\pi_{\mathbf{x}}(p_{\mathbf{x}_c} - p_{\mathbf{x}})Z}{1-(p_{\mathbf{x}} - p_{\mathbf{x_c}})z^{\mathbf{x}}}, \tag{5.163}$$

where $p_{\mathbf{x}}$ and $p_{\mathbf{x}_c}$ are the two rows corresponding to the states $\mathbf{x}$ and $\mathbf{x}_c$ in P, $z^{\mathbf{x}}$ is a column vector corresponding to the state $\mathbf{x}$ in Z, $\pi_{\mathbf{x}}$ is the stationary mass for $\mathbf{x}$, and $\tilde{\pi}(\mathbf{x})$ denotes the steady-state distribution after we apply gene flipping at the state $\mathbf{x}$. This formula enables fast computation of the total stationary mass for undesirable states π_U and $\tilde{\pi}_U(\mathbf{x})$, and therefore the shifted mass after the possible controls to each state. Given that, we can formulate a *steady-state distribution (SSD) control policy* based on a procedure similar to deriving the MFPT control policy by comparing the total stationary mass of undesirable states after applying control to $\mathbf{x}$ and $\mathbf{x}_c$: $\tilde{\pi}_U(\mathbf{x})$ and $\tilde{\pi}_U(\mathbf{x}_c)$. If both are larger than the original stationary mass, π_U, of undesirable states, then we do not apply any control; otherwise, we adopt the control on the state which leads to less stationary mass of the undesirable states. Like the MFPT policy, the computational complexity for finding this new control policy is $\mathbf{O}(2^n)$, while the complexity for each iteration in the algorithm increases a bit by the vector-matrix multiplications involved in Eq. (5.163).

The SSD control policy depends on effects caused by perturbations to the pairs of states $\mathbf{x}$ and $\mathbf{x}_c$. Considering the perturbation to the original transition matrix P, for each network state $\mathbf{x}$, we compare the steady-state distributions π and $\tilde{\pi}(\mathbf{x})$ that correspond to P and the controlled transition matrix $P^{u_g(\mathbf{x})=1}$ by replacing the row in P corresponding to the state $\mathbf{x}$ by the row that corresponds to the flipped state $\mathbf{x}_c$. If the undesirable stationary mass based on this one-row perturbation caused by flipping the control gene g at $\mathbf{x}$ reduces the undesirable stationary mass, then we decide the control policy for the state $\mathbf{x}$: $u_g(\mathbf{x}) = 1$. The full stationary control policy is derived by independently doing this for all states in the network, that is, by characterizing the changes from P to the corresponding controlled transition matrices, $P \to P^{u_g(\mathbf{x})=1}$, by one-row perturbations at all states $\mathbf{x}$. The final control policy $\mathbf{u}_g$ for the network leads to a multirow perturbation to P by combining all the beneficial one row perturbations determined in the algorithm: $P^{\mathbf{u}_g}$. Generally, $P^{\mathbf{u}_g}$ is different from all the controlled transition matrices by one-row perturbations considered during the derivation of $\mathbf{u}_g$. Intuitively, one might conjecture that this combination of beneficial one-row perturbations should reduce the total undesirable stationary mass. Although no proof of this conjecture has been given, over the course of the large amount of simulations performed in Qian *et al.* (2009b), the derived SSD control policy always reduces the stationary mass for undesirable states.

5.9.2 Conservative steady-state-distribution control policy

We now present a control policy, the *conservative steady-state distribution (CSSD) control policy*, for which there is a theoretical guarantee that the undesirable mass of the steady-state distribution after intervention will be less than or equal to the stationary mass of the undesirable states in the original network.

Flipping one control gene at any given state $\mathbf{x}$ replaces the row in the original transition matrix corresponding to the state $\mathbf{x}$ by the row that corresponds to the flipped state $\mathbf{x}_c$. Hence, in Eq. (4.34) we have $a = e_{\mathbf{x}}$, which has 1 for the element corresponding to the state $\mathbf{x}$ and all 0's for the remaining elements in the vector; and $b = p_{\mathbf{x}_c} - p_{\mathbf{x}}$, which allows us to compute the updated fundamental matrix. This result can be used to construct a sequential algorithm that iteratively chooses states to control in such a way that we can prove that the control policy reduces the stationary mass of undesirable states.

At each iteration, all states for which the control policy has not been decided are checked to see which state to control in order to achieve the largest reduction of the undesirable stationary mass. Similar to the SSD policy, for each state $\mathbf{x}$, the steady-state distributions π and $\tilde{\pi}(\mathbf{x})$ that correspond to P and the controlled transition matrix $P^{u_g(\mathbf{x})=1}$ are compared by replacing the row in P corresponding to state $\mathbf{x}$ by the row corresponding to the flipped state $\mathbf{x}_c$. Unlike deriving $u_g(\mathbf{x})$ independently for all the states, as in the SSD control policy, for the CSSD policy, we do not directly combine all the beneficial one-row perturbations into the new transition matrix $P^{\mathbf{u}_g}$ decided by the derived SSD control policy. The CSSD algorithm checks all possible one-row perturbations and finds the one-row perturbation that results in the largest reduction of undesirable mass. Hence, only one state, $\mathbf{x}^{best}$, is selected to control at each iteration if there is a reduction of undesirable mass. Denoting the controlled transition matrix resulting from flipping the control gene g at that state by $P_k^{u_g(\mathbf{x}_k^{best})=1}$ at the kth iteration, the sequential algorithm generates a sequence of controlled transition matrices:

$$P \rightarrow P_1^{u_g(\mathbf{x}_1^{best})=1} \rightarrow P_2^{u_g(\mathbf{x}_2^{best})=1} \rightarrow \cdots \rightarrow P_K^{u_g(\mathbf{x}_K^{best})=1}, \tag{5.164}$$

where K is the total number of iterations and each pair of neighboring transition matrices differs by only one row. The derived control policy, $\mathbf{u}_g$, possesses the controlled transition matrix

$$P^{\mathbf{u}_g} = P_K^{u_g(\mathbf{x}_K^{best})=1}. \tag{5.165}$$

Because the controlled transition matrix is obtained at each iteration by a one-row perturbation to the previously computed controlled transition matrix, we can keep updating the exact steady-state distribution $\tilde{\pi}(\mathbf{x}_k^{best})$ and the fundamental matrix Z_k^{best} using Eqs. (5.163) and (4.27), respectively. Thus, at each iteration, we can directly compute the true stationary mass for undesirable states after intervention and also make the decision about the control policy for the selected state. We let the algorithm run iteratively until intervention to any state will actually increase the stationary mass of undesirable states from the previous iteration. This guarantees that the derived control policy will always have undesirable stationary mass less than or equal to that of the undesirable states in the original network. The CSSD algorithm is computationally more expensive than the MFPT and SSD algorithms because the search space is $\mathbf{O}(2^n)$ at each iteration and the number of iterations K depends on the controllability of the networks. The advantage here is that the CSSD control policy is guaranteed to decrease undesirable stationary mass after intervention (as proven in the next theorem) and, as simulations have shown, tends to outperform the SSD and MFPT policies.

Theorem 5.12. *The derived CSSD stationary control policy cannot increase the total undesirable stationary mass:*

$$\tilde{\pi}_U(\mathbf{x}_K^{best}) \leq \pi_U, \tag{5.166}$$

where K is the number of total iterations of the sequential algorithm.

Proof. The theorem is proven by induction. Starting with the first iteration, $k = 1$, $\tilde{\pi}_U(\mathbf{x}_1^{best}) \leq \pi_U$ since control is applied only when $\Delta^{best} = \pi_U - \tilde{\pi}_U(\mathbf{x}_1^{best}) > 0$. Now at the kth

Table 5.3. *Average* $\sum_{x_1=0}\pi_{\mathbf{x}}$ *over* 1000 *randomly generated PBNs before and after applying all of four control policies: ORG – original stationary mass for undesirable states before control; MFPT – mean-first-passage-time control policy; SSD – steady-state distribution control policy; CSSD – conservative SSD control policy with* x_{10} *as the control gene.*

Control policies	p_b				
	0.1	0.3	0.5	0.7	0.9
ORG	0.8939	0.6934	0.4997	0.2967	0.1063
MFPT	0.8567	0.5807	0.3484	0.1912	0.0747
SSD	0.8547	0.5637	0.3349	0.1822	0.0728
CSSD	0.8525	0.5439	0.2971	0.1670	0.0723

iteration, assuming $\tilde{\pi}_U(\mathbf{x}_{k-1}^{best}) \leq \pi_U$, we want to show that $\tilde{\pi}_U(\mathbf{x}_K^{best}) \leq \pi_U$. Indeed, at each iteration the control is applied only when $\Delta^{best} = \tilde{\pi}_U(\mathbf{x}_{k-1}^{best}) - \tilde{\pi}_U(\mathbf{x}_k^{best}) > 0$. Hence, $\tilde{\pi}_U(\mathbf{x}_k^{best}) \leq \tilde{\pi}_U(\mathbf{x}_{k-1}^{best}) \leq \pi_U$. □

Although we have we have focused on instantaneously random PBNs, where the GAP space corresponds to the Markov-chain space, the MFPT, SSD, and CSSD algorithms can be extended to intervene in context-sensitive PBNs with no theoretical obstacles. Computationally, however, as the state space changes from the space of GAPs to the space of (context, GAP) pairs in context-sensitive PBNs, the computational complexity of these algorithms increases. Moreover, for the algorithms directly based on steady-state distributions, one must apply iterative update schemes to compute the shifted steady-state distributions since the perturbations to the transition matrix become multiple-row perturbations.

5.9.3 Performance comparison

The performances of the MFPT, SSD, and CSSD policies have been compared in Qian *et al.* (2009b) based on a large number of randomly generated networks with similar network properties. Here we consider randomly generated instantaneously random 10-gene, 2-context PBNs. The perturbation probability is $p = 0.01$ and the selection probabilities are $c_1 = c_2 = 0.5$. States with $x_1 = 0$ are undesirable and states with $x_1 = 1$ are desirable. The control policies are applied with x_{10} as the control gene. In each constituent BN, the input degree (number of predictors) is set to 3. The bias p_b of each BN is randomly selected from a beta distribution. The mean of the beta distribution varies from 0.1 to 0.9, with step-size 0.2, and the variance of the beta distribution is 0.000064.

Table 5.3 gives the average stationary mass for the undesirable states before control (ORG) and after applying the three different control policies, with $\gamma = 0$ for the MFPT policy, and different p_b's, with 1000 randomly generated PBNs being used in each case. The CSSD policy has the best average performance and the SSD policy achieves better average performance than the MFPT policy, the advantages depending on p_b. Table 5.4 shows the percentages of random PBNs with stationary mass shift $\Delta = \pi_U - \tilde{\pi}_U \geq 0$ for the different control policies and different p_b's. 100% performance is guaranteed for the

Table 5.4. *Percentages of random PBNs with* $\Delta = \pi_U - \tilde{\pi}_U \geq 0$ *within* 1000 *random PBNs after applying three control policies with* x_{10} *as the control gene.*

Control policies	p_b 0.1	0.3	0.5	0.7	0.9
MFPT	94.4%	91.3%	94.4%	92.5%	95.6%
SSD	100.0%	100.0%	100.0%	100.0%	100.0%
CSSD	100.0%	100.0%	100.0%	100.0%	100.0%

CSSD policy, but we also see 100% performance for the SSD policy. On the other hand, Δ can be negative for the MFPT policy, although it exceeds 0 in over 90% of the generated PBNs. As noted previously, no case is known in which the SSD policy does not yield $\Delta \geq 0$. Owing to the superior computational performance of the SSD algorithm in comparison to the CSSD policy, it would be nice to have a theorem showing this or a counterexample to settle the issue.

Chapter 6

Asynchronous Networks

To this point we have assumed synchronicity in network transitioning. In effect, time is not part of the model because transitions are from state to state, with the "time" t merely serving as a transition index. The assumption of synchronous updating facilitates mathematical tractability and reduces the amount of data required for network inference, since substantial time-course data are required to characterize asynchronism. As with any simplification, a price must be paid; however, given the severe impediments to inferring asynchronous networks, this price is acceptable if synchronicity leads to successful applications that could not otherwise be accomplished.

Numerous issues regarding the assumption of synchronicity have been noted. For instance, a synchronous model might result in spurious attractors (Deng *et al.*, 2007). In the same vein, deviation from synchronous updating modifies the attractor structure of BNs (Greil and Drossel, 2005) and can change their long-run behavior (Gershenson, 2002). From a biological perspective, interactions among genes causing transcription, translation, and degradation occur over a wide range of time scales (Monk, 2003; Yu *et al.*, 2006). While present technology may make application of asynchronous models impractical, these observations suggest that asynchronous regulatory networks have the potential to provide a better description of underlying biological processes and, therefore, their theoretical investigation is warranted. Thus, in this chapter we will introduce asynchronous PBNs and develop the corresponding intervention theory. The presentation is based on the developments in Faryabi *et al.* (2008a) and Faryabi (2009).

6.1 Asynchronous PBNs

We consider two asynchronous models. The first model considers asynchronism at the gene level; the second considers asynchronism at the state level.

6.1.1 Deterministic asynchronous PBNs

Historically, asynchronism has been introduced into BNs by updating each gene based on its period, which may differ from one gene to another. Studies of this kind of asynchronicity generally try to understand generic characteristics of asynchronous updating schemes in randomly generated BNs. To accomplish this aim, a wide range of artificial asynchronous updating protocols with different degrees of freedom in the selection of the updating period for each gene has been postulated.

Proposed asynchronous protocols can be categorized into two groups. With *stochastic asynchronous* protocols, the updating period of each gene is randomly selected based on a given distribution (Gershenson, 2002; Greil and Drossel, 2005; Di Paolo, 2000). With *deterministic asynchronous* protocols, the updating period of each gene is fixed but can differ from one gene to another (Gershenson, 2002; Harvey and Bossomaier, 1997; Cornforth *et al.*, 2001). There have also been studies that consider both stochastic and deterministic asynchronous protocols in an effort to investigate the predictability of BNs when asynchronous updating schemes are used instead of synchronous schemes (Chaves *et al.*, 2005; Fauré *et al.*, 2006).

Studies of both randomly generated and experimentally validated BNs reveal that stochastic asynchronism has some limitations. Stochastic asynchronous updating methods can significantly change the properties of state transition graphs (Cornforth *et al.*, 2001; Harvey and Bossomaier, 1997). Starting from wild-type gene expressions, neither the BNs of Chaves *et al.* (2006) nor Fauré *et al.* (2006) successfully predict the anticipated long-run attractors of their networks. Earlier studies indicate that constraining the degrees of freedom in the asynchronous protocols can improve the predictability of BNs. More structured asynchronous protocols predict the long-run behavior of BNs more effectively by representing their cyclic attractors (Di Paolo, 2000; Fauré *et al.*, 2006). A caveat to the existing study of asynchronism is that the timing protocols have been modeled mathematically without biological verification.

Focusing on deterministic gene-level asynchronicity, we can define a *deterministic-asynchronous Boolean network* (*DA-BN*) as a BN in which each gene has its own fixed updating period. This means that the order of updating among all genes is deterministic over time. From an inference perspective, even if it were possible to measure the level of each gene in isolation while the other genes remain constant, owing to the effects of measurement noise and the existence of latent variables, one could, at best, produce estimates for updating periods. In particular, on account of latent variables, in practice we can estimate only a set consisting of the most probable updating periods for each gene in the network, each set depending on the status of latent variables. Hence, there would be a finite collection of DA-BNs defining the dynamics of the system.

In the framework of PBNs, if we were to consider DA-BNs instead of BNs as the constituent networks, having defined the probabilities of selecting each context, the model would select a constituent DA-BN at each updating instant and the system would evolve according to the selected constituent DA-BN until the constituent network changes. We refer to this asynchronous extension of a PBN as a *deterministic-asynchronous probabilistic Boolean network* (*DA-PBN*), the terms "probabilistic" and "deterministic" referring to the random selection of a context and the asynchronous protocol within each context of the regulatory network, respectively.

In providing a rigorous definition of a DA-PBN, for consistency with context-sensitive PBNs, we use the same notation to define a DA-PBN. A DA-PBN consists of a set $V = \{x_i\}_{i=1}^{n}$, of n nodes, where $x_i \in \{0, 1, \ldots, d\}$. Each x_i represents the expression value of a gene. For simplicity and to remain in the framework of binary logic, we will assume $d = 1$, so that each gene is ON or OFF, and we will have a binary DA-PBN. Under this assumption, a DA-PBN is composed of a collection of m constituent DA-BNs. Analogous to a

PBN, in a DA-PBN, the active DA-BN changes at updating instants selected by a binary switching random variable. A DA-PBN acts like one of its constituent DA-BNs, each being referred to as a *context*, between two switching instants.

The lth DA-BN $(V, \mathbf{f}_l, \Theta_l)$ is defined by two vector-valued functions. The vector-valued function $\mathbf{f}_l$ consists of n predictors, $\mathbf{f}_l = (f_{l1}, \ldots, f_{ln})$, where $f_{li} : \{0,1\}^n \to \{0,1\}$ denotes the predictor of gene i, whenever context l is selected. The vector-valued function Θ_l consists of n updating components, $\Theta_l = (\theta_{l1}, \ldots, \theta_{ln})$. Each function $\theta_{li} : \mathbb{N} \to \{0,1\}$ is defined by a pair (a_{li}, b_{li}). The parameter $a_{li} \in \mathbb{N}$ specifies the updating period of gene i, relative to the moment of selection of context l. The parameter $b_{li} \in \{0, \ldots, a_{li}-1\}$ further differentiates the exact updating instant of each gene within its updating period. The two degrees of freedom in θ_{li} are sufficient to assign any instant of time $t \in \mathbb{N}$ as the updating epoch of gene i:

$$\theta_{li}(t) = \begin{cases} 1 & \text{if} \quad t \equiv b_{li} \pmod{a_{li}}, \\ 0 & \text{otherwise.} \end{cases} \tag{6.1}$$

We focus on DA-PBNs with perturbation, in which each gene may change its value with probability p at each epoch. At each updating instant a decision is made whether to switch the current constituent DA-BN. The switching probability q is a system parameter. If the current DA-BN is not switched, then the DA-PBN behaves as a fixed DA-BN and genes are updated synchronously according to the current constituent network:

$$x_i(t+1) = \begin{cases} \mathbf{1}(\gamma(t+1) \neq \mathbf{0})(x_i(t) \oplus \gamma_i(t+1)) & \\ \qquad + \mathbf{1}(\gamma(t+1) = \mathbf{0}) f_{li}(x_1(t), \ldots, x_n(t)) & \text{if } \theta_{li}(t+1) = 1, \\ x_i(t) & \text{if } \theta_{li}(t+1) = 0, \end{cases} \tag{6.2}$$

where $\gamma_i(t)$ is a Bernoulli random variable with parameter p and the random vector γ at instant t is defined as $\gamma(t) = (\gamma_1(t), \gamma_2(t), \ldots, \gamma_n(t))$. The operator $\oplus$ is componentwise addition in modulo 2 and f_{li} is the predictor of gene i according to the DA-BN l.

If a switch occurs, then a new constituent network is randomly selected according to a selection probability measure $\{c_l\}_{l=1}^k$. After selecting the new constituent network $\hat{l}$, the values of the genes are updated using Eq. (6.2), but with $\mathbf{f}_{\hat{l}}$ and $\Theta_{\hat{l}}$ instead.

6.1.2 Semi-Markov asynchronous PBNs

Assuming asynchronism at the gene level for Markovian networks has practical and theoretical impediments that may prevent independent gene updating to serve as a basis for system-based treatment discovery (Faryabi *et al.* 2008a). In particular, the delay and the updating order of a given gene are observable only with respect to the activity levels of other genes and proteins involved in the regulation process. Thus, it is impractical to study the alteration of one specific gene over time while keeping the levels of all other genes in the model constant. Practically, we can measure the gene-expression profile (state) at each observation instant. The intertransition interval between two states can then be modeled by a random variable.

Asynchronously updating the genes changes the global behavior of regulatory networks on account of it changing the state transition graph, which models the dynamics of the system (Faryabi *et al.*, 2008a). Along this line, it has been shown that small perturbations do not settle down in a random BN with gene-level asynchronism. Consequently, the asynchronous network is in the chaotic regime while its synchronous counterpart is in the critical regime (Mesot and Teuscher, 2003). Studies of experimentally validated BNs (Chaves *et al.*, 2005; Fauré *et al.*, 2006) suggest that the state transition graphs of given BNs provide accurate predictability, whereas the state transition graphs of networks utilizing the same Boolean rules with asynchronously updated genes are very complex and possess many incompatible or unrealistic pathways.

These observations motivate the assumption that an asynchronous regulatory model should maintain the topology of the state transition graph as specified by the logical rules governing the interactions between genes. In other words, regulatory models should accurately translate the logical relationships governing the interactions of genes to the state transition graph specifying the dynamics of the model. Moreover, they should enable analysis of the temporal behaviors of biological systems. Since our ultimate objective is to alter the long-run behavior of biological systems via effective intervention strategies, our regulatory models should not only possess the previous two characteristics, but they should also be inferable from empirical data.

Based upon these considerations, we define the *semi-Markov asynchronous probabilistic Boolean network* (*SMA-PBN*) model, in which asynchronism is at the state level, thereby facilitating the incorporation of empirically measurable timing information of biological systems into the model. This temporal information determines the typical time delay between transitions from one state to another. Since the order of updating genes and their relative time delays depend on the levels of other regulatory components, estimating the updating time of each gene in isolation, and independent of the values of other genes, is highly problematic, if not impossible. Time-course data enable the estimation of intertransition times between states, not the updating time of each gene.

A SMA-PBN is specified with two sets of information. The first set determines the rule-based multivariate interactions between genes. Considering simultaneous updating, we can specify the state transition graph of the model based on this information. In other words, the first set of information specifies a PBN generated from a given set of Boolean functions for updating each gene. The generated state transition graph guarantees the predictability of the rule-based topology. The second set of information consists of the distributions of intertransition intervals between any two directly connected states. These values can be empirically inferred from time-course data.

Rigorously, a SMA-PBN consists of a set, $V = \{x_i\}_{i=1}^{n}$, of n nodes, representing gene-expression values. Continuing with the binary assumption, with extension to any finite quantization being immediate, the expression value, $x_i(t)$, of gene i at time t belongs to $\{0,1\}$. The states of a SMA-PBN are defined as ordered pairs consisting of a constituent network κ and a gene-activity profile $\mathbf{x}(t) = (x_1(t),\ldots,x_n(t))$ giving the values of the genes at time t, where $\mathbf{x}(t) \in \mathcal{X} = \{0,1\}^n$. At each time $t \in \mathbb{R}^+$, the state $\mathbf{z}(t)$ of a SMA-PBN is selected from the set

$$\mathcal{Z} = \{(\kappa,\mathbf{x}) : \kappa \in \{1,2,\ldots,m\}, \mathbf{x} \in \mathcal{X}\} \tag{6.3}$$

of all possible states.

Considering two consecutive epochs t_k and t_{k+1} per Fig. 6.1, the state of the SMA-PBN for all the times $t \in [t_k, t_{k+1})$ is given by $\mathbf{z}(t_k) = \mathbf{z}_k$. At t_{k+1}, the model enters a new state $\mathbf{z}(t_{k+1}) = \mathbf{z}_{k+1}$. If τ_{k+1} is the time spent in state $\mathbf{z}_k$ prior to transition to state $\mathbf{z}_{k+1}$, then $\tau_{k+1} = t_{k+1} - t_k$. In the SMA-PBN model, this intertransition interval is modeled with a nonnegative random variable with probability distribution

$$\lambda_{\mathbf{z}_k,\mathbf{z}_{k+1}}(\tau) = P\left(\tau_{k+1} \le \tau \,|\, \mathbf{z}(t_k) = \mathbf{z}_k, \mathbf{z}(t_{k+1}) = \mathbf{z}_{k+1}\right), \tag{6.4}$$

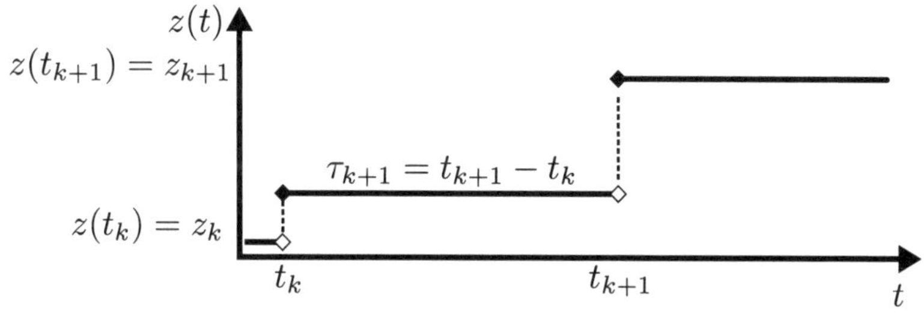

Figure 6.1. *A schematic of transition in a SMA-PBN with two consecutive epoch times t_k and t_{k+1}. The intertransition interval, τ_{k+1}, is the sojourn time in state $\mathbf{z}_k$ prior to the transition to state $\mathbf{z}_{k+1}$.*

where $\mathbf{z}_k, \mathbf{z}_{k+1} \in \mathcal{Z}$. According to Eq. (6.4), the probability distribution of sojourn time in the current state $\mathbf{z}_k$ prior to transition to the successive state $\mathbf{z}_{k+1}$ could depend on both states. We require the set of intertransition interval distributions, $\lambda_{\mathbf{z}_k,\mathbf{z}_{k+1}}(\tau)$, for any two directly connected states as one of the two sets of information needed to define a SMA-PBN.

Transitions in a SMA-PBN are determined by its *embedded PBN,* which models the probabilistic rule-based connections of gene interactions and constitutes the other set of information required for specification of a SMA-PBN. The embedded PBN determines the state transition graph of the SMA-PBN based on the predictors of the genes. Specifically, let $\{\mathbf{f}_l\}_{l=1}^m$ be the set of m realizations of the embedded PBN. If the genes are coupled, then at each simultaneous updating instant, one of the m possible realizations of the embedded PBN is selected. Each vector-valued function $\mathbf{f}_l$ has the form $\mathbf{f}_l = (f_{l1}, \ldots, f_{ln})$. Each function $f_{li} : \{0,1\}^n \rightarrow \{0,1\}$ denotes the predictor of gene i when the realization l is selected. At each simultaneous updating instant a decision is made whether to switch the context of the network. The switching probability q is a system parameter. If at a particular updating instant it is decided that the realization of the network should not be switched, then the embedded PBN behaves as a fixed BN and simultaneously updates the values of all the genes according to their current predictors. If it is decided that the network should be switched, then a realization of the embedded PBN is randomly selected according to a selection distribution $\{c_l\}_{l=1}^m$. After selecting the vector-valued function $\mathbf{f}_l$, the values of the genes are updated according to the predictors determined by $\mathbf{f}_l$. In addition, we allow perturbations in the embedded PBN, whereby each gene may change its value with a small probability p at each updating instant.

The SMA-PBN and its embedded PBN share the same state transition graph, but the state updates in the state transition graph of a SMA-PBN occur on various time scales according to intertransition interval distributions. Thus, whereas the state transitions associated with a PBN constitute a Markov chain (discrete-space, discrete-time Markov process), the state transitions associated with a SMA-PBN constitute a semi-Markov process (discrete-space, continuous-time Markov process). Ignoring transition timing, the embedded PBN shares its Markov chain structure with the SMA-PBN. Originating from a state $\mathbf{z}(t_k) = \mathbf{z}_k$, the successor state $\mathbf{z}(t_{k+1}) = \mathbf{z}_{k+1}$ is selected randomly within the set $\mathcal{Z}$ according to the transition probability

$$p(\mathbf{z}_k, \mathbf{z}_{k+1}) = P\left(\mathbf{z}(t_{k+1}) = \mathbf{z}_{k+1} | \mathbf{z}(t_k) = \mathbf{z}_k\right) \tag{6.5}$$

for all $\mathbf{z}_k, \mathbf{z}_{k+1} \in \mathcal{Z}$.

Gene perturbation ensures that all the states of the SMA-PBN communicate. Hence, the fraction of time that the SMA-PBN spends in state $\mathbf{z}$ in the long run can be computed:

$$p_{\tau}(\mathbf{z}) = \frac{\pi(\mathbf{z})\overline{\tau}(\mathbf{z})}{\sum_{\mathbf{z}\in\mathcal{Z}} \pi(\mathbf{z})\overline{\tau}(\mathbf{z})}, \tag{6.6}$$

with probability 1, where $\pi(\mathbf{z})$ is the steady-state probability of state $\mathbf{z}$ and $\overline{\tau}(\mathbf{z})$ is the expected sojourn time in state $\mathbf{z}$, which can be computed from the information in Eqs. (6.4) and (6.5) (Gallager, 1996). One can easily verify that $p_{\tau}(\mathbf{z}) = \pi(\mathbf{z})$ if all nodes are synchronously updated.

6.2 Intervention in Asynchronous Networks

Having defined two asynchronous PBN models, we now turn to developing optimal intervention strategies in each.

6.2.1 Intervention in DA-PBNs

In Faryabi *et al.* (2008a), a synchronization method is used to provide the synchronous version of a DA-PBN's state transition graph, which can then be utilized for designing optimal intervention strategies to effectively alter the dynamical behavior of the DA-PBN.

To synchronize the state transition graph of a DA-PBN, we encode all the dynamical steps within an interval of duration equal to the least common multiple (LCM) of all the updating periods. The LCM of all the updating periods a_{li} in $\{\Theta_l\}_{l=1}^{m}$, for $l \in \{1,2,\ldots,m\}$ and $i \in \{1,2,\ldots,n\}$,

$$\zeta = LCM\,(a_{11},\ldots,a_{1n},\ldots,a_{m1},\ldots,a_{mn}) \tag{6.7}$$

defines the number, r, of new components added to the state $\mathbf{z}$ to yield the *augmented logical state* $\hat{\mathbf{z}}(t)$. The integer r is the smallest integer larger than the logarithm to the base 2 of ζ:

$$r = \lceil \log_2(\zeta) \rceil . \tag{6.8}$$

This value r is a nonoptimal number of elements required to distinguish all the time steps within one ζ. For state $\mathbf{z}(t) = (\kappa(t), x_1(t),\ldots,x_n(t))$ at time t, with context $\kappa(t)$ and GAP $(x_1(t),\ldots,x_n(t))$, the augmented logical state $\hat{\mathbf{z}}(t)$ is created by adjoining r new elements, $x_{n+1}(t),\ldots,x_{n+r}(t)$:

$$\hat{\mathbf{z}}(t) = (\kappa(t), x_1(t),\ldots,x_n(t), x_{n+1}(t),\ldots,x_{n+r}(t)). \tag{6.9}$$

Figure 6.2 shows the time instants at which the genes of a hypothetical 3-gene DA-PBN are updated. The updating function θ_{l1} of x_1 has the parameters $(a_{l1} = 2, b_{l1} = 1)$. The parameters of the updating functions of genes x_2 and x_3 are $(a_{l2} = 2, b_{l2} = 0)$ and $(a_{l3} = 3, b_{l3} = 0)$, respectively. The pattern of updates is repeated after each 6 updating instants. We can use 3 extra elements to encode all the instants in the duration of $\zeta = 6$.

The evolution of a synchronized dynamical system with its augmented logical state space can be modeled by a stationary discrete-time equation:

$$\hat{\mathbf{z}}(t+1) = f\,(\hat{\mathbf{z}}(t), w(t)) \tag{6.10}$$

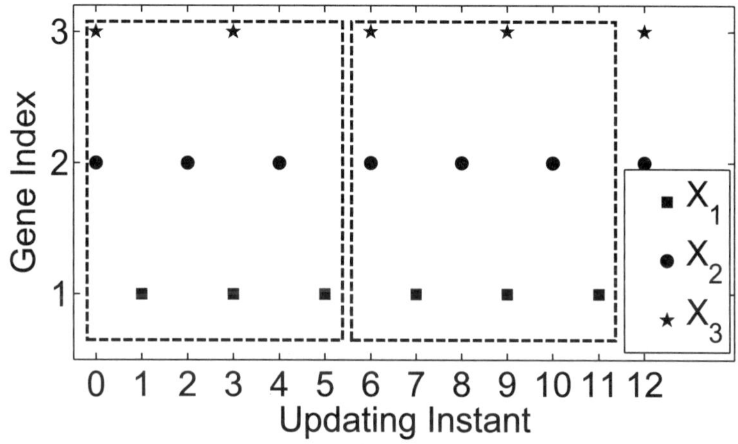

Figure 6.2. *Schematic of updating instants of genes of a DA-PBN with* $(a_{l1} = 2, b_{l1} = 1)$, $(a_{l2} = 2, b_{l2} = 0)$, *and* $(a_{l3} = 3, b_{l3} = 0)$. *The pattern of updates is repeated at each LCM* ξ *shown with a dashed-line box. Each marker indicates the instant in which the corresponding gene updates its value.*

for $t = 0, 1, \ldots,$ where the augmented logical state $\hat{\mathbf{z}}(t)$ is an element of the state space

$$\hat{\mathcal{Z}} = \left\{ (\kappa, \hat{\mathbf{x}}) : \kappa \in \{1, 2, \ldots, m\}, \hat{\mathbf{x}} \in \{0, 1\}^{n+r} \right\}. \tag{6.11}$$

The disturbance $w(t)$ is the manifestation of uncertainties in the DA-PBN. It is assumed that both the gene perturbation distribution and the network switching distribution are independent and identical for all time steps t.

The n-gene DA-PBN is now modeled as a synchronous context-sensitive PBN with augmented state space. This new PBN has $n+r$ (augmented logical) states and an associated Markov chain with $m2^{n+r}$ states. Originating from an augmented logical state $\hat{\mathbf{z}}_1 \in \hat{\mathcal{Z}}$, the successor augmented logical state $\hat{\mathbf{z}}_2 \in \hat{\mathcal{Z}}$ is selected randomly within the set $\hat{\mathcal{Z}}$ according to the transition probability

$$p(\hat{\mathbf{z}}_1, \hat{\mathbf{z}}_2) = P\left(\hat{\mathbf{z}}(t+1) = \hat{\mathbf{z}}_2 | \hat{\mathbf{z}}(t) = \hat{\mathbf{z}}_1\right) \tag{6.12}$$

for all $\hat{\mathbf{z}}_1, \hat{\mathbf{z}}_2 \in \hat{\mathcal{Z}}$. Gene perturbation ensures that all the states in the Markov chain communicate with one another. Hence, the finite-state Markov chain is ergodic and possesses a steady-state distribution.

Now that the dynamical behavior of a DA-PBN is described by a Markov chain, the theory of Markov decision processes can be utilized to find an optimal sequence of actions in the same way as in the previous chapter.

We suppose that the DA-PBN has an external control input $u \in \mathcal{C}$ at each updating instant t. In the presence of external control, the system evolution in Eq. (6.10) can be modeled by a discrete-time equation:

$$\hat{\mathbf{z}}(t+1) = f\left(\hat{\mathbf{z}}(t), u(t), w(t)\right) \tag{6.13}$$

for $t = 0, 1, \ldots$. Optimal intervention in the DA-PBN is then modeled as a classical intervention with $m2^{n+r}$ states, the state $\hat{\mathbf{z}}(t)$ at any time step t being an augmented logical state. Originating from state $\hat{\mathbf{z}}_1$, the successor state $\hat{\mathbf{z}}_2$ is selected randomly within the set $\hat{\mathcal{Z}}$ according to the transition probability

$$p(\hat{\mathbf{z}}_1, \hat{\mathbf{z}}_2, u) = P\left(\hat{\mathbf{z}}(t+1) = \hat{\mathbf{z}}_2 | \hat{\mathbf{z}}(t) = \hat{\mathbf{z}}_1, u(t) = u\right). \tag{6.14}$$

We associate a cost per stage, $c(\hat{\mathbf{z}}_1, \hat{\mathbf{z}}_2, u)$, with each intervention in the system. The cost per stage could depend on the origin state $\hat{\mathbf{z}}_1$, the successor state $\hat{\mathbf{z}}_2$, and the control input u. We assume that the cost per stage is stationary and bounded for all states $\hat{\mathbf{z}}_1, \hat{\mathbf{z}}_2 \in \hat{\mathcal{Z}}$ and all controls $u \in \mathcal{C}$. The cost of a transition from a desirable state to an undesirable state is the highest, and the cost of a transition from an undesirable state to a desirable state is the lowest. An optimal intervention strategy is now determined in the same manner as described in Section 5.3, the objective of intervention in the discrete-time problem being to reduce the likelihood of visiting undesirable states. Since the time between two consecutive epochs of a DA-PBN is fixed, the effect of intervention is equivalent to the reduction of the time spent in undesirable states.

6.2.2 Intervention in SMA-PBNs

In considering the stochastic control of a SMA-PBN, we suppose that the SMA-PBN has a single control input $u(t) \in \mathcal{C}$ at time t. In the presence of external control, the SMA-PBN is modeled as a semi-Markov decision process. Originating from state $\mathbf{z}_1$, the successor state $\mathbf{z}_2$ is selected randomly within the set $\mathcal{Z}$ according to the transition probability

$$p(\mathbf{z}_1, \mathbf{z}_2, u) = P\left(\mathbf{z}(t_{k+1}) = \mathbf{z}_2 | \mathbf{z}(t_k) = \mathbf{z}_1, u(t_k) = u\right) \tag{6.15}$$

for all $\mathbf{z}_1, \mathbf{z}_2 \in \mathcal{Z}$ and $u \in \mathcal{C}$. Moreover, the intertransition interval distribution is also a function of control $u(t)$:

$$\lambda_{\mathbf{z}_1,\mathbf{z}_2}(\tau, u) = P\left(\tau_{k+1} \le \tau | \mathbf{z}(t_k) = \mathbf{z}_1, \mathbf{z}(t_{k+1}) = \mathbf{z}_2, u(t_k) = u\right) \tag{6.16}$$

for all $\mathbf{z}_1, \mathbf{z}_2 \in \mathcal{Z}$ and $u \in \mathcal{C}$.

We associate a rate-of-cost $c(\mathbf{z}(t), u(t))$ for sojourning in state $\mathbf{z}(t)$ per unit of time while the action $u(t)$ is selected. Considering consecutive epoch times t_k and t_{k+1}, $c(\mathbf{z}(t), u(t))$ is constant for $t_k \le t < t_{k+1}$. It is equal to $c(\mathbf{z}, u)$, whenever $\mathbf{z}(t_k) = \mathbf{z}$ and $u(t_k) = u$. It is higher for undesirable states than for desirable states. We also consider the cost of applying a control action, which increases the rate-of-cost of each state.

Figure 6.3 shows several epoch times of a hypothetical 3-gene SMA-PBN. We assume that the undesirable states are those with an up-regulated gene in the most significant position in the GAP. We then assign lower rates-of-cost to desirable states 0 through 3 compared to the undesirable states 4 through 7. Given that r_1 and r_2 are the rates when the model is in the undesirable and desirable state, respectively, the cost $(t_2 - t_1)r_2$ gained between the two epoch times t_1 and t_2 is lower than the cost $(t_6 - t_5)r_2$ gained between the two epoch times t_5 and t_6.

We desire an effective intervention policy that minimizes the accumulated cost over time, that is, an intervention strategy to reduce the time spent in undesirable states with higher rate-of-cost compared to desirable states with lower rate-of-cost. In practice, the rate-of-cost has to capture the relative preferences for the different states. For instance, the cost gained between the two epoch times t_4 and t_5 may need to be greater than the cost

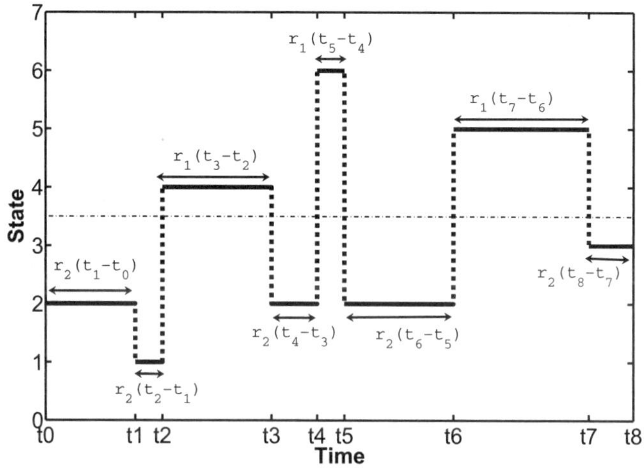

Figure 6.3. *Schematic of transitions in a hypothetical 3-gene SMA-PBN along with their epoch times and cost during each sojourn interval. The total cost between two epoch times t_1 and t_2 is less than the total cost between two epoch times t_5 and t_6.*

gained between the two epoch times t_5 and t_6. In order to penalize the sojourn time in undesirable states, the ratio of r_1 to r_2 should be large enough to capture the relative preferences for the desirable states. If the intervals between any two epoch times in Fig. 6.3 were equal, then the problem of intervention in a SMA-PBN would reduce to the intervention problem in PBNs. In this intervention problem, the objective is to reduce the number of visits to undesirable states, because the sojourn times in all states are the same, so that reducing the number of visits to undesirable states is equivalent to reducing the amount of time spent in these states.

We consider a discounted cost formulation. If $\alpha \in (0,1)$ is the discount factor per unit of time and we divide the time unit to small intervals δ, then at each interval the discount is α/δ, given the initial interval is 1. Hence, $(1-\frac{\alpha}{\delta})^{\delta t}$ represents the discount over t units of time. As $\delta \to 0$, the discount goes to $e^{-\alpha t}$.

Among the set Ω of all admissible deterministic strategies, the decision maker finds a strategy $\omega = (\mu, \mu, \ldots)$, where $\mu : \mathcal{Z} \to \mathcal{C}$ is the decision rule at time t that minimizes the expected total discounted cost. The expected total discounted cost, given the policy ω and the initial state $\mathbf{z}_0$, is

$$J_\omega(\mathbf{z}_0) = \lim_{N\to\infty} E\left[\int_0^{t_N} e^{-\alpha t} c(\mathbf{z}(t), \mu(\mathbf{z}(t))dt | \mathbf{x}(t_0) = \mathbf{z}_0\right], \tag{6.17}$$

where t_N is the Nth time epoch. We seek a strategy ω^* that minimizes $J_\omega(\mathbf{z}_0)$ for each state $\mathbf{z}_0$. An optimal intervention strategy is a solution of the continuous-time decision making problem

$$\omega^*(\mathbf{z}_0) = \arg\min_{\omega\in\Omega} J_\omega(\mathbf{z}_0) \tag{6.18}$$

for any $\mathbf{z}_0 \in \mathcal{Z}$. Intervention using the strategy ω^* increases the time spent in desirable states determined through appropriate assignment of rates-of-cost $c(\mathbf{z}(t), u(t))$.

For solving the optimization problem of Eq. (6.18), we can use the intertransition interval distributions in Eq. (6.16) and the transition probability distributions in Eq. (6.15) to define the joint transition distribution of an intertransition interval and the successor state, given the current state and control:

$$Q_{\mathbf{z}_1,\mathbf{z}_2}(\tau,u) = P\left(\tau_{k+1} \le \tau, \mathbf{z}(t_{k+1}) = \mathbf{z}_2 | \mathbf{z}(t_k) = \mathbf{z}_1, u(t_k) = u\right). \tag{6.19}$$

Consequently, the expected cost of a single transition from state $z(t_k) = \mathbf{z}_1$ and control $u(t_k) = u$, given by

$$C(\mathbf{z}_1,u) \triangleq E\left[\int_0^\tau e^{-\alpha t} c(\mathbf{z}_1,u)dt\right], \tag{6.20}$$

can be computed. Noting that

$$C(\mathbf{z}_1,u) = c(\mathbf{z}_1,u) E_{\mathbf{z}_2}\left[E_\tau\left[\int_0^\tau e^{-\alpha t}dt | \mathbf{z}_2\right]\right], \tag{6.21}$$

we obtain

$$C(\mathbf{z}_1,u) = c(\mathbf{z}_1,u) \sum_{\mathbf{z}_2 \in \mathcal{Z}} p(\mathbf{z}_1,\mathbf{z}_2,u) \int_0^\infty \left(\int_0^\tau e^{-\alpha t} dt\right) \frac{dQ_{\mathbf{z}_1,\mathbf{z}_2}(\tau,u)}{p(\mathbf{z}_1,\mathbf{z}_2,u)}, \tag{6.22}$$

which reduces to

$$C(\mathbf{z}_1,u) = c(\mathbf{z}_1,u) \sum_{\mathbf{z}_2 \in \mathcal{Z}} \int_0^\infty \frac{1-e^{-\alpha\tau}}{\alpha} dQ_{\mathbf{z}_1,\mathbf{z}_2}(\tau,u). \tag{6.23}$$

A recursive relation exists between the value function $J^N_{\omega_N}$ of stage N,

$$J^N_{\omega_N}(\mathbf{z}_0) = \sum_{k=0}^{N-1} E\left[\int_{t_k}^{t_{k+1}} e^{-\alpha t} c\big(\mathbf{z}_k, \mu(\mathbf{z}(t_k))\big)\, dt | \mathbf{z}(t_0) = \mathbf{z}_0\right], \tag{6.24}$$

and the value function $J^{N-1}_{\omega_{N-1}}$ of stage $(N-1)$ based on the definition of Eq. (6.17), given the $(N-1)$-stage policy ω_{N-1} is the subset of the N-stage policy

$$\omega_N = (\mu(\cdot,0), \mu(\cdot,1), \ldots, \mu(\cdot,N-1)) \tag{6.25}$$

when $\mu(\cdot,0)$ is excluded (Faryabi, 2009). Using Eq. (6.23), we can express Eq. (6.24) as

$$J^N_{\omega_N}(\mathbf{z}_0) = C(\mathbf{z}_0, \mu(\mathbf{z}_0)) + E\left[e^{-\alpha\tau} J^{N-1}_{\omega_{N-1}}(\mathbf{z}_1) | \mathbf{z}(t_0) = \mathbf{z}_0, \mu(\mathbf{z}(t_0)) = \mu(\mathbf{z}_0)\right]. \tag{6.26}$$

This expression can be further modified to

$$J^N_{\omega_N}(\mathbf{z}_0) = C(\mathbf{z}_0, \mu(\mathbf{z}_0)) + E_{\mathbf{z}_1}\left[E_\tau\left[e^{-\alpha\tau}|\mathbf{z}_1\right] J^{N-1}_{\omega_{N-1}}(\mathbf{z}_1)|\mathbf{z}(t_0) = \mathbf{z}_0, \mu(\mathbf{z}(t_0)) = \mu(\mathbf{z}_0)\right], \tag{6.27}$$

given τ is the first random intertransition interval and $\mathbf{z}_1$ is the successor state, which can be further expressed as

$$J^N_{\omega_N}(\mathbf{z}_0) = C(\mathbf{z}_0, \mu(\mathbf{z}_0)) + \sum_{\mathbf{z}_1 \in \mathcal{Z}} p(\mathbf{z}_0, \mathbf{z}_1, \mu(\mathbf{z}_0)) \left(\int_0^\infty e^{-\alpha\tau} \frac{dQ_{\mathbf{z}_0,\mathbf{z}_1}(\tau, \mu(\mathbf{z}_0))}{p(\mathbf{z}_0, \mathbf{z}_1, \mu(\mathbf{z}_0))} \right) J^{N-1}_{\omega_{N-1}}(\mathbf{z}_1). \tag{6.28}$$

Simplification yields the desired recursive relation between the value function $J^N_{\omega_N}$ of stage N and the value function $J^{N-1}_{\omega_{N-1}}$ of stage $(N-1)$:

$$J^N_{\omega_N}(\mathbf{z}_0) = C(\mathbf{z}_0, \mu(\mathbf{z}_0)) + \sum_{\mathbf{z}_1 \in \mathcal{Z}} \int_0^\infty e^{-\alpha\tau} dQ_{\mathbf{z}_0,\mathbf{z}_1}(\tau, \mu(\mathbf{z}_0)) J^{N-1}_{\omega_{N-1}}(\mathbf{z}_1), \tag{6.29}$$

where $\mathbf{z}(t_k) = \mathbf{z}_k$.

Equation (6.29) can be rewritten as

$$J^N_{\omega_N}(\mathbf{z}_0) = C(\mathbf{z}_0, \mu(\mathbf{z}_0)) + \sum_{\mathbf{z}_1 \in \mathcal{Z}} m(\mathbf{z}_0, \mathbf{z}_1, \mu(\mathbf{z}_0)) J^{N-1}_{\omega_{N-1}}(\mathbf{z}_1), \tag{6.30}$$

where $m(\mathbf{z}_0, \mathbf{z}_1, u)$ is defined as

$$m(\mathbf{z}_0, \mathbf{z}_1, u) \triangleq \int_0^\infty e^{-\alpha\tau} dQ_{\mathbf{z}_0,\mathbf{z}_1}(\tau, u). \tag{6.31}$$

Equation (6.30) is similar to the Bellman optimality equation in dynamic programming algorithms, in which the expected immediate cost is replaced by $C(\mathbf{z}_0, u)$, which is the expected cost of a single transition from state $\mathbf{z}_0$ under control $\mu(\mathbf{z}_0, 0) = u$, and $\alpha \times p(\mathbf{z}_0, \mathbf{z}_1, u)$ is replaced by $m(\mathbf{z}_0, \mathbf{z}_1, u)$. Hence, the optimal cost is the unique fixed point of the Bellman optimality equation:

$$J^*(\mathbf{z}_0) = \min_{u \in \mathcal{C}} \left[C(\mathbf{z}_0, u) + \sum_{\mathbf{z}_1 \in \mathcal{Z}} m(\mathbf{z}_0, \mathbf{z}_1, u) J^*(\mathbf{z}_1) \right]. \tag{6.32}$$

Any numerical method that solves the classical intervention optimization, for instance, value iteration, can be used to find the fixed point of Eq. (6.32) and also provide an optimal intervention strategy that is a solution to the optimization of Eq. (6.18). Here, we consider three hypothetical cases for the intertransition interval distribution.

6.2.3 Solution for three intertransition interval distributions

Following Faryabi *et al.* (2008a), we formulate the Bellman optimality equation, Eq. (6.32), for three intertransition interval distributions.

Suppose that the duration of the transcription of a specific gene is approximately fixed, given the expression status of other genes in the network. Owing to latent variables, we assume that this value is drawn from a discrete set, $\{\tau_{\mathbf{z}_0,\mathbf{z}_1}(k,u)\}_{k=1,\ldots,m}$, of possible values, with probabilities $\{\varrho_{\mathbf{z}_0,\mathbf{z}_1}(k,u)\}_{k=1,\ldots,m}$. According to Eq. (6.23),

$$C(\mathbf{z}_0,u) = c(\mathbf{z}_0,u) \sum_{\mathbf{z}_1\in\mathcal{Z}} \sum_{k=1}^{m} \frac{1-\exp\left(-\alpha\,\tau_{\mathbf{z}_0,\mathbf{z}_1}(k,u)\right)}{\alpha} p(\mathbf{z}_0,\mathbf{z}_1,u)\,\varrho_{\mathbf{z}_0,\mathbf{z}_1}(k,u). \tag{6.33}$$

According to Eq. (6.31),

$$m(\mathbf{z}_0,\mathbf{z}_1,u) = \sum_{k=1}^{m} p(\mathbf{z}_0,\mathbf{z}_1,u)\,\varrho_{\mathbf{z}_0,\mathbf{z}_1}(k,u)\exp(-\lambda\,\tau_{\mathbf{z}_0,\mathbf{z}_1}(k,u)). \tag{6.34}$$

With these two equations we formulate the Bellman optimality equation, Eq. (6.32), for the intertransition interval with a discrete distribution.

Next assume that, given the expression status of other genes in the network, we can measure the maximum and minimum durations of transcription of a specific gene. If we postulate that the intertransition interval between two states can take any value within the range from the maximum value to the minimum value with an equal probability, then the intertransition interval between two states $\mathbf{z}_0$ and $\mathbf{z}_1$ has a uniform distribution in the interval $[c_{\mathbf{z}_0,\mathbf{z}_1}(u), d_{\mathbf{z}_0,\mathbf{z}_1}(u)]$. In this case,

$$C(\mathbf{z}_0,u) = \frac{c(\mathbf{z}_0,u)}{\alpha} \sum_{\mathbf{z}_1\in\mathcal{Z}} \left(1 - \frac{\exp(-\alpha\, c_{\mathbf{z}_0,\mathbf{z}_1}(u)) - \exp(-\alpha\, d_{\mathbf{z}_0,\mathbf{z}_1}(u))}{\alpha\,(d_{\mathbf{z}_0,\mathbf{z}_1}(u) - c_{\mathbf{z}_0,\mathbf{z}_1}(u))}\right) p(\mathbf{z}_0,\mathbf{z}_1,u) \tag{6.35}$$

and

$$m(\mathbf{z}_0,\mathbf{z}_1,u) = \frac{\exp(-\alpha\, c_{\mathbf{z}_0,\mathbf{z}_1}(u)) - \exp(-\alpha\, d_{\mathbf{z}_0,\mathbf{z}_1}(u))}{\alpha\,(d_{\mathbf{z}_0,\mathbf{z}_1}(u) - c_{\mathbf{z}_0,\mathbf{z}_1}(u))} p(\mathbf{z}_0,\mathbf{z}_1,u). \tag{6.36}$$

Finally, it is commonplace that the amount of data observed from a biological system is limited. Hence, instead of using the data to estimate an arbitrary intertransition interval distribution, we can postulate a class of parametric distributions whose members can be defined with a small number of moments, for instance, the expected value. Suppose the distribution of the intertransition interval follows an exponential distribution. If all the intertransition intervals of state $\mathbf{z}$ are exponentially distributed, then the sojourn time of state $\mathbf{z}$ possesses an exponential distribution:

$$p_{\mathbf{z}}(\tau,u) = 1 - e^{-\nu_{\mathbf{z}}(u)\tau} \tag{6.37}$$

for $\tau \geq 0$, where $\nu_{\mathbf{z}}(u)$ is the rate of transition from state $\mathbf{z}$ whenever the action is u. Practically, the rates $\nu_{\mathbf{z}}(u)$ are bounded for all states $\mathbf{z} \in \mathcal{Z}$ and all controls $u \in \mathcal{C}$. In Faryabi (2009), it is shown that if

$$\nu = \max_{\mathbf{z}\in\mathcal{Z},u\in\mathcal{C}} \nu_{\mathbf{z}}(u), \tag{6.38}$$

then

$$C(\mathbf{z},u) = \frac{c(\mathbf{z},u)}{\alpha+\nu} \tag{6.39}$$

and

$$m(\mathbf{z}_1,\mathbf{z}_2,u) = \frac{\nu}{\alpha+\nu}\tilde{p}(\mathbf{z}_1,\mathbf{z}_2,u), \tag{6.40}$$

where

$$\tilde{p}(\mathbf{z}_1,\mathbf{z}_2,u) = \frac{\nu_{\mathbf{z}_1}(u)}{\nu}p(\mathbf{z}_1,\mathbf{z}_2,u) \tag{6.41}$$

if $\mathbf{z}_1 \neq \mathbf{z}_2$ and

$$\tilde{p}(\mathbf{z}_1,\mathbf{z}_2,u) = \frac{\nu_{\mathbf{z}_1}(u)}{\nu}p(\mathbf{z}_1,\mathbf{z}_2,u) + 1 - \frac{\nu_{\mathbf{z}_1}(u)}{\nu} \tag{6.42}$$

if $\mathbf{z}_1 = \mathbf{z}_2$.

6.2.4 Intervention in a mutated mammalian cell-cycle SMA-PBN

Once again we consider the mammalian cell-cycle network with mutated p27 (always OFF) introduced in Section 4.5.8, where, in this cancerous scenario, the permanent inactivation of p27 creates a situation where both CycD and Rb might be inactive, so that the states in which both Rb and CycD are down-regulated are undesirable. Recall that Table 4.7 summarizes the mutated Boolean functions and, in the case of a SMA-PBN, these are used to construct the embedded PBN. As in Section 4.5.8, the first constituent BN is determined from Table 4.7 when CycD = 0 and the second when CycD = 1. The mutated cell-cycle's PBN consists of nine genes: CycD, Rb, E2F, CycE, CycA, Cdc20, Cdh1, UbcH10, and CycB, with $q = 0.01$ and $p = 0.001$. This order of the genes is used in the binary representation of the logical states, with CycD as the most significant bit and CycB as the least significant bit. This order of the genes in the logical states facilitates the presentation of the results and does not affect the derived control policy.

In the case of a SMA-PBN, we need to specify the intertransition interval distribution. We assume that all intertransition intervals are exponentially distributed. Letting $\tau(\mathbf{z}_1,\mathbf{z}_2)$ be the sojourn time in state $\mathbf{z}_1$ before transition to state $\mathbf{z}_2$, we need the rate of transition from state $\mathbf{z}_1$ to $\mathbf{z}_2$ to specify its distribution. Letting $p_{\mathbf{z}_1,\mathbf{z}_2}$ denote the probability of the transition from $\mathbf{z}_1$ to $\mathbf{z}_2$ in the embedded PBN, we assume that the rate of the transition from $\mathbf{z}_1$ to $\mathbf{z}_2$ is assigned such that

$$P\left[\tau(\mathbf{z}_1,\mathbf{z}_2) < \min_{\mathbf{z}\in\mathcal{Z},\mathbf{z}\neq\mathbf{z}_1} \tau(\mathbf{z}_1,\mathbf{z})\right] = p(\mathbf{z}_1,\mathbf{z}_2). \tag{6.43}$$

Under this assumption, the probability of the first transition out of state $\mathbf{z}_1$ to $\mathbf{z}_2$ is equal to the transition probability $p_{\mathbf{z}_1,\mathbf{z}_2}$. The left-hand side of Eq. (6.43) can be determined from exponentially distributed sojourn times.

The rate of penalizing the states with down-regulated Rb and CycD is set to be higher than those for the states in which these two genes are not simultaneously down-regulated. We postulate the following rate-of-cost function:

$$c(\mathbf{z},u) = \begin{cases} 0 & \text{if } u = 0 \text{ and } \mathbf{z} \in \mathcal{D}, \\ 5 & \text{if } u = 0 \text{ and } \mathbf{z} \in \mathcal{U}, \\ 1 & \text{if } u = 1 \text{ and } \mathbf{z} \in \mathcal{D}, \\ 6 & \text{if } u = 1 \text{ and } \mathbf{z} \in \mathcal{U}, \end{cases} \tag{6.44}$$

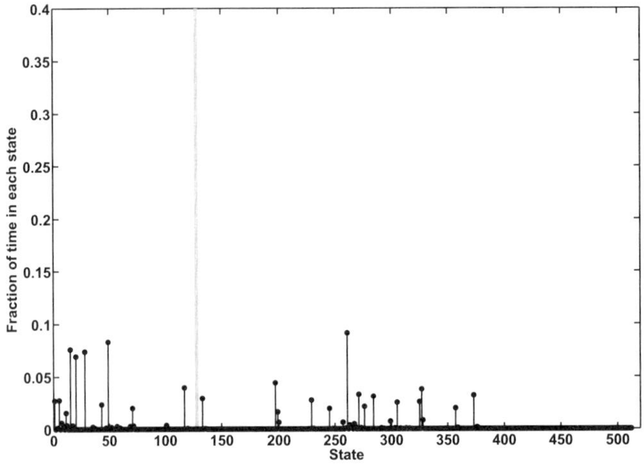

Figure 6.4. *The fraction of time that the SMA-PBN of a mammalian cell cycle spends in each state prior to intervention. The vertical line separates the undesirable states in $\mathcal{U}$ from the desirable states in $\mathcal{D}$.*

Table 6.1. *The Δp_g for the intervention strategy based on various control genes.*

Control Gene g	*Rb*	*E2F*	*CycE*	*CycA*	*Cdc20*	*Cdh1*	*UbcH10*	*CycB*
Δp_g	94.2%	89.1%	71.1%	62.1%	63.5%	68.4%	59.7%	75.2%

where $\mathcal{U}$ and $\mathcal{D}$ are the sets of undesirable and desirable states, respectively. State $\mathbf{z} \in \mathcal{D}$ if (CycD, Rb) $\neq (0,0)$ and $\mathbf{z} \in \mathcal{U}$ if (CycD, Rb) $= (0,0)$.

Figure 6.4 shows the fraction of time spent in each state prior to intervention, the vertical line separating the undesirable from desirable states. The aggregated fraction of time that the mutated cell-cycle SMA-PBN spends in the states with simultaneously down-regulated CycD and Rb is 49%.

We define ΔP to be the percentage of the change in the fraction of time that the SMA-PBN spends in the states with simultaneously down-regulated CycD and Rb before and after the intervention:

$$\Delta P = \frac{\sum_{\mathbf{z}\in\mathcal{U}} \varphi(\mathbf{z}) - \sum_{\mathbf{z}\in\mathcal{U}} \varphi_{\mu^*}(\mathbf{z})}{\sum_{\mathbf{z}\in\mathcal{U}} \varphi(\mathbf{z})}, \tag{6.45}$$

where $\varphi(\mathbf{z})$ and $\varphi_{\mu^*}(\mathbf{z})$ are the fractions of time that the SMA-PBN spends in state $\mathbf{z}$ in the long run when no control is applied and after intervention with an optimal strategy, μ^*. As a performance measure, ΔP indicates the percentage of the reduction in the fraction of time that the model spends in undesirable states in the long run.

Letting each gene in the network act as the control gene, Table 6.1 lists the value of ΔP corresponding to each. Among all the genes, Rb and E2F have the best performance. The fraction of time that the SMA-PBN spends in states after direct Rb-based intervention is shown in Fig. 6.5. Using Rb as the control gene, the fraction of time that the model spends

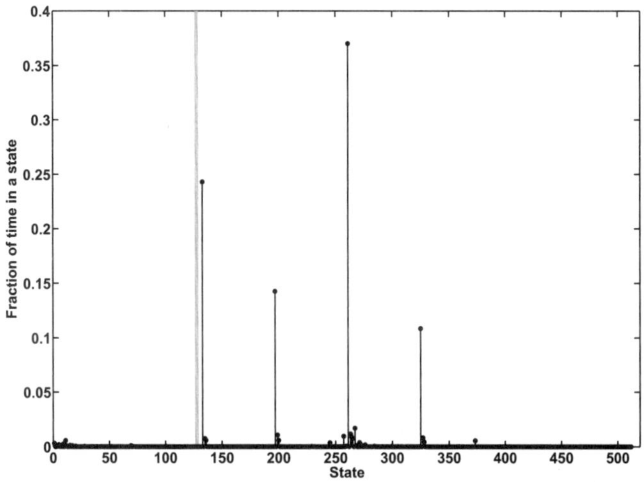

Figure 6.5. *The fraction of time that the SMA-PBN of a mammalian cell cycle spends in states after intervention using Rb as the control gene. The vertical line separates the undesirable states in $\mathcal{U}$ from the desirable states in $\mathcal{D}$.*

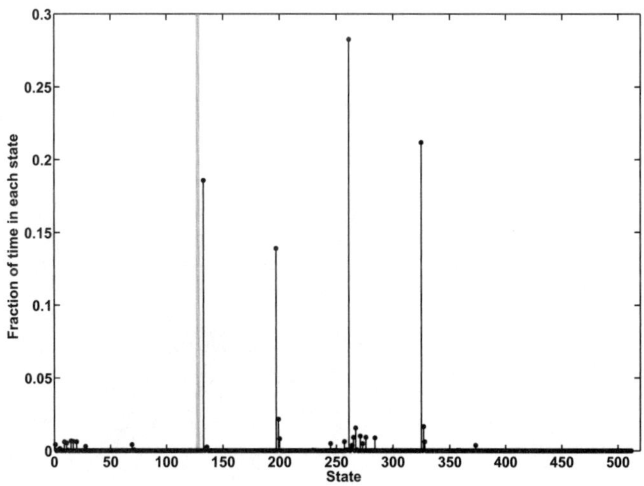

Figure 6.6. *The fraction of time that the SMA-PBN of a mammalian cell cycle spends in states after intervention using E2F as the control gene. The vertical line separates the undesirable states in $\mathcal{U}$ from the desirable states in $\mathcal{D}$.*

in the undesirable states is reduced to less than 2%. If direct intervention based on Rb is not feasible, then one can use E2F as the control gene. In this case the system spends slightly more time in the undesirable states, but even this value is still less than 4.5% (see Fig. 6.6).

Bibliography

Akutsu T, Miyano S, Kuhara S. (1999) Identification of genetic networks from a small number of gene expression patterns under the Boolean network model. *Pac Symp Biocomput.* 4:17-28.

Akutsu T, Hayashida M, Ching WK, Ng MK. (2007) Control of Boolean networks: Hardness results and algorithms for tree structured networks. *J Theor Biol.* 244(4):670-9.

Albert R, Barabási AL. (2002) Statistical mechanics of complex networks. *Rev. Mod. Phys.* 74:47-97.

Albert R, Jeong H, Barabási AL. (2000) Error and attack tolerance of complex networks. *Nature.* 406:378-82.

Albert R, Othmer HG. (2003) The topology of the regulatory interactions predicts the expression pattern of the Drosophila segment polarity genes. *J. Theor. Biol.* 223:1-18.

Aldana M, Cluzel P. (2003) A natural class of robust networks. *Proc Natl Acad Sci USA.* 100(15):8710-4.

Aldana M. (2003) Boolean dynamics of networks with scale-free topology. *Physica D.* 185:45-66.

Aldana M, Coppersmith S, Kadanoff LP. (2002) Boolean dynamics with random couplings. in *Perspectives and Problems in Nonlinear Science*, eds. Kaplan, E., Marsden, J. E. & Sreenivasan, K. R., Springer, New York, 23-89.

Altman E. (1999) *Constrained Markov Decision Processes*. Chapman and Hall/CRC, Boca Raton, FL.

Alvarez-Buylla ER, Chaos A, Aldana M, Benítez M, Cortes-Poza Y, Espinosa-Soto C, Hartasánchez DA, Lotto RB, Malkin D, Escalera Santos GJ, Padilla-Longoria P. (2008) Floral morphogenesis: Stochastic explorations of a gene network epigenetic landscape. *PLoS ONE.* 2008;3(11):e3626.

Ananko EA, Podkolodny NL, Stepanenko IL, Ignatieva EV, Podkolodnaya OA, Kolchanov NA. (2002) GeneNet: A database on structure and functional organisation of gene networks. *Nucleic Acids Res.* 30(1):398-401.

Arsura M, Wu M, Sonenshein GE. (1996) TGF beta 1 inhibits NF-kappa B/Rel activity inducing apoptosis of B cells: Transcriptional activation of I kappa B alpha. *Immunity*. 5(1):31-40.

Attoor S, Dougherty ER, Chen Y, Bittner ML, Trent JM. (2004) Which is better for cDNA-microarray-based classification: Ratios or direct intensities. *Bioinformatics*. 20(16):2513-20.

Bang-Jensen J, Gutin G. (2001) *Digraphs: Theory, Algorithms and Applications*. Springer-Verlag, London.

Barabási AL, Albert R. (1999) Emergence of scaling in random networks. *Science*. 286(5439):509-12.

Barron A, Rissanen J, Yu B. (1998) The minimum description length principle in coding and modeling. *IEEE Transactions on Information Theory*. 44(6):2743-60.

Bastolla U, Parisi G. (1998a) Relevant elements, magnetization and dynamical properties in Kauffman networks: A numerical study. *Physica D*. 115:203-218.

Bastolla U, Parisi G. (1998b) The modular structure of Kauffman networks. *Physica D*. 115:219-233.

Bellman R. (1957) *Dynamic Programming*. Princeton University Press, Princeton.

Bernard A, Hartemink AJ. (2005) Informative structure priors: Joint learning of dynamic regulatory networks from multiple types of data. *Pac Symp Biocomput*. 2005:459-70.

Bertsekas DP. (1976) *Dynamic Programming and Stochastic Control*. Academic Press, New York.

Bertsekas DP, Tsitsiklis JN. (1996) *Neuro-Dynamic Programming*. Athena Scientific, Belmont, MA.

Bertsekas DP. (2005) *Dynamic Programming and Optimal Control*. Athena Scientific, Belmont, MA, 1995, 3rd ed., 2005.

Bittner M, Meltzer P, Chen Y, Jiang Y, Seftor E, Hendrix M, Radmacher M, Simon R, Yakhini Z, Ben-Dor A, Sampas N, Dougherty E, Wang E, Marincola F, Gooden C, Lueders J, Glatfelter A, Pollock P, Carpten J, Gillanders E, Leja D, Dietrich K, Beaudry C, Berens M, Alberts D, Sondak V. (2000) Molecular classification of cutaneous malignant melanoma by gene expression profiling. *Nature*. 406(6795):536-40.

Blondel VD, Tsitsiklis JN. (2000) A survey of computational complexity results in systems and control. *Automatica*. 36(9):1249-74.

Bornholdt S. (2008) Boolean network models of cellular regulation: Prospects and limitations. *J R Soc Interface*. 5 Suppl 1:S85-94.

Boros E, Ibaraki T, Makino K. (1998) Error-free and best-fit extensions of partially defined Boolean functions. *Information and Computation*. 140:254-83.

Boyd S, Vandenberghe L. (2004) *Convex Optimization*. Cambridge University Press, Cambridge, UK.

Braunewell S, Bornholdt S. (2006) Superstability of the yeast cell-cycle dynamics: Ensuring causality in the presence of biochemical stochasticity. *J Theor Biol.* 245(4):638-643.

Brun M, Dougherty ER, Shmulevich I. (2005) Steady-state probabilities for attractors in probabilistic Boolean networks. *Signal Processing*. 85(4):1993-2013.

Bu D, Zhao Y, Cai L, Xue H, Zhu X, Lu H, Zhang J, Sun S, Ling L, Zhang N, Li G, Chen R. (2003) Topological structure analysis of the protein-protein interaction network in budding yeast. *Nucleic Acids Res*. 31(9):2443-50.

Chang HH, Hemberg M, Barahona M, Ingber DE, Huang S. (2008) Transcriptome-wide noise controls lineage choice in mammalian progenitor cells. *Nature*. 453(7194):544-7.

Chaves M, Albert R, Sontag D. (2005) Robustness and fragility of Boolean models for genetic regulatory networks. *J Theor Biol.* 235:431-449.

Chaves M, Sontag ED, Albert R. (2006) Methods of robustness analysis for Boolean models of gene control networks. *IEE Proceedings in Systems Biology*. 153:154-167.

Chen Y, Dougherty ER, Bittner ML. (1997) Ratio-based decisions and the quantitative analysis of cDNA microarray images. *J Biomed Opt.* 2(4):364-74.

Cheng SY, Huang HJ, Nagane M, Ji XD, Wang D, Shih CC, Arap W, Huang CM, Cavenee WK. (1996) Suppression of glioblastoma angiogenicity and tumorigenicity by inhibition of endogenous expression of vascular endothelial growth factor. *Proc Natl Acad Sci USA*. 93(16):8502-7.

Ching WK, Zhang S, Ng MK, Akutsu T. (2007) An approximation method for solving the steady-state probability distribution of probabilistic Boolean networks. *Bioinformatics*. 23(12):1511-8.

Ching WK, Ng MK, Fung ES, Akutsu T. (2005) On construction of stochastic genetic networks based on gene expression sequences. *Int J Neural Syst.* 15(4):297-310.

Ching WK, Zhang SQ, Jiao Y, Akutsu T, Tsing NK, Wong AS. (2009) Optimal control policy for probabilistic Boolean networks with hard constraints. *IET Syst Biol.* 3(2):90-9.

Cho GE, Meyer CD. (2001) Comparison of perturbation bounds for the stationary distribution of a Markov chain. *Linear Algebra and Its Applications*. 335:137-150.

Choudhary A, Datta A, Bittner ML, Dougherty ER. (2006) Intervention in a family of Boolean networks. *Bioinformatics*. 22 (2): 226-32.

Çınlar, E. (1997) *Introduction to Stochastic Processes*. Prentice-Hall, Englewood Cliffs, NJ.

Cook S, Dwork C, Reischuk R. (1986) Upper and lower time bounds for parallel random access machines without simultaneous writes. *SIAM J. Comput.* 15, 87-97.

Cornforth D, Green DG, Newth D, Kirley MR. (2001) Ordered asynchronous processes in natural and artificial systems. in *Proceeding of the 5th Australia-Japan Joint Workshop on Intelligent and Evolutionary Systems*, New Zealand, November 2001, 105-112.

Cowles MK, Carlin, BP. (1996) Markov chain Monte Carlo convergence diagnostics: A comparative study. *Journal of the American Statistical Association*. 91:883-904.

Dassow G, Meir E, Munro EM, Odell GM. (2000) The segment polarity network is a robust developmental module. *Nature*. 406:188-92.

Datta A, Choudhary A, Bittner M, Dougherty ER. (2003) External control in Markovian genetic regulatory networks. *Machine Learning*. 52:169-81.

Datta A, Choudhary A, Bittner M, Dougherty ER. (2004) External control in Markovian genetic regulatory networks: The imperfect information case. *Bioinformatics*. 20(6):924-30.

Davidich M, Bornholdt S. (2008a) The transition from differential equations to Boolean networks: A case study in simplifying a regulatory network model. *J Theor Biol.* 255(3):269-77.

Davidich MI, Bornholdt S. (2008b) Boolean network model predicts cell cycle sequence of fission yeast. *PloS One*. 3(2): e1672.

Davidson EH. (2006) *The Regulatory Genome: Gene Regulatory Networks in Development and Evolution*. Academic Press, New York.

de Jong H. (2002) Modeling and simulation of genetic regulatory systems: A literature review. *J. Comput. Biol.* 9:69-105.

Deng X, Geng H, Matache MT. (2007) Dynamics of asynchronous random Boolean networks with asynchrony generated by stochastic processes. *Biosystems*. 88(1-2):16-34.

Deo N. (1974) *Graph Theory with Applications to Engineering and Computer Science*. Prentice-Hall, Englewood Cliffs, NJ.

Derman C. (1970) *Finite State Markovian Decision Processes*. Academic Press, New York.

Derrida B, Pomeau Y. (1986) Random networks of automata: A simple annealed approximation. *Europhys. Lett.* 1:45-49.

Di Paolo EA. (2000) Searching for rhythms in asynchronous Boolean networks. in *Proceedings of the Seventh International Conference on Artificial Life (Artificial Life* VII*)*, Portland, OR, August 2000, 1-6.

Dougherty J, Kohavi R, Sahami M. (1995) Supervised and unsupervised discretization of continuous features. in *Proceedings of the Twelfth International Conference on Machine Learning*, Tahoe City, CA: Morgan Kaufmann, San Francisco, 194-202.

Dougherty ER, Shmulevich I. (2003) Mappings between probabilistic Boolean networks. *Signal Processing*. 83(4):799-809.

Dougherty ER, Hua J, Xiong Z, Chen Y. (2005) Optimal robust classifiers. *Pattern Recognition*. 38(10):1520-32.

Dougherty ER, Xiao Y. (2006) Design of probabilistic Boolean networks under the requirement of contextual data consistency. *IEEE Transactions on Signal Processing*. 54(9):3603-3613.

Dougherty ER. (2007) Validation of inference procedures for gene regulatory networks. *Curr Genomics*. 8(6):351-9.

Dougherty J, Tabus I, Astola J. (2008) Inference of gene regulatory networks based on a universal minimum description length. *EURASIP Journal on Bioinformatics and Systems Biology*. 2008:482090.

Dougherty ER, Kim S, Chen Y. (2000) Coefficient of determination in nonlinear signal processing. *Signal Processing*. 80:2219-35.

Drossel B. (2007) Random Boolean networks. in *Annual Review of Nonlinear Dynamics and Complexity*, 1, ed. Schuster HG, Wiley, New York.

Faryabi B, Datta A, Dougherty ER. (2007) On approximate stochastic control in genetic regulatory networks. *IET Syst Biol.* 1(6):361-8.

Faryabi B, Chamberland J-F, Vahedi G, Datta A, Dougherty ER. (2008a) Optimal intervention in asynchronous genetic regulatory networks. *IEEE Journal of Selected Topics in Signal Processing*. 2(3):412-23.

Faryabi B, Chamberland J-F, Vahedi G, Datta A, Dougherty ER. (2008b) Optimal constrained stationary intervention in gene regulatory networks. *EURASIP Journal on Bioinformatics and Systems Biology*. 2008:620767.

Faryabi B, Vahedi G, Chamberland JF, Datta A, Dougherty ER. (2009) Intervention in context-sensitive probabilistic Boolean networks revisited. *EURASIP Journal on Bioinformatics Systems Biology*. 2009:360864.

Faryabi B. (2009) Systems medicine: An integrated approach with decision making perspective. Ph.D. Dissertation, *Department of Electrical and Computer Engineering*, Texas A&M University, College Station, 2009.

Fauré A, Naldi A, Chaouiya C, Thieffry D. (2006) Dynamical analysis of a generic Boolean model for the control of the mammalian cell cycle. *Bioinformatics*. 22(14):e124-31.

Feinberg E, Shwartz A. (2002) *Handbook of Markov Decision Processes*. Kluwer Academic Publishers, Boston.

Flyvbjerg H. (1988) An order parameter for networks of automata. *J. Phys. A: Math. Gen.* 21:L955-L960.

Fraser AG, Marcotte EM. (2004) A probabilistic view of gene function. *Nat Genet.* 36(6):559-64.

Friedman N, Linial M, Nachman I, Pe'er D. (2000) Using Bayesian network to analyze expression data. *J Comput Biol.* 7(3-4):601-20.

Fuller GN, Rhee CH, Hess KR, Caskey LS, Wang R, Bruner JM, Yung WK, Zhang W. (1999) Reactivation of insulin-like growth factor binding protein 2 expression in glioblastoma multiforme: A revelation by parallel gene expression profiling. *Cancer Res.* 59(17):4228-32.

Gallager RG. (1996) *Discrete Stochastic Processes.* Kluwer Academic Publishers, Boston.

Gershenson C. (2002) Classification of random Boolean networks. in *Proceedings of the Eighth International Conference on Artificial Life (Artificial Life* VIII*)*, Sydney, Australia, December 2002, 1-8.

Gersho A, Gray RM. (1992) *Vector Quantization and Signal Compression.* Kluwer Academic Publishers, Boston.

Glass L. (1975) Classification of biological networks by their qualitative dynamics. *J Theor Biol.* 54:85-107.

Glass L, Kauffman SA. (1973) The logical analysis of continuous, nonlinear biochemical control networks. *J Theor Biol.* 39:103-29.

Goodwin BC. (1963) *Temporal Organization in Cells.* Academic Press, New York.

Goodwin BC. (1965). Oscillatory behavior in enzymatic control processes. in *Advances in Enzyme Regulation*, ed. Weber, G., Pergamon Press, Oxford, 425-438.

Goryanin I, Hodgman TC, Selkov E. (1999) Mathematical simulation and analysis of cellular metabolism and regulation. *Bioinformatics.* 15(9):749-758.

Greil F, Drossel B. (2005) Dynamics of critical Kauffman networks under asynchronous stochastic update. *Phys Rev Lett.* 95(4):048701.

Grigoryan AM, Dougherty ER. (2001a) Design and analysis of robust optimal binary filters in the context of a prior distribution for the states of nature. *Mathematical Imaging and Vision.* 11:239-54.

Grigoryan AM, Dougherty ER. (2001b) Bayesian robust optimal linear filters. *EURASIP J. Signal Processing.* 81:2503-21.

Han K, Ju BH, Jung H. (2004) WebInterViewer: Visualizing and analyzing molecular interaction networks. *Nucleic Acids Res.* 32(Web Server Issue):W89-W95.

Harris SE, Sawhill BK, Wuensche A, Kauffman, SA. (2002) A model of transcriptional regulatory networks based on biases in the observed regulation rules. *Complexity.* 7(4), 23-40.

Harvey I, Bossomaier T. (1997) Time out of joint: Attractors in asynchronous random Boolean networks. in *Proceeding of the 4th European Conference on Artificial Life (ECAL97)*, MIT Press, Cambridge, MA, July 1997, 67-75.

Hashimoto RF, Kim S, Shmulevich I, Zhang W, Bittner ML, Dougherty ER. (2004) Growing genetic regulatory networks from seed genes. *Bioinformatics.* 20(8):1241-7.

Hayashi S, Yamamoto M, Ueno Y, Ikeda K, Ohshima K, Soma G, Fukushima T. (2001) Expression of nuclear factor-kappa B, tumor necrosis factor receptor type 1, and c-Myc in human astrocytomas. *Neurol Med Chir (Tokyo)*. 41(4):187-95.

Holland JH. (1995) *Hidden Order: How Adaptation Builds Complexity*. Helix Books, Reading, MA.

Huang S. (1999) Gene expression profiling, genetic networks, and cellular states: An integrating concept for tumorigenesis and drug discovery. *J Mol Med*. 77(6):469-80.

Huang S. (2001) Genomics, complexity and drug discovery: Insights from Boolean network models of cellular regulation. *Pharmacogenomics*. 2(3):203-22.

Huang S, Ingber DE. (2000) Shape-dependent control of cell growth, differentiation, and apoptosis: Switching between attractors in cell regulatory networks. *Exp Cell Res*. 261(1):91-103.

Huang S, Eichler G, Bar-Yam Y, Ingber DE. (2005) Cell fates as high-dimensional attractor states of a complex gene regulatory network. *Phys Rev Lett*. 94(12):128701.

Hunter JJ. (1986) Stationary distributions of perturbed Markov chains. *Linear Algebra and its Applications*. 82:201-214, 1986.

Hunter JJ. (1988) Characterizations of generalized inverses associated with Markovian kernels. *Linear Algebra and its Applications*. 102:121-142.

Hunter JJ. (1990) Parametric forms for generalized inverse of Markovian kernels and their applications. *Linear Algebra and its Applications*. 127:71-84.

Hunter JJ. (2005) Stationary distributions and mean first passage times of perturbed Markov chains. *Linear Algebra and its Applications*. 410:217-243.

Hunter JJ. (2006) Mixing times with applications to perturbed Markov chains. *Linear Algebra and its Applications*. 417:108-123.

Ideker TE, Thorsson V, Karp RM. (2000) Discovery of regulatory interactions through perturbation: Inference and experimental design. *Pac Symp Biocomput*. 5:305-16.

Ioannou PA, Sun J. (1996) *Robust Adaptive Control: A Unified Approach*. Prentice-Hall, Englewood Cliffs, NJ.

Ito T, Chiba T, Ozawa R, Yoshida M, Hattori M, Sakaki Y. (2001) A comprehensive two-hybrid analysis to explore the yeast protein interactome. *Proc Natl Acad Sci USA*. 98(8):4569-74.

Ivanov I, Dougherty ER. (2004) Reduction mappings between probabilistic Boolean networks. *EURASIP Journal on Applied Signal Processing*. 2004(1):125-131.

Ivanov I, Pal R, Dougherty ER. (2007) Dynamics preserving size reduction mappings for probabilistic Boolean networks. *IEEE Transactions on Signal Processing*. 55(5):2310-2322.

Iyengar GN. (2005) Robust dynamic programming. *Mathematics of Operations Research.* 30(2):257-80.

Jacob F, Monod J. (1961) On the regulation of gene activity. Cold Spring Harbor Symposia on Quantitative Biology.

Jensen FV. (2001) *Bayesian Networks and Decision Graphs.* Springer-Verlag, New York.

Jeong H, Tombor B, Albert R, Oltvai ZN, Barabasi AL. (2000) The large-scale organization of metabolic networks. *Nature.* 407(6804):651-4.

Jeong H, Mason SP, Barabasi AL, Oltvai ZN. (2001) Lethality and centrality in protein networks. *Nature.* 411(6833):41-2.

Kaelbling LP, Littman ML, Cassandra AR. (1998) Planning and acting in partially observable stochastic domains. *Artificial Intelligence.* 101:99-134.

Kahn J, Kalai G, Linial N. (1988) The influence of variables on Boolean functions. in *Proceedings of the 29th Annual Symposium on Foundations of Computer Science*, IEEE, Washington, DC, 68-80.

Kamath AP, Karmarkar NK, Ramakrishnan KG, Resende MGC. (1992) A continuous approach to inductive inference. *Mathematical Programming.* 57(2): 215-238.

Kamath AP, Karmarkar NK, Ramakrishnan KG, Resende MGC. (1993) An interior point approach to Boolean vector function synthesis. in *Proceedings of the 36th Midwest Symposium on Circuits and Systems*, 1:185-189.

Kanehisa M, Goto S. (2000) KEGG: Kyoto encyclopedia of genes and genomes. *Nucleic Acids Res.* 28(1):27-30.

Kauffman SA. (1969a) Metabolic stability and epigenesis in randomly constructed genetic nets. *J Theor Biol.* 22:437-67.

Kauffman SA. (1969b) Homeostasis and differentiation in random genetic control networks. *Nature.* 224:177-8.

Kauffman SA. (1974) The large scale structure and dynamics of genetic control circuits: An ensemble approach. *J Theor Biol.* 44:167-90.

Kauffman SA. (1993) *The Origins of Order: Self-Organization and Selection in Evolution.* Oxford University Press, New York.

Kauffman SA. (1995) *At Home in the Universe.* Oxford University Press, New York.

Kauffman SA. (2000) *Investigations.* Oxford University Press, New York.

Kauffman S, Peterson C, Samuelsson B, Troein C. (2004) Genetic networks with canalyzing Boolean rules are always stable. *Proc Natl Acad Sci USA.* 101(49):17102-7.

Kemeny JG, Snell JL. (1960) *Finite Markov Chains.* Van Nostrand, New York.

Kim S, Dougherty ER, Chen Y, Sivakumar K, Meltzer P, Trent JM, Bittner M. (2000a) Multivariate measurement of gene expression relationships. *Genomics.* 67(2):201-9.

Kim S, Dougherty ER, Bittner ML, Chen Y, Sivakumar K, Meltzer P, Trent JM. (2000b) General nonlinear framework for the analysis of gene interaction via multivariate expression arrays. *J Biomed Opt*. 5(4):411-24.

Kim S, Li H, Dougherty ER, Cao N, Chen Y, Bittner ML, Suh EB. (2002) Can Markov chain models mimic biological regulation? *Journal of Biological Systems*. 10(4):431-45.

Kim SY, Imoto S, Miyano S. (2003) Inferring gene networks from time series microarray data using dynamic Bayesian networks. *Brief Bioinform*. 4(3):228-35.

Kirienko GI. (1964) On self-correcting schemes from functional elements. *Problemy Kibernet*. 12:29-37 (in Russian).

Krawitz P, Shmulevich I. (2007) Basin entropy in Boolean network ensembles. *Physical Review Letters*. 98:158701.

Kumar P, Lin W. (1982) Optimal adaptive controller for unknown Markov chains. *IEEE Trans. Automatic Control*. 27(4):765-774.

Kumar P, Varaiya P. (1986) *Stochastic Systems Estimation Identification and Adaptive Control*. Prentice-Hall, Englewood Cliffs.

Kuznetsov VP. (1976). Stable detection when the signal and spectrum of normal noise are inaccurately known. *Telecommunications and radio engineering*. 30-31:58-64.

Lähdesmäki H, Shmulevich I, Yli-Harja O. (2003) On learning gene regulatory networks under the Boolean network model. *Machine Learning*. 52:147-67.

Lähdesmäki H, Hautaniemi S, Shmulevich I, Yli-Harja O. (2006) Relationships between probabilistic Boolean networks and dynamic Bayesian networks as models of gene regulatory networks. *Signal Processing*. 86(4):814-834.

Lähdesmäki H, Shmulevich I. (2008) Learning the structure of dynamic Bayesian networks from time series and steady state measurements. *Machine Learning*. 71:185-217.

Lambert JD. (1991) *Numerical Methods for Ordinary Differential Equations*. Wiley, Chichester.

Langholz G, Kandel A, Mott JL. (1998) *Foundations of Digital Logic Design*. World Scientific, Singapore.

Lauritzen S. (1996) *Graphical Models*. Oxford University Press, Oxford.

Layek R, Datta A, Pal R, Dougherty ER. (2009) Adaptive intervention in probabilistic Boolean networks. *Bioinformatics*. 25(16): 2042-48.

Li F, Long T, Lu Y, Quyang Q, Tang C. (2004) The yeast cell-cycle network is robustly designed. *Proc Natl Acad Sci USA*. 101(14):4781-4786.

Li P, Zhang C, Perkins EJ, Gong P, Deng Y. (2007) Comparison of probabilistic Boolean network and dynamic Bayesian network approaches for inferring gene regulatory networks. *BMC Bioinformatics*. 8 Suppl 7:S13.

Liang S, Fuhrman S, Somogyi R. (1998) REVEAL, a general reverse engineering algorithm for inference of genetic network architectures. *Pac Symp Biocomput.* 3:18-29.

Liu W, Lähdesmäki H, Dougherty ER, Shmulevich I. (2008) Inference of Boolean networks using sensitivity regularization. *EURASIP Journal on Bioinformatics and Systems Biology*. 2008:780541.

Luque B, Solé RV. (1997) Phase transitions in random networks: Simple analytic determination of critical points. *Phys. Rev. E*. 55:257-260.

Luque B, Solé RV. (2000) Lyapunov exponents in random Boolean networks. *Physica A*. 284:33-45.

Lynch JF. (1995) On the threshold of chaos in random Boolean cellular automata. *Random Structures and Algorithms*. 6:239-260.

Ma HW, Zeng AP. (2003) The connectivity structure, giant strong component and centrality of metabolic networks. *Bioinformatics*. 19(11):1423-30.

MacLeod MC. (1996) A possible role in chemical carcinogenesis for epigenetic, heritable changes in gene expression. *Mol Carcinog*. 15(4):241-50.

Madani O. (2000) Complexity Results for Infinite-Horizon Markov Decision Processes. Ph.D. Dissertation, Dept. Comp. Sci. Eng., University of Washington, Seattle.

Maki Y, Tominaga D, Okamoto M, Watanabe S, Eguchi Y. (2001) Development of a system for the inference of large scale genetic networks. *Pac Symp Biocomput*. 6:446-58.

Marshall S, Yu L, Xiao Y, Dougherty ER. (2007) Inference of a probabilistic boolean network from a single observed temporal sequence. *EURASIP Journal on Bioinformatics and Systems Biology*. 2007:32454.

McAdams HH, Arkin A. (1999) It's a noisy business! Genetic regulation at the nanomolar scale. *Trends Genet*. 15(2):65-9.

McCluskey EJ. (1956) Minimization of Boolean functions. *Bell System Technical Journal*. 35:1417-1444.

Mendes P. (1993) GEPASI: A software package for modelling the dynamics, steady states and control of biochemical and other systems. *Comput Appl Biosci*. 9(5):563-71.

Mendoza L, Thieffry D, Alvarez-Buylla ER. (1999) Genetic control of flower morphogenesis in *Arabidopsis thaliana*: A logical analysis. *Bioinformatics*. 15(7-8):593-606.

Mesot B, Teuscher C. (2003) Critical values in asynchronous random Boolean networks. in *Proceeding of the 7th European Conference on Artificial Life (ECAL03)*, MIT Press, Cambridge, MA, September 2003, 367-377.

Mestl T, Plahte E, Omholt SW. (1995) A mathematical framework for describing and analysing gene regulatory networks. *J Theor Biol.* 176(2):291-300.

Milgram S. (1967) The small world problem. *Psychology Today*. May 1967, 60-67.

Mircean C, Tabus I, Astola J, Kobayashi T, Shiku H, Yamaguchi M, Shmulevich I, Zhang W. (2004) Quantization and similarity measure selection for discrimination of lymphoma subtypes under k-nearest neighbor classification. *BIOS* 2004, San Jose, January 2004.

Monk NA. (2003) Oscillatory expression of *Hes*1, *p*53, and *NF-κB* driven by transcriptional time delays. *Curr Biol.* 13(16):1409-13.

Moriyama T, Shinohara A, Takeda M, Maruyama O, Goto T, Miyano S, Kuhara S. (1999) A system to find genetic networks using weighted network model. *Genome Inform Ser Workshop Genome Inform.* 10:186-195.

Muchnik AA, Gindikin SG. (1962) On the completeness of systems of unreliable elements which realize functions of the algebra of logic. *Dokl. Akad. Nauk SSSR.* 144:1007-1010 (in Russian).

Murphy K, Mian S. (1999) Modelling Gene Expression Data Using Dynamic Bayesian Networks. *Technical Report, University of California, Berkeley.*

Ng MK, Zhang SQ, Ching WK, Akutsu T. (2006) A control model for Markovian genetic regulatory networks. Lecture Notes in Computer Science. 4070:36-48.

Nilim A, El Ghaoui L. (2004) Robust Markov Decision Processes with Uncertain Transition Matrices. EECS Department, University of California, Berkeley, UCB/ERL M04/28.

Noda K, Shinohara A, Takeda M, Matsumoto S, Miyano S, Kuhara S. (1998) Finding genetic network from experiments by weighted network model. *Genome Inform Ser Workshop Genome Inform.* 9:141-150.

Norris JR. (1998) *Markov Chains*, Cambridge University Press, Cambridge, UK.

Novak B, Tyson JJ. (1997) Modeling the control of DNA replication in fission yeast. Cell biology. *Proc Natl Acad Sci USA.* 94:9147-9152.

Nykter M, Price ND, Aldana M, Ramsey SA, Kauffman SA, Hood LE, Yli-Harja O, Shmulevich I. (2008) Gene expression dynamics in the macrophage exhibit criticality. *Proc Natl Acad Sci USA.* 105(6):1897-900.

Öktem H, Pearson R, Egiazarian K. (2003) An adjustable aperiodic model class of genomic interactions using continuous time Boolean networks (Boolean delay equations). *Chaos.* 13(4):1167-74.

Oosawa C, Savageau MA. (2002) Effects of alternative connectivity on behavior of randomly constructed Boolean networks. *Phys. D.* 170:143-161.

Pal R, Datta A, Fornace AJ Jr, Bittner ML, Dougherty ER. (2005a) Boolean relationships among genes responsive to ionizing radiation in the NCI 60 ACDS. *Bioinformatics.* 21(8):1542-9.

Pal R, Datta A, Bittner ML, Dougherty ER. (2005c) Intervention in context-sensitive probabilistic Boolean networks. *Bioinformatics.* 21(7):1211-8.

Pal R, Datta A, Dougherty ER. (2006) Optimal infinite horizon control for probabilistic Boolean networks. *IEEE Transactions on Signal Processing*. 54(6-2):2375-2387.

Pal R, Datta A, Dougherty ER. (2008) Robust intervention in probabilistic Boolean networks. *IEEE Transactions on Signal Processing*. 56(3):1280-94.

Pal R, Datta A, Dougherty ER. (2009) Bayesian robustness in the control of gene regulatory networks. *IEEE Transactions on Signal Processing*. 57(9):3667-78.

Pal R, Ivanov I, Datta A, Bittner ML, Dougherty ER. (2005b) Generating Boolean networks with a prescribed attractor structure. *Bioinformatics*. 21(21):4021-5.

Pearl J. (1988) *Probabilistic Reasoning in Intelligent Systems: Networks of Plausible Inference*. Morgan Kaufmann, San Mateo, CA.

Perrin BE, Ralaivola L, Mazurie A, Bottani S, Mallet J, D'Alche-Buc F. (2003) Gene networks inference using dynamic Bayesian networks. *Bioinformatics*. 19 Suppl 2:II138-II148.

Pfahringer B. (1995) Compression-based discretization of continuous attributes. in *Machine Learning: Proceedings of the Twelfth International Conference,* eds. Prieditis, A. & Russell, S., Morgan Kaufmann, Los Altos/Palo Alto/San Francisco, 456-63.

Pitt L, Valiant LG. (1988) Computational limitations on learning from examples. *Journal of the ACM*. 35:965-84.

Poor HV. (1980) On robust Wiener filtering. *IEEE Trans. Automatic Control*. 26:531-36.

Post E. (1921) Introduction to a general theory of elementary propositions. *Amer. J. Math.* 43:163-185.

Post E. (1941) *Two-Valued Iterative Systems of Mathematical Logic*. Princeton University Press, Princeton.

Przulj N, Wigle DA, Jurisica I. (2004) Functional topology in a network of protein interactions. *Bioinformatics*. 20(3):340-8.

Qian X, Dougherty ER. (2008) Effect of function perturbation on the steady-state distribution of genetic regulatory networks: Optimal structural intervention. *IEEE Transactions on Signal Processing*. 56(10-1):4966-4975.

Qian X, Dougherty ER. (2009a) On the long-run sensitivity of probabilistic Boolean networks. *J Theor Biol.* 257(4):560-77.

Qian X, Ivanov, I, Ghaffari N, Dougherty ER. (2009b) Intervention in gene regulatory networks via greedy control policies based on long-run behavior. *BMC Systems Biology*. 3:61.

Raftery AE, Lewis S. (1992) How many iterations in the Gibbs sampler? in *Bayesian Statistics* 4, eds. Berger, J. O. *et al.*, Oxford University Press, Oxford, UK, 763-773.

Ramö P, Kesseli J, Yli-Harja O. (2006) Perturbation avalanches and criticality in gene regulatory networks. *J Theoret Biol.* 242:164-170.

Ramsey S, Orrell D, Bolouri H. (2005) Dizzy: Stochastic simulations of large-scale genetic regulatory networks. *J Bioinform Comput Biol.* 3(2):1-21.

Richmond, CS, Glasner, JD, Mau, R, Jin, H & Blattner, FR. (1999) Genome-wide expression profiling in Escherichia coli K-12. *Nucleic Acids Res.* 27:3821-3835.

Rissanen J. (1978) Modelling by shortest data description. *Automatica.* 14:465-71.

Rissanen J. (1986) Stochastic complexity and modelling. *Ann. Statist.* 14:1080-100.

Rissanen J. (2007) *Information and Complexity in Statistical Modeling.* Springer, New York.

Robert CP. (1995) Convergence control techniques for Markov chain Monte Carlo algorithms. *Statistical Science.* 10(3):231-53.

Robert CP, Casella G. (1999) *Monte Carlo Statistical Methods.* Springer, New York.

Rosenthal JS. (1995) Minorization conditions and convergence rates for Markov Chain Monte Carlo. *Journal of the American Statistical Association.* 90(430):558-66.

Saez-Rodriguez J, Simeoni L, Lindquist JA, Hemenway R, Bommhardt U, Arndt B, Haus UU, Weismantel R, Gilles ED, Klamt S, Schraven B. (2007) A logical model provides insights into T cell receptor signaling. *PloS Comput Biol.* 3(8):e163.

Samsonova MG, Savostyanova EG, Serov VN, Spirov AV, Reinitz J. (1998) GeNet, a database of genetic networks. in *Proceedings of First Int. Conf. Bioinformatics Genome Regul. Struct., BGRS'*98, Novosibirsk, 105-110.

Sánchez L, Thieffry D. (2001) A logical analysis of the Drosophila gap-gene system. *J Theor Biol.* 211(2):115-41.

Sánchez L, Thieffry D. (2003) Segmenting the fly embryo: A logical analysis of the pair-rule cross-regulatory module. *J Theor Biol.* 224(4):517-37.

Sato TN, Qin Y, Kozak CA, Audus KL. (1993) Tie-1 and tie-2 define another class of putative receptor tyrosine kinase genes expressed in early embryonic vascular system. *Proc Natl Acad Sci USA.* 90(20):9355-8.

Scharf LL. (1991) *Statistical Signal Processing.* Addison-Wesley, Reading, MA.

Schweitzer PJ. (1968) Perturbation theory and finite Markov chains. *Journal of Applied Probability.* 5:401-13.

Seneta E. (1988) Perturbation of the stationary distribution measured by ergodicity coefficients. *Appl. Process.* 20:228-30.

Seneta E. (1993) Sensitivity of finite Markov chains under perturbation. *Statist. Probab. Lett.* 17:163-8.

Serov VN, Spirov AV, Samsonova MG. (1998) Graphical interface to the genetic network database GeNet. *Bioinformatics.* 14(6):546-7.

Serra R, Villani M, Semeria A. (2004) Genetic network models and statistical properties of gene expression data in knock-out experiments. *J Theor Biol.* 227(1):149-57.

Shmulevich I, Gabbouj M, Astola J. (2001) Complexity of the consistency problem for certain Post classes. *IEEE Transactions on Systems, Man, and Cybernetics, Part B.* 31(2):251-3.

Shmulevich I, Zhang W. (2002) Binary analysis and optimization-based normalization of gene expression data. *Bioinformatics.* 18(4):555-65.

Shmulevich I, Dougherty ER, Kim S, Zhang W. (2002a) Probabilistic Boolean networks: A rule-based uncertainty model for gene regulatory networks. *Bioinformatics.* 18(2):261-74.

Shmulevich I, Dougherty ER, Zhang W. (2002b) Gene perturbation and intervention in probabilistic Boolean networks. *Bioinformatics.* 18(10):1319-31.

Shmulevich I, Dougherty ER, Zhang W. (2002c) Control of stationary behavior in probabilistic Boolean networks by means of structural intervention. *Journal of Biological Systems.* 10(4):431-45.

Shmulevich I, Dougherty ER, Zhang W. (2002d) From Boolean to probabilistic Boolean networks as models of genetic regulatory networks. *Proceedings of the IEEE.* 90(11):1778-92.

Shmulevich I, Saarinen A, Yli-Harja O, Astola J. (2002e). Inference of genetic regulatory networks under the Best-Fit Extension paradigm. in *Computational and Statistical Approaches to Genomics*, eds. Zhang W. and Shmulevich I., Kluwer Academic Publishers, Boston, 197-210.

Shmulevich I, Gluhovsky I, Hashimoto R, Dougherty ER, Zhang W. (2003) Steady-state analysis of probabilistic Boolean networks. *Comparative and Functional Genomics.* 4(6):601-8.

Shmulevich I. (2003) Model selection in genomics. *Environmental Health Perspectives.* 111(6):A328-9.

Shmulevich I, Kauffman SA. (2004) Activities and sensitivities in Boolean network models. *Phys Rev Lett.* 93(4):048701.

Shmulevich I, Kauffman SA, Aldana M. (2005) Eukaryotic cells are dynamically ordered or critical but not chaotic. *Proc Natl Acad Sci USA.* 102(38):13439-44.

Shmulevich I, Dougherty ER. (2007) *Genomic Signal Processing.* Princeton University Press, Princeton.

Silvescu A, Honavar V. (1997) Temporal Boolean network models of genetic networks and their inference from gene expression time series. *Complex Syst.* 13:54-75.

Simon I, Barnett J, Hannett N, Harbison CT, Rinaldi NJ, Volkert TL, Wyrick JJ, Zeitlinger J, Gifford DK, Jaakkola TS, Young RA. (2001) Serial regulation of transcriptional regulators in the yeast cell cycle. *Cell.* 106(6):697-708.

Smallwood RD, Sondik EJ. (1973) Optimal control of partially observable Markov processes over a finite horizon. *Operations Research.* 21:1071-88.

Smolen P, Baxter DA, Byrne JH. (2000) Modeling transcriptional control in gene networks–methods, recent results, and future directions. *Bull Math Biol.* 62(2):247-92.

Snoussi EH. (1989) Qualitative dynamics of piecewise-linear differential equations: A discrete mapping approach. *Dyn. Stability Syst.* 4:189-207.

Socolar JES, Kauffman SA. (2003) Scaling in ordered and critical random Boolean networks. *Physical Review Letters.* 90:068702.

Spellman PT, Sherlock G, Zhang MQ, Iyer VR, Anders K, Eisen MB, Brown PO, Botstein D, Futcher B. (1998) Comprehensive identification of cell cycle-regulated genes of the yeast Saccharomyces cerevisiae by microarray hybridization. *Mol Biol Cell.* 9(12):3273-97.

Stauffer D. (1987) On forcing functions in Kauffman's random Boolean networks. *J. Stat. Phys.* 46:789-794.

Stewart WJ. (1994) *Introduction to the Numerical Solution of Markov Chains.* Princeton University Press, Princeton.

Stolovitzky G, Prill RJ, Califano A. (2009) Lessons from the DREAM2 Challenges. *Ann NY Acad Sci.* 1158:159-95.

Strauch ED, Yamaguchi J, Bass BL, Wang JY. (2003) Bile salts regulate intestinal epithelial cell migration by nuclear factor-κB-induced expression of transforming growth factor-β. *J Am Coll Surg.*197(6):974-84.

Suh EB, Dougherty ER, Kim S, Bittner ML, Chen Y, Russ DE, Martino R. (2002) Parallel computation and visualization tools for codetermination analysis of multivariate gene-expression relations. in *Computational and Statistical Approaches to Genomics*, Zhang W. and Shmulevich I., eds., Kluwer Academic Publishers, Boston, 227-240.

Tabus I, Astola J. (2001) On the use of MDL principle in gene expression prediction. *EURASIP Journal of Applied Signal Processing.* 2001(4):297-303.

Tabus I, Rissanen J, Astola A. (2002) Normalized maximum likelihood models for Boolean regression with applications to prediction and classification in genomics. in *Computational and Statistical Approaches to Genomics*, Zhang W. and Shmulevich I., eds., Kluwer Academic Publishers, Boston, 173-196.

Thieffry D, Colet M, Thomas R. (1993) Formalisation of regulatory nets: A logical method and its automatization. *Math. Modelling Sci. Computing.* 2:144-151.

Thieffry D, Thomas R. (1995) Dynamical behaviour of biological networks: II. Immunity control in bacteriophage lambda. *Bull. Math. Biol.* 57(2):277-297.

Thomas R, D'Ari R. (1990) *Biological Feedback*. CRC Press, Boca Raton, FL.

Thomas R, Kaufman M. (2001a) Multistationarity, the basis of cell differentiation and memory. I. Structural conditions of multistationarity and other nontrivial behavior. *Chaos*.11(1):170-179.

Thomas R, Kaufman M. (2001b) Multistationarity, the basis of cell differentiation and memory. II. Logical analysis of regulatory networks in terms of feedback circuits. *Chaos*. 11(1):180-195.

Tseng P. (1990) Solving H-horizon, stationary Markov decision problems in time proportional to log(H). *Operations Research Letters*. 9(5):287-97.

Uetz P, Giot L, Cagney G, Mansfield TA, Judson RS, Knight JR, Lockshon D, Narayan V, Srinivasan M, Pochart P, Qureshi-Emili A, Li Y, Godwin B, Conover D, Kalbfleisch T, Vijayadamodar G, Yang M, Johnston M, Fields S, Rothberg JM. (2000) A comprehensive analysis of protein-protein interactions in Saccharomyces cerevisiae. *Nature*. 403(6770):623-7.

Vahedi G, Faryabi B, Chamberland JF, Datta A, Dougherty ER. (2008) Intervention in gene regulatory networks via a stationary mean-first-passage-time control policy. *IEEE Trans Biomed Eng*. 55(10):2319-31.

Vahedi G, Ivanov IV, Dougherty ER. (2009a) Inference of Boolean networks under constraint on bidirectional gene relationships. *IET Syst. Biol.* 3(3): 191-202.

Vahedi G, Faryabi B, Chamberland JF, Datta A, Dougherty ER. (2009b) Optimal intervention strategies for cyclic therapeutic methods. *IEEE Trans Biomed Eng*. 56(2):281-91.

Vasudevan N, Zhu YS, Daniel S, Koibuchi N, Chin WW, Pfaff D. (2001) Crosstalk between oestrogen receptors and thyroid hormone receptor isoforms results in differential regulation of the preproenkephalin gene. *Neuroendocrinology*. 13:779-790.

Verdu S, Poor HV. (1984) On minimax robustness: A general approach and applications. *IEEE Trans. Information Theory*. 30:328-40.

von Neumann J. (1956) Probabilistic logics and the synthesis of reliable organisms from unreliable components. in *Automata Studies*, eds. C. E. Shannon and J. McCarthy, Princeton University Press, Princeton, NJ, pp. 43-98.

Wagner A. (2001) The yeast protein interaction network evolves rapidly and contains few redundant duplicate genes. *Mol Biol Evol*.18(7):1283-92.

Watts DJ, Strogatz SH. (1998) Collective dynamics of "small-world" networks. *Nature*. 393(6684):440-2.

Weeraratna AT, Jiang Y, Hostetter G, Rosenblatt K, Duray P, Bittner M, Trent JM. (2002) Wnt5a signalling directly affects cell motility and invasion of metastatic melanoma. *Cancer Cell*. 1(3):279-88.

Whitfield ML, Sherlock G, Saldanha AJ, Murray JI, Ball CA, Alexander KE, Matese JC, Perou CM, Hurt MM, Brown PO, Botstein D. (2002) Identification of genes periodically expressed in the human cell cycle and their expression in tumors. *Mol Biol Cell.* 13(6):1977-2000.

Winkler G. (1995) *Image Analysis, Random Fields and Dynamic Monte Carlo Methods: A Mathematical Introduction*. Springer-Verlag, New York.

Wolf DM, Eeckman FH. (1998) On the relationship between genomic regulatory element organization and gene regulatory dynamics. *J Theor Biol.* 195(2):167-86.

Xiao Y, Dougherty ER. (2006) Optimizing consistency-based design of context-sensitive gene regulatory networks. *IEEE Transactions on Circuits and Systems.* 53(11):2431-2437.

Xiao Y, Dougherty ER. (2007) The impact of function perturbations in Boolean networks. *Bioinformatics.* 23(10):1265-73.

Yablonsky SV, Gavrilov GP, Kudryavtsev VB. (1966) *Functions of the Algebra of Logic and Post Classes*. Nauka, Moscow (in Russian).

Yook SH, Oltvai ZN, Barabasi AL. (2004) Functional and topological characterization of protein interaction networks. *Proteomics.* 4(4):928-42.

Yu J, Xiao J, Ren X, Lao K, Xie XS. (2006) Probing gene expression in live cells, one protein molecule at a time. *Science.* 311(5767):1600-3.

Zhang Z, Teng CT. (2001) Estrogen receptor alpha and estrogen receptor-related receptor alpha1 compete for binding and coactivator. *Mol Cell Endocrinol.* 172(1-2):223-33.

Zhang W, Shmulevich I, Astola J. (2004) *Microarray Quality Control*, John Wiley & Sons, Hoboken, NJ.

Zhang S-Q, Ching WK, Ng MK, Akutsu T. (2007) Simulation study in Probabilistic Boolean Network models for genetic regulatory networks. *International Journal on Data Mining and Bioinformatics.* 1(3):217-240.

Zhao W, Serpedin E, Dougherty ER. (2006) Inferring gene regulatory networks from time series data using the minimum description length principle. *Bioinformatics.* 22(17):2129-35.

Zhao W, Serpedin E, Dougherty ER. (2008) Inferring connectivity of genetic regulatory networks using information-theoretic criteria. *IEEE/ACM Transactions on Computational Biology and Bioinformatics.* 5(2):262-74.

Zhou X, Kao MC, Wong WH. (2002) Transitive functional annotation by shortest-path analysis of gene expression data. *Proc Natl Acad Sci USA.* 99(20):12783-8.

Zhou X, Wang X, Dougherty ER. (2003a) Binarization of microarray data on the basis of a mixture model. *Mol Cancer Ther.* 2(7):679-84.

Zhou X, Wang X, Dougherty ER. (2003b) Construction of genomic networks using mutual-information clustering and reversible-jump Markov-chain-Monte-Carlo predictor design. *Signal Processing*. 83(4):745-61.

Zhou X, Wang X, Pal R, Ivanov I, Bittner M, Dougherty ER. (2004) A Bayesian connectivity-based approach to constructing probabilistic gene regulatory networks. *Bioinformatics*. 20(17):2918-27.

Zou M, Conzen SD. (2005) A new dynamic Bayesian network (DBN) approach for identifying gene regulatory networks from time course microarray data. *Bioinformatics*. 21(1):71-9.

Index